T0173068

GEOHUMANITIES

In the past decade, there has been a convergence of transdisciplinary thought characterized by geography's engagement with the humanities, and the humanities' integration of place and the tools of geography into its studies.

GeoHumanities maps this emerging intellectual terrain with 30 cutting-edge contributions from internationally renowned scholars, architects, artists, activists, and scientists. This book explores the humanities' rapidly expanding engagement with geography and the multimethodological inquiries that analyze the meanings of place, and then reconstructs those meanings to provoke new knowledge as well as the possibility of altered political practices. It is no coincidence that the geohumanities are forcefully emerging at a time of immense intellectual and social change. This book focuses on a range of topics to address urgent contemporary imperatives, such as the link between creativity and place; altered practices of spatial literacy; the increasing complexity of visual representation in art, culture, and science; and the ubiquitous presence of geospatial technologies in the Information Age.

GeoHumanities is essential reading for students wishing to understand the intellectual trends and forces driving scholarship and research at the intersections of geography and the humanities disciplines. These trends hold far-reaching implications for future work in these disciplines, and for understanding the changes gripping our societies and our globalizing world.

Michael Dear is Professor of City and Regional Planning at the University of California, Berkeley.

Jim Ketchum is Special Projects Coordinator and Newsletter Editor for the Association of American Geographers (AAG), Washington, D.C.

Sarah Luria is Associate Professor of English at College of the Holy Cross in Worcester, Massachusetts.

Douglas Richardson is Executive Director of the Association of American Geographers (AAG), Washington, D.C.

"This volume stands at the forefront of one of the most exciting new fields of cross-disciplinary work. The editors have assembled a spectacular array of original contributions from an impressive group of authors, whose work opens new routes into the emerging field known as the geohumanities. It is bound to become a landmark book."

Anthony J. Cascardi, Director, Townsend Center for the Humanities, U.C. Berkeley, USA.

"Making a compelling case for re-aligning geography with the humanities, *GeoHumanities* provides a series of richly-interwoven textual, visual and cartographic essays to demonstrate the creative potential of new forms of artistic, literary and historical engagement with place. Issuing a challenge to transcend disciplinary boundaries, to forge novel connections between past and present, and to re-imagine the world in novel ways, the contributors to *GeoHumanities* invite us to explore afresh the politics and poetics of place."

Professor Peter Jackson, University of Sheffield, UK.

GEOHUMANITIES

Art, history, text at the edge of place

Edited by

Michael Dear, Jim Ketchum,
Sarah Luria, and Douglas Richardson

LONDON AND NEW YORK

First published 2011
by Routledge
2 Park Square, Milton Park, Abingdon, Oxon, OX14 4RN

Simultaneously published in the USA and Canada
by Routledge
711 Third Avenue, New York, NY 10017

Routledge is an imprint of the Taylor & Francis Group, an informa business

© 2011 Selection and editorial matter, Association of American
Geographers; individual chapters, the contributors.

The right of the Association of American Geographers to be identified
as authors of the editorial material and of the authors for their
individual chapters has been asserted in accordance with the
Copyright, Designs and Patents Act 1988.

Typeset in Garamond 3 by Glyph International Ltd.

British Library Cataloguing in Publication Data
A catalogue record for this book is available from the British Library

Library of Congress Cataloging in Publication Data
A catalog record for this book has been requested

ISBN 13: 978-0-415-58979-6 (hbk)
ISBN 13: 978-0-415-58980-2 (pbk)
ISBN 13: 978-0-203-83927-0 (ebk)

Contents

List of figures

Acknowledgments

The editors would like to thank the National Endowment for the Humanities and the Virginia Foundation for the Humanities for providing financial support for this book project and related activities. The editors also thank Ed Ayers, formerly of the University of Virginia (and now President of the University of Richmond), who co-hosted the AAG's 2007 Geography & Humanities Symposium at the University of Virginia. We are also grateful for the support of our editors Andrew Mould and Faye Leerink of Routledge and also to Candida Mannozzi of the Association of American Geographers, for guiding the book through the process of compilation and proofing, assisted by colleagues Miranda Lecea, Megan Overbey, and Marcela Zeballos. Our special thanks go to artist Nick Schiller, for generously granting permission to use his artwork "The Modern Geographer" for the cover of this book.

Contributors

Stuart C. Aitken is Professor and Chair of Geography at San Diego State University, and Director of the Center for Interdisciplinary Studies of Young People and Space (ISYS). His research interests include film and media, critical social theory, qualitative methods, children, families and communities. Stuart's recent books include *Qualitative Geographies* (Sage 2009), *The Awkward Spaces of Fathering* (Ashgate, 2009) and *Global Childhoods* (2008).

Edward L. Ayers is President and Professor of History at the University of Richmond. Winner of the Bancroft Prize and a National Professor of the Year, Ayers pioneered in digital history with *The Valley of the Shadow* and works on geography and visualization with the Digital Scholarship Lab at Richmond.

Susan Bergeron, Assistant Professor of Geography at Coastal Carolina University, has research interests in geovisualization and immersive simulation, historical geography and GIS. Her current work is focused on implementing an advanced virtual world application that integrates humanities information within an immersive and interactive virtual landscape.

Ursula Biemann is an artist, curator, and videomaker whose work focuses on gender, migration, and the global economy. She exhibits her work internationally, has published several books on art practice in the field, geography, and mobility, and is a Researcher at the Institute for Theory of Art and Design, Zurich. www.geobodies.org.

Peter K. Bol is Charles H. Carswell Professor East Asian Languages and Civilizations and Director of the Center for Geographic Analysis at Harvard University. He is Director of the China Historical Geographic Information Systems project, which is aimed at integrating geographical knowledge and analysis into historical research, and the China Biographical Database project, which aims eventually to include all biographical records in China's history.

Michael Dear is Professor of City and Regional Planning at the University of California, Berkeley.

Deborah P. Dixon is Senior Lecturer in the Institute of Geography and Earth Sciences, Aberystwyth University, and editor of the journal *Gender, Place and Culture*. Her research focuses on science/art collaborative endeavors, particularly with respect to the production of the monstrous, and spaces of representation more broadly.

Barbara Eckstein is the author of *Sustaining New Orleans* (Routledge 2006) and co-editor, with James Throgmorton, of *Story and Sustainability* (MIT 2003). She teaches in the English Department at the University of Iowa.

Robbert Flick is a Southern California artist who uses photography as his primary medium. He has exhibited his photographs for over 30 years and his work has been shown and collected at numerous private and public venues both nationally and internationally. He has been the recipient of two NEA Fellowships, a Guggenheim Fellowship and a Flintridge Visual Artist Award, and was a Getty Scholar at the Getty Center for the History of Art and the Humanities.

Philip Govedare is an artist and Professor at the University of Washington School of Art. He has had more than 20 solo exhibitions nationally and internationally and received numerous awards including an individual artist fellowship from the NEA and a Pollack-Krasner Foundation Award.

Ian N. Gregory is a Senior Lecturer in Digital Humanities at Lancaster University. His work centers on applying GIS approaches to the humanities. He has authored several books in this field including (with P. Ell) *Historical GIS: Technologies, methodologies and scholarship* (CUP) and numerous journal articles.

Matthews M. Hamabata is at The Kohala Center (available at www.kohalacenter.org).

Trevor Harris is Eberly Professor of Geography and Chair of the Department of Geology and Geography at West Virginia University. He co-directs the West Virginia GIS Technical Center and the Virtual Center for Spatial Humanities. Harris specializes in Geographic Information Science and in particular humanities GIS, participatory GIS, and virtual geographic environments.

Jordi Marti-Henneberg is Professor of Human Geography at Lleida University, Spain. He leads a European Science Foundation project "The Development of European Waterways, Road and Rail Infrastructures: A Geographical Information System for the History of European Integration (1825–2005)," a collaborative partnership involving nine countries.

Amy Hillier is an Assistant Professor in the Department of City and Regional Planning in the School of Design at the University of Pennsylvania where she teaches courses in GIS and conducts research about geographic disparities in urban history and public health. Hillier holds an MSW and a PhD in social welfare.

Martin Hogue is a licensed architect and has taught in architecture and landscape architecture at several schools across the United States and Canada. His research is informed by a rich and diverse body of ideas and influences ranging from architecture to conceptual art, land art, landscape, geography, and film.

Howard Horowitz is a Professor of Geography at Ramapo College. Always drawn to the edge of the field, the former tree planter creates wordmaps and other geographical poems, fights herbicide overuse on public lands and waters, and leads students on field programs to the American West.

Norma Iglesias-Prieto is Professor and Chair of the Department of Chicana and Chicano Studies at San Diego State University. Previously she was a Researcher-Professor in the Department of Cultural Studies at El Colegio de la Frontera Norte (Tijuana). She has developed research projects for Centro Cultural Tijuana and Instituto Nacional de Bellas Artes (Mexico). She is a member of Sistema Nacional de Investigadores (level II) in Mexico. She has published many books, worked in video, film and radio production, and been curator in many art events. Her current work is focused on Tijuana's art scene and its re-definition as a transborder urban space.

Janelle Jenstad is Associate Professor of English at the University of Victoria, where she teaches Shakespeare, Early Modern Studies, and Bibliography. She is General Editor of *The Map of Early Modern London*, editor of The Merchant of Venice for Broadview/Internet Shakespeare Editions, and author of articles and book chapters on London's merchants and livery companies, childbirth and christening, usury, Shakespeare in Performance, and teaching Shakespeare.

Ian Johnson is Director of the Archaeological Computing Laboratory at the University of Sydney. He has been developing tools for digital humanities since the 1980s, with a particular emphasis on collaborative data bases, mapping and the temporal dimension. His major software projects include TimeMap (time-enabled distributed-data mapping) and Heurist (collaborative database).

Catherine D'Ignazio, a.k.a. **kanarinka**, is an artist and educator. Her artwork is invested in the idea that small, everyday actions (such as cooking, singing, or running) can have a poetic, transformative resonance. She has invited the public to rename the city of Cambridge, measured post-9/11 fear by running Boston's evacuation route system, and tracked her body's weather systems. She teaches at the Rhode Island School of Design and lives in Waltham, Massachusetts. Her website is www.kanarinka.com.

Caren Kaplan is Professor of Women and Gender Studies at the University of California, Davis. She has published work on post-colonial travel, cultural geography, and feminist theory. Her current research focuses on aerial views, militarization, and modern visual culture.

Kekuhi Keali'ikanaka'oleohaililani is at the Edith Kanaka'ole Foundation (www.edithkanakaolefoundation.org).

Karen Kemp is a Senior Consultant to the Redlands Institute in Redlands, California, and senior scientist at the Kohala Center in Waimea, Hawaii (www.kohalacenter.org). Her research has worked across disciplines to focus on formalizing conceptual models of space acquired by scientists and humanities scholars. She is currently working on the GIS framework for the Ala Kahaki National Historic Trail and with the Kohala Center on a vision for an Island-wide GIS infrastructure integrating Hawaiian and Western sciences.

Jim Ketchum is Special Projects Coordinator and newsletter editor for the Association of American Geographers in Washington, DC. A cultural geographer with interests in contemporary art, landscape, and visual culture, his research examines the ways artists use geographic perspectives and technologies in responding to contemporary issues, including war and globalization.

Gustavo Leclerc is an architect and artist currently pursuing a doctorate in architecture at UCLA. He is a partner and founding member of the multidisciplinary collective ADOBE LA (Artists, Architects, and Designers Opening the Border of Los Angeles). He was a Loeb Fellow at Harvard University during the 1998–99 academic year, and lectures throughout the USA and Mexico on art, architecture, border culture, immigration and cultural criticism.

Sarah Luria is Associate Professor of English at College of the Holy Cross in Worcester, Massachusetts. The author of *Capital Speculations: Writing and Building Washington, D.C.* (University of New Hampshire Press, 2006), her current book project is a study of land surveying and property making in the work of Thomas Jefferson, Henry David Thoreau, and Robert Moses.

Timothy Mennel, a former editor at *Artforum*, the Andy Warhol Foundation for the Visual Arts, and Random House, is senior editor and acquisitions manager for Planners Press and the PAS Reports series at the American Planning Association in Chicago. He was the coeditor of *Block by Block: Jane Jacobs and the Future of New York* and, most recently, *Green Community*.

Peta Mitchell is Lecturer in the School of English, Media Studies, and Art History at The University of Queensland. Her research focuses on spatial metaphor and literary geography, and she is author of *Cartographic Strategies of Postmodernity* (Routledge 2008). She gratefully acknowledges assistance provided by the Centre for Critical and Cultural Studies at The University of Queensland in the preparation of her chapter.

Lize Mogel is an interdisciplinary artist who works with the interstices between art and cultural geography. Her work has been exhibited and published internationally.

With Alexis Bhagat, she is editor and curator of the book and exhibition project *An Atlas of Radical Cartography*.

Wolfgang Moschek studied History and Geography (at the Technical University Darmstadt), and completed his PhD (Dr. phil.) at the University Erfurt on the topic of the cognitive and cultural backgrounds of the Roman Limes. He works at a private high school near Frankfurt, Germany.

Trevor Paglen is an artist and geographer. He has written four books and exhibits his work internationally.

René Peralta was educated at the New School of Architecture in San Diego and at the Architectural Association in London. He is currently Visiting Professor at Washington University, St. Louis, and is on the faculty of Woodbury University School of Architecture in San Diego. His research publications include *World View Cities Tijuana*, a web-based report on urbanism for The Architectural League of New York; he is co-author of *Here is Tijuana*, Black Dog Publishing, London, 2006.

Douglas Richardson is Executive Director of the Association of American Geographers. He previously founded and was President of the firm GeoResearch, Inc., which invented, developed, and patented the first real-time interactive GPS/GIS technology, leading to major advances in the ways geographic information is collected, mapped, integrated, and used within geography and in society at large. He is co-editor of *Envisioning Landscapes, Making Worlds: Geography and the Humanities* (Routledge, 2011).

L. Jesse Rouse, Assistant Professor of Geography at the University of North Carolina at Pembroke, is currently focusing on research exploring the intersection of Geographic Information Science and phenomenology to address social science and humanities research questions.

Robert M. Schwartz is E. Nevius Rodman Professor of History at Mount Holyoke College. His most recent publication applying historical GIS is "Rail Transport, Agrarian Crisis, and the Restructuring of Agriculture: France and Great Britain Confront Globalization, 1860–1900," *Social Science History* 34:2 (June 2010): 229–55.

Emily Eliza Scott is an artist and scholar whose work explores intersections between contemporary art, geography, and environmental sciences and politics. In 2010, she completed a PhD in Art History at UCLA, with her dissertation, *Wasteland: American Landscapes in/and 1960s Art*. In 2004, she founded the Los Angeles Urban Rangers art collective. She is currently based in Zurich, Switzerland.

Alexander von Lünen holds a 'Diplom' in 'Informatik' (equivalent to an MSc in Computer Science) and a Dr. phil. (or PhD) in history, both from the Technical

University Darmstadt, Germany. Since 2007, he has worked with the Great Britain Historical GIS team at the University of Portsmouth, UK.

Stephen S. Young is Chair of the Department of Geography at Salem State College and has been a member of the Department since 1995. He has had over a dozen gallery exhibitions of his Art and Science work, including shows at the headquarters of the National Science Foundation as well as international shows in Tunisia and Australia.

INTRODUCTION

Introducing the geohumanities

Douglas Richardson, Sarah Luria, Jim Ketchum,
and Michael Dear

The term "geohumanities" refers to the rapidly growing zone of creative interaction between geography and the humanities. The traditions of these various disciplines are being actively breached by a profusion of intellectually and artistically challenging scholarship and practice, itself propelled by social, technological, and political change. Just as geography, with its focus on space and place, is now actively engaging with the humanities, so are the discourses of the humanities increasingly incorporating place and the spatial dimension. The fortuitous convergence of this intellectual traffic outlines a distinctive scholarly terrain and emerging zone of practice that is the focus of this book.

As geography and the humanities converge, new topics are being suggested that require a transdisciplinary perspective and a combination of methodologies. From such fusions, a kaleidoscope of intellectual and artistic outputs is currently emanating. You can see it in the hybrid maps of radical cartographers and the artistic creations of experimental geography; or when philosophers and literary theorists concerned to examine life's fundamental spatiality engage with constructs of place and landscape; or when urbanists turn to art for visions of the future city. In this book, we capture some of the excitement breaking out across the intellectual landscape of the geohumanities. We try to nudge the diversity of contributions toward a more collaborative and creative group-awareness that explores the meaning of place, and attempts to reconstruct those meanings in ways that produce new knowledge and better-informed scholarly or political practices.

Efforts to introduce new ideas often dwell on their differences from past and present traditions. By emphatically denying the legacies of predecessors, advocates call attention to the radical break implied by their novel approaches. Such is not our strategy. Our first sketch of a "geohumanities" is resonant with the histories of our twin origins. As editors, we represent four "spatial minds" with significantly different training in, respectively, urbanism, literature, the visual arts, and science. Our goals are to give voice to these separate traditions while bringing them into the same discursive space, and to illuminate the new opportunities thereby revealed. We also understand our responsibility to demonstrate something of the form and utility of these uncharted territories, even as we readily concede that our first mappings are necessarily tentative and incomplete.

Hence, this volume may be properly regarded as an exercise in becoming. We do not aim to define a comprehensive new "field" or "discipline." The works collected in this book provide an opening to a project that is just beginning. In it, academics, architects, artists, scientists, activists, and writers reveal an astonishing degree of common ground when explaining the production and meanings of place. But they also uncover new and insightful terrains that advance our understanding of knowledge and action.

The structure and content of this volume are themselves an active part of our investigation. While the essays themselves for the most part are rooted in conventional disciplinary heritages, each deliberately seeks to transgress disciplinary boundaries in order to encompass a wider explanation of the production and meaning of place. Our contributors' objects of inquiry are text-based (e.g., novels, histories), visual (photography, film), and cartographic (maps, spatial ecologies). In what follows, we maintain this traditional categorization – text, image, map – but are keenly aware that it is precisely this divide that we hope to span via our conversations. Such ontological self-awareness is an important strategy, even an imperative in any transdisciplinary work.

Another aspect of the volume's transdisciplinarity collaboration are the "vignettes" that are peppered throughout the text. These are meant to convey emerging forms of the geohumanities; they are examples of unfettered practice in the geohumanities. These short interventions appear irregularly and without ceremony, as emotional voices, colors, non-sequiturs, contradictions, and un-categories. They call attention to diverse practices of place-making: to the object that is being created, a change of perspective being sought, or to what is going on around the practice itself. The rough edges of these place-practices spill wittily, uncomfortably, even dazzlingly into the adjacent spaces of academic discourse. They raise questions about convention, dissent, difference, altered identity, and social action. They exist in a flux of exceptions, at the burgeoning edges of a geohumanities' discourse. As such, they are beacons signaling new directions.

Any cartography of the emerging geohumanities conveyed in the essays and vignettes that follow necessarily remains time- and place-contingent. We do not intend our initial mapping to spell out an immutable, comprehensive definition of the project, and it certainly should not be read as such. But while this book is an experiment, it is also an invitation, opening up new spaces for a geohumanities and inviting readers to advance its agenda.

CREATIVE PLACES
Geocreativity

Michael Dear

Conventional social theory makes an important distinction between *structure* and *agency*, that is, between the enduring, deep-seated practices that undergird society and the everyday voluntaristic behavior of individuals. This dichotomy is important for what it can tell us about the relative importance of social constraint and free will, and how these opposing forces become articulated in the practices of everyday life. Approaches to the study of the production of place tend to fall into these two categories. Most of the structuralist visions of place production derive out of a Marxist-inspired political economy, including Lefebvre's concept of the production of space, and Harvey's treatment of the capitalist urban process. An agency-oriented view of place production has a diverse theoretical heritage that includes material concerns but also cognitive, cultural, and social dimensions (such as attitudes and beliefs, feelings and emotions).

Social theory is also concerned to articulate the relation between social process and spatial structure, that is, how social forces become manifest in geographies, and how geography is constitutive of social relations – a problematic sometimes referred to as the "socio-spatial dialectic." Needless to say, the relationships articulating society and space relate to more than pure theory; they also have consequences for the work of practicing professionals such as architects and urban planners who are charged with forging new geographies of the built environment.

The spaces of cities are of special concern in this book. While consensus has it that we have entered a global urban age, there is little understanding of, much less agreement about, what this trend entails. The proliferation of urban sprawl has caused investigators to look more closely at the forms of emerging urbanism, but the rash of neologisms describing these forms is more indicative of intellectual confusion rather than understanding. These terms include such descriptors as *polycentric*, *postmodern*, *patchwork*, *splintered*, and *post-(sub)urban*. The places produced by these altered processes are variously labeled as *city-region*, *micropolitan region*, *exopolis*, *edge city*, or *metroburbia*. Despite the profusion of terminology, there is a growing sense that the geography of cities is changing; no longer are cities being built from the inside out (from core to periphery) but from the outside in (from hinterlands to what remains of the core). This decentered urbanism has the effect of shifting the traditional bases of power in the city. Power lies less in the center than at the edges, and is correspondingly more dispersed, even hidden; but such arrangements also offer greater opportunities for widespread local autonomies. At the same time, other dynamics of the Information Age, such as globalization, domestic and international migration, and so on, underscore how local outcomes in city-regions are being buffeted by forces operating at different scales, including the national and international.

Each of the essays in this section addresses the question of place production in cities, with special emphasis on the creative process in urban places. Cities have always been regarded as the locus of innovation and social change in all dimensions of human activity. In this section's first essay, geographer Michael Dear considers some fundamentals in the relationship between place and creativity. He distinguishes between creativity *in* place and creativity *of* place: the former refers to the role that a particular location has in facilitating the creative process; the latter to the ways in which place becomes an artifact in the creative output, be it a dance movement or photographic frame. Dear describes a two-year collaborative project among an international group of academics, artists, critics, and curators charged to imagine reconstructed places along the controversial and rapidly changing US–Mexico borderlands.

The urban outcomes that characterize the Information Age are dramatically described in architect René Peralta's account of the border city of Tijuana, in Baja California. One of North America's fastest-growing and most dynamic cities, Tijuana is globally engaged through the presence of the *maquiladora* (assembly-plant) industry, a hemispheric trade in drug and human trafficking, and its locus as the busiest international boundary crossing in the world. Peralta's reading of Tijuana's urban ecologies rewrites many conventions of urban theory, and reveals some of the startling material conditions of Information Age urbanism.

Cities also have a "soft" dimension, that is, they comprise an infinite number of mental maps lodged in the minds of their inhabitants. To see how these cognitive maps are formed, artist kanarinka invited residents of the city of Cambridge, Massachusetts, to rename their favorite streets and places. The soft city she discovered is full of humanity, invention, and fun, totally unlike the "hard" city with its ponderous monuments, commemorative namings, patrolled spaces, and formal geometries. Produced with friends from the Institute for Infinitely Small Things – itself worthy of attention – kanarinka's hypothetical map of a city formerly known as Cambridge shows just how deep is the fissure between the formalities of the hard city and the spontaneities of the soft city.

Similar lessons about cognitive mapping come out of architect/artist Gustavo Leclerc's visual and textual reflection on his migrant experience. Born in Veracruz, Mexico, but now a long-term resident of Los Angeles, California, Leclerc presents a selection from notebooks of his experience of transition to the USA. His drawings were accompanied by textual annotations, also reproduced here, including a grandmother's recipe and quotes from Mesoamerican manuscripts. In addition, Leclerc has added a present-day commentary to the elements of drawing and annotation. This triple-layered narrative stretches backwards and forwards through time and space in a continuous unfolding reminiscent of the ancient codices (scrolls) of Mayan and other indigenous populations. The syncopated texts bleed into one another, echoing and fusing, making tangible the elusive workings of memory in our lives.

Keeping in mind these introductions to the hard city, soft city, and memory, the following essays turn to the ways creative people work with and in creative places. First, geographer Trevor Paglen outlines what he calls "experimental geography." Careful to avoid precise definitions and always welcoming nuance, Paglen is nevertheless clear about what he does: he deliberately works in many fields simultaneously; is passionately transdisciplinary and collaborative; communicates through popular media and academic publications; is self-consciously aware of the political in his work; and possesses a keen sense of public responsibility regarding his work. His testimony may amount to a kind of manifesto of experimental geography, but Paglen resolutely refuses this label as contrary to the open-ended, participatory spirit of what he advocates and practices.

Emily Scott's offers a kind of "undisciplined geography." She is an art historian and artist, and self-described "long-time interloper" attracted by geography's breadth and interdisciplinarity. From a base in contemporary art, she asserts that geographers and artists should break boundaries with their undisciplined geographies, drawing on three examples to make her case. Scott is a founder of another activist collective (called the Los Angeles Urban Rangers), which seems to be a common feature of many geo-humanities projects.

The centrality of the *map* as an analytical focus and inspiration is strongly evident in an essay by architect Martin Hogue. He reports on a painstaking documentation of the state of properties and sites in the borough of Queens, New York. Hogue is most interested in what he terms the "agency of the map," that is, what the map gives to us. He shows the possibilities offered by a map for contemplation and taking stock, as well as the satisfaction he derived from its comprehensive accounting of place. Hogue's absorption in the map is driven by many desires: a taxonomic urge that seems almost akin to rational and scientific reasoning; but also by flights of imagination. As in Leclerc's essay, the multiple layers of accumulated meaning dissolve the boundaries between categories and offer opportunities for original insight into both object and observer.

In 1999, Emily Scott recalled cultural geographer Denis Cosgrove's comment on the "startling explosion" of interest in cartography, the cartographic trope and the map within the humanities and cultural studies. This observation remains current, but a decade later we are witnessing a more general "spatial turn" that is emblematic of the geohumanities project as a whole. In this opening section on geocreativity and the production of place, our contributors have already illuminated some of the geo-humanities' strategies: a proclivity to transgress disciplinary boundaries; to accumulate layer upon layer of transdisciplinary data, and then make connections; to imagine the world as well as describe it; and to produce scholarship, art, poetry, community, and politics (often simultaneously) from their works. The inventive, edgy settings inhabited by these practitioners occur in many places: five of the essays in this first selection focus in some way on California and the US–Mexico border, the latter a place of great turmoil and potential; but creativity also occurs in older more established urban places, sometimes driven by small collaboratives dedicated to their communities. Another preliminary observation relates to the social-theoretic question of structure and agency: even at this early stage in our exegesis, a geohumanities approach appears to promise greater insight into the question of human agency in its myriad forms and dynamics.

Figure 1.1 Walls, Muros. Source: Marcos Ramirez. Used by permission.

Figure 1.2 What the River Gave Me. Source: Amalia Mesa-Bains. Used by permission.

Creativity and place

Michael Dear

Place

Creativity takes place. That is to say, it occurs in specific locations, but also it requires space for its realization. Creativity *in* place refers to the role that a particular location, or time-space conjunction, has in facilitating the creative process: think for example of *fin-de-siècle* Vienna, Silicon Valley, or Hollywood. Creativity *of* place refers to the ways in which space itself is an artifact in the creative practice or output, as when a dancer moves an arm through an arc or a photographer crops an image to create a representational space. The simplest creative acts are fraught with geographies that instruct the spectator how to see, but also hide things from us.

Most artists readily concede the significance of place *in* the creative process. For instance, director Peter Sellars found in California a delightful sense of playful experimentation and an important freedom from overbearing tradition:

> This is the place where people have come to try new things. It's the newest edge of America … New York and Chicago remind you of what America was. In California, the question is, what will America be?[1]

Amalia Mesa-Bains, a Chicana scholar and artist, reminds us of the complications of identity:

> all the time California has been California, it's always been Mexico. There is a Mexico within the memory, the practices, the politics, the economy, the spirituality of California.[2]

And novelist Carolyn See reveals how the specificity of space adds universality to her fiction:

> When I talk to my students about writing, I talk about characters and plot, space and time, but, most important, geography. Place has always been where you start. The more specific you can get about a particular place, the better chance you have of making it be universal and of really grounding it.[3]

The inspiration provided by geography is, in short, a rich and robust wellspring that operates in myriad ways.[4] The city itself may become a *crucible* of artistic innovation whenever a critical mass of artists is assembled. Perhaps the most famous example of this is *fin-de-siècle* Vienna. In a widely admired study, Carl Schorske argued that Vienna, "with its acutely felt tremors of social and political disintegration," proved to be a fertile breeding ground for contemporaneous shifts in painting, music, philosophy, psychoanalysis, economics, architecture, and urban planning—even though the protagonists in each field might scarcely have known one another.[5] Place acts too as an *inspiration* for art. In his landmark study of visual arts and poetry in California from 1925 to 1975, Richard Cándida Smith accords great weight to the State's relative isolation as a stimulus to creativity, linking remoteness to an anticipation of a renaissance, of bringing a "new culture into the world."[6] It was not that California artists desired isolation, but that the *tabula rasa* represented by California offered the opportunity to strike out against the primacy of New York and Europe.

The specifics of place can become the *subject* of art. For instance, Williams and Rutkoff revealed how early twentieth-century realists embraced New York City, "making its variety and life their subject."[7] Photographer Alfred Stieglitz saw New York as "a great modern machine, which, if carefully observed and recorded, offered artists a unique opportunity to comprehend the modern."[8] A remarkable case of city as subject is Walther Ruttman's documentary film *Berlin: Symphony of a Great City* (1927). The camera brilliantly perceives the diurnal rhythms of the hard (material) city and soft (cognitive) city at varying scales: in the early morning, a speeding train approaches Berlin, portrayed as massive clumps of streets, residential neighborhoods, and industries; inside the city, the pre-dawn boulevards are empty, still dormant; but later, as the city awakes, every corner is abuzz with minutely detailed life—a brief flirtation, the raised eyebrow of a disdainful child, a forlorn dog. The onset of evening cloaks Berlin in a veil of anonymity, and the camera lens adjusts to reveal the revels of the demi-monde, before blinking its tired eye as the city settles to sleep.

The place where art is consumed influences the *reception* of an art object. This is not solely a matter of reception theory, with its emphasis on who reads books, views art, and so on. It's also a mater of where. As Azar Nafisi's memoir of Iran points out, Nabokov's *Lolita* becomes a different book if it is read by a Muslim who does not regard the novel's eponymous heroine as an underage girl because, in other eyes, she had reached marriageable age much earlier.[9]

The synergies of *geographical ensembles of cultural producers, consumers, and markets* that congeal in a particular place and time create what Sharon Zukin has called the "artistic mode of production."[10] Harvey Molotch demonstrates such synergies among LA's industrial/cultural producers, including movie-making, tourism, food, architecture, furniture, and automobiles.[11] Important in the emergence of a California counter-culture during the latter half of the twentieth century was the proliferation of art schools. Such conditions proved significant in the rise of women artists in the latter half of the twentieth century.[12]

The creativity *of* place is under-theorized in most studies of the creative process. Place is everywhere in photography, for example, from the choice of subject, through the placement of camera, to the composition of an image. Yet space/place is peculiarly absent as an explicit component in the theory and practice of photography.[13] In his concise history of American photography, Miles Orvell lists seven cultural determinants that give meaning to a photographic image: the photographer's personal history; the social and commercial matrix, including photographer–client relationship; the semiotic context, that is, the codes of meaning incorporated into the image; the aesthetic composition (balance, lighting, etc.); interpretive codes (the symbolic or literal implications); presentation format (gallery, newspaper, etc.); and the history of the photograph's reception.[14] Yet while geography is inherent in each of these facets, it nowhere enters Orvell's conceptualization in any systematic way. The role of space/place in film studies is equally underspecified. According to David Clarke: "Today by far the dominant paradigm for the study of film covers a theoretical terrain triangulated by semiotics, psychoanalysis and historical materialism."[15] Or to put it more simply: the image, how it is received, and the social context.

In my judgment, the entire panoply of place-based contingencies involved in photographic production, reproduction, distribution, and consumption incorporates:

- *the place of production*, which incorporates both the specific site of photography, but also the circumstances of the ambient artistic mode of production (particularly the market for photographic images);
- *the production of place*, including the narrative and compositional aspects of the image, as well as the spatial techniques employed by the image-maker (camera angle, depth of focus, etc.);
- *the place of presentation*, referring to the image and its mode of presentation (on a website, in an album, or on a gallery wall); and
- *reception in place*, what happens to an image when it is released to the world of consumption (including purchase).[16]

This characterization emphasizes the dynamics of the place/art dialectic: as the material conditions of society alter so inevitably will the production and reception of cultural products; but equally, art will challenge perceptions of social change and alter its meaning. In our present globalizing, technologically dynamic society, the analytical onus is on understanding how material changes are influencing the creative process, and how art uses spaces to inform us about these changes.

The creative academy

The academy is a locus of innovation where creativity is universally prized and regarded as a good thing. Its outputs are measured in terms of prize-winning discoveries of basic sciences, technological innovation, contributions to human wellbeing, as well as the production of a skilled workforce and an informed citizenry. Yet pathological forms of creativity are also present: some individuals fraudulently misrepresent their findings and deliberately suppress the contradictory ideas of competitors. Other pathologies are more systemic: researchers strive for the new, but research training tends to institutionalize the old. Students bear some responsibility for this conservativism because their concern with job market prospects act to the detriment of an innovative education (cf. Chapter 2). But perhaps the greatest enemy to academic creativity is disciplinary boundaries because they favor marginal advancements appreciated by relatively few specialists locked in disciplinary silos. It is therefore unsurprising that the loudest voices in today's research communities call for an interdisciplinarity based in the ability to speak simultaneously in many intellectual tongues – the visual and mathematical, spiritual and scientific, the fragmented and formalistic.

Creative interdisciplinary exchanges occur when people venture beyond their disciplines and speak across boundaries in their respective tongues about matters of common concern. This takes courage (to risk rebuff or humiliation) and hard work (to become fluent in many representational modes). I want briefly to describe three salutary efforts in creative interdisciplinarity: a scientist approaching the humanities; an art historian embracing neuroscience; and a self-conscious practitioner of "artscience."

Antonio Damasio is equally at home with Spinoza as well as synapses. Years ago, puzzled by the age-old problem of the mind/body relation, he began to "map the geography of the feeling brain."[17] Damasio's wide-ranging research (with Hanna Damasio and colleagues) distinguishes between *feelings*, or perceptions of arousal in the brain's body map, and truly primal *emotions* encoded within the evolved subcortical networks of the nervous system. Damasio expresses his conviction this way:

> We have emotions first and feelings after because evolution came up with emotions first and feelings later. Emotions are built from simple reactions that easily promote survival of an organism and thus could easily prevail in evolution.[18]

What interests me is the way Damasio's neurobiological formulation is being stretched into humanistic dimensions. Writing with exemplary clarity, he has entered adjacent disciplines not to do battle but instead to invite a conversation with others who know better the realms he is engaging.

Barbara Maria Stafford, an art historian, also journeys from the realm of feelings to a place where "organic intensity … exceeds mere physiochemical processes."[19] Pointing to the reluctance of historians to consider seriously the biological underpinnings of art and place, Stafford concedes that "concepts do not travel well from one context to the next."[20] Nevertheless, she describes culture and science as a joint project, suggesting that "a broader range of visual materials has to be mobilized in the discussion of complex phenomena like attention or consciousness."[21] Her starting point is the image itself and how our reactions make visible the invisible ordering of human consciousness through expressions and representations of an object: "Art enables us to observe the space inside our bodies. It gives a face to the secret life of consciousness."[22] This complicated interaction avoids reducing human feelings to "a galaxy of neurons, awash in neurotransmitters, and dispersed in synaptic circuitry."[23] When brain science and cultural configurations meet, Stafford claims, our work gains an "added bounce."[24] In her own brilliant performances, she creates the perfect foil for her liminal thinking – presenting a deliberate bricolage of unfamiliar images and colliding perspectives that overwhelm the listener/viewer, undermine existing categories and logic, as well as resisting easy reconstruction simply because of the sheer number of images. On the overdetermined cognitive field she has opened, Stafford is then able to begin a kind of post-disorientation reconstruction.

David Edwards calls his fusion of scientific and aesthetic methods "artscience."[25] He is a bridge-builder in the normative dimension who considers what it takes to mold creative institutions that successfully link science and humanities. Edwards excoriates scholars for their tendency to remain within comfortable disciplinary confines and for rarely sticking with an idea long enough to realize its full potential. The follow-up work of "idea translation" moves beyond the conceptual toward a general process of realization. This might involve taking an intellectual idea and deliberately nurturing it as an economic, social, or cultural practice. Crucial to the success of this translation are two conditions: the existence of "idea translators," passionate people willing to endure stiff resistance over extended periods in order to arrive at an original artistic or scientific expression; and the "impact space," a terrain on which innovation and creativity is marshaled and put to work. Edwards suggests that universities possess knowledge banks that facilitate creativity, but that the institution's inherent inertia may be assisted by an integrative "lab" (or terrain) designed to accelerate idea translation and intellectual fecundity.[26]

It is precisely this normative dimension that Richard Florida and Elizabeth Currid have considered in their work.[27] For instance, Florida suggests that the energies of so-called "creative classes" in the regeneration of inner cities depend on the confluence of three factors: the availability of technology and talent, plus a tolerance for innovation and the innovators. Currid explains the roles of fashion, art, and music in upholding the vitality of New York City. The notion that place promotes creativity is substantiated by the case of academic "truth-spots." This term refers to the tendency to promote the findings of one's own laboratory, team, or location over others'. For example, the precepts of the "Chicago School" have dominated urban studies since their introduction in the mid-1920s. More recently, however, the hegemony of the Chicago School has been challenged by the "LA School of Urbanism," based in the accumulation of theoretical/empirical knowledge and a deliberative process of knowledge creation and transmission in that urban place.

Altered places

If, as I have argued, creativity is time- and space-specific, what happens during eras and geographies of exceptional social fluidity? Our own era is characterized by the rise of the Information Age: the emergence of network society, globalization, social polarization, hybridization, and crises of sustainability and governance (discussed more fully in the Afterword to this volume). What have been the impacts of such altered social conditions on creativity in the spaces of my research focus, the US–Mexico borderlands?

A few years ago, with co-curator Gustavo Leclerc I convened an international group of academics and artists to consider the future of the US–Mexico borderlands. This two-year collaborative venture involved a group of "idea translators" – multiethnic artists and multidisciplinary cultural scholars from both sides of the border – who met on several occasions in Tijuana and Los Angeles to produce an exhibition and book on what we termed the emerging "postborder condition." The project participants set out to explore how the conjugations of the new world order are being concretized in the borderlands and how border art represents those changes.[28]

The artworks in the exhibition, entitled *Mixed Feelings*, grappled with identity, space, and representation (including forms of memory, aspiration, hybridity, and resistance). The concurrent book, *Postborder City*, included critical essays that focused on the historical development of a regional consciousness in the borderland art and how the artists' groundedness was being challenged by the global present. The essayists were inevitably inflected by the artworks but they also prompted the artists toward a more expansive, theory-inflected visioning. Perhaps most significantly, the borderland became jointly construed as a place where mental and material cartographies are being recast as a new *local cosmopolitanism*. This term was not used in its customary sense, referring to identification with the fundamental interests of humanity as a whole, free from parochial or national affiliations. Rather, it referred to a congealing in place of many varieties of transnational experience. No longer universal and privileged, cosmopolitanisms are now plural and particular, occurring from the bottom up and grounded in the experiences of migration and diaspora in space. The borderlands' burgeoning cosmopolitanisms do not uphold traditional structures of power and knowledge; instead they daily reinvent unanticipated pathways for living and original means for personal and collective visioning whose significance is extending far beyond the region of their inception. In short, they are part of a process of inventing transborder identities and affiliations. In a world where immigration hysteria has granted new currency to nativism and racism, borderland art and society are revealing how creativity and place come together to imagine and build better ways of realizing our collective futures.

Tijuana-based Marcos Ramirez (aka ERRE) uses everyday construction materials to reveal the uneven development of the borderlands. His work is thick with place and echoes of his many years of working in the construction industry on both sides of the border.

> All my life I have felt that I live in a space where two worlds overlap, clash, touch, attract and repel each other. The mutual fascination they awaken can only be contained by the fear they provoke in each other. However, united by this wound like the edges of a single scar, my worlds of numbed, calloused, aching skin still support, seduce and penetrate each other.[29]

In *Walls/Muros*, ERRE's installation consists of two walls back-to-back, one made of wood, the other of brick. On each wall is mounted a video screen that portrays house construction on either side of the border. The dual images drive home the differing perceptions of work and community on either side of the border. On the US side, the construction is largely mechanized; on the Mexican side, workers use traditional labor-intensive, trowel-and-mortar methods. In Mexico, the sound of work is the scrape of trowel against brick, a quiet tapping to ensure a brick is in place. In the USA, the mostly Mexican workers use loud powertools at breakneck speed.

Their soundtrack is the repetitive bang of nails being shot into wood framing, and the constant scream of electric saws. At mealtimes, workers in Mexico erect a makeshift grill, then collectively cook and share food brought from home. In the USA, workers purchase individual prepared meals from itinerant truck vendors. ERRE's presentation emphasizes that it is the Mexican worker who is being emasculated by development's embrace on the US side of the border.

What the River Gave to Me is part of a continuing series on "Spiritual Geographies" begun in 1998 by California-based Chicana artist Amalia Mesa-Bains, who explains:

> The landscape in my work is a revelation of my own family history, the human geography of my ancestors and my own gendered life experience and the allegory of woman and nature. In the deepest sense the violation of the land is a scar on Mother Nature and the evidence of our displacement that can only be healed by a politicizing spirituality.

Portraying a canyon etched into a forbidding landscape, Mesa-Bains explores the natural markers of division in the terrain and the spirits residing therein alongside the subversions of forced separation due to human-induced boundaries. The arid landscape evokes the hard journeys of migrants. Yet the terrain is dramatically bisected by the life-giving flow of a river, brilliantly illuminated and studded with multicolored glass boulders bearing the names of migrants past and present, and etched with names of their origins and destinations. The river's flow transcends natural and human barriers, bearing its glittering treasures to a place where dreams and spirit may be fulfilled, while at the same time allowing the travelers to recall their past identities:

> The river in this geography is the natural demarcation of separation and simultaneously the element of transformation, the flow of memory and the space of crossing and conflict.

Issues of identity and hybridity come to the foreground in the serio-comic works by Einar and Jamex de la Torre, who are based in San Diego, California, and Ensenada, Baja California. In a deliberate assault on a modernist aesthetic, the brothers reveal a sensibility akin to a cluttered suburban lawn, or when a magnet has been dragged through the streets of Tijuana and Los Angeles and its detritus dumped for our delectation. Their vivid blown and molded glass creations combine elements of popular culture from both sides of the border and infuse them with a strong dose of pre-Columbian iconography.

In *Colonial Atmosphere*, the de la Torres created a massive lunar module in the style of an Olmec head, accompanied by an astronaut who resembles a Mesoamerican *tule* (statue figure) with a spacesuit helmet atop its head and life-support system on its back. The colossal prehistoric Olmec heads discovered in Veracruz on the Gulf of Mexico have African facial features that the artists transcribed onto the face of the lunar module, deliberately confounding the question of origins. A flood of indigenous, contemporary, and extraterrestrial references and representations clutter the lunar module's interior, mocking our need to know. The brothers offer a version of their own origins:

> We're very fortunate to live and work on both sides of the border, a circumstance that has fed our work over the years with a continuing flip-flop of insider-outsider perspectives that have compelled us to become additive in our work. ... For us, the border is a rather organic entity. Even though one thinks differently on either side, it is often difficult to know where the line is.

Creative places, people, and products

Artists are visionaries better able than most to imagine possible futures.

ERRE's walls mimic the enormous fences dividing the USA from Mexico. They reveal in precise detail how everyday lives are slashed through by the disruptions of the boundary line. Yet the lives of ERRE's workers are pasted back to back on the walls, suggesting an intimacy and an inevitable merging across the international border.

Amalia Mesa-Bains is transfixed by the spirituality and physicality of landscape. Each glass boulder/migrant she portrays possesses a vital individuality and inner beauty; together they create a cascading flow of riches. Her river metaphor links the migrants' past (origins), present (the journey), and future (unknown, but guided by the candles that spread out before the migrants' flow like a delta of hope).

The de la Torre brothers invoke the confusions of cosmopolitanism, overflowing with deep histories and daily detritus, and created by extraterrestrials who have landed among us from places unknown. These transgressive transborder "aliens" mock our obsession with origins and authenticity, pointing out the inevitability of a postborder world.

Scholar critics are a vital component part of the artistic mode of production in the creative places of the borderlands. They can explain why the future is taking place, even though no-one knows the shape of that future. Artists and scholar critics are both idea translators; that is, their works can lead from concept to altered practices. For better or worse, their actions are implicated in the creation of a *mestizaje* world.

Notes

1 P. Sellars, quoted in B. Isenberg, *State of the Arts*, New York, NY: Morrow, 2002, p. 151.
2 A. Mesa-Bains, quoted in Isenberg, *op. cit.*, p. 152.
3 C. See, quoted in Isenberg, *op. cit.*, p. 35.
4 The relationship between "regional groundedness" and "regional imagination" is examined in depth in M. Dear and G. Leclerc (eds), *Postborder City: Cultural Spaces of Bajalta California*, New York, NY: Routledge, 2003.

5 C. E. Schorske, *Fin-de-Siècle Vienna: Politics and Culture*, New York, NY: Alfred Knopf, 1979, p. xxvii.
6 R. C. Smith, *Utopia and Dissent: Art, Poetry, and Politics in California*, Berkeley, CA: University of California Press, 1995, p. 5.
7 S. Williams and P. M. Rutkoff, *New York Modern: The Arts and the City*, Baltimore, MD: The Johns Hopkins University Press, 1999, p. 18.
8 Williams and Rutkoff, *op. cit.*, p. 50.
9 A. Nafisi, *Reading Lolita in Tehran*, New York, NY: Random House, 2004, p. 43. Also see Peter Brook's wonderful account of the reception accorded his production of *King Lear* in various countries: P. Brook, *The Empty Space*, New York, NY: Atheneum, 1968, Ch. 1.
10 S. Zukin, "How to Create a Culture Capital: Reflections on Urban Markets and Places," in Blazwick, I. (ed.), *Century City: Art and Culture in the Modern Metropolis*, London: Tate Gallery, 2001, p. 260.
11 H. Molotch, 'L.A. as Design Product: how art works in a regional economy,' in A. J. Scott and E. W. Soja, (eds), *The City: Los Angeles and Urban Theory at the End of the Twentieth Century*, Berkeley, CA: University of California Press, 1996, pp. 225–75.
12 D. B. Fuller and D. Salvioni, *Art/Women/California: Parallels and Intersections, 1950–2000*, Berkeley, CA: University of California Press, and San Jose Museum of Art, 2002.
13 Important exceptions to this rule include: G. Bruno, *Streetwalking on a Ruined Map*, Princeton, NJ: Princeton University Press, 1993; T. Conley, *Cartographic Cinema*, Minneapolis: University of Minnesota Press, 2007; and R. Deutsche, *Evictions: Art and Spatial Politics*, Cambridge, MA: MIT Press, 1997.
14 M. Orvell, *American Photography*, Oxford: Oxford University Press, 2003, p. 15.
15 D. B. Clarke, (ed.), *The Cinematic City*, New York, NY: Routledge, 1997, p. 7.
16 This formulation is adapted from the theory of filmspace outlined in M. Dear, *The Postmodern Urban Condition*, Ch. 9.
17 A. Damasio, *Looking for Spinoza: Joy, Sorrow and the Feeling Brain*, New York, NY: Harcourt, 2003, p. 6.
18 Damasio, *op. cit.*, p. 30.
19 B. M. Stafford, *Echo Objects: the cognitive work of images*, Chicago, IL: University of Chicago Press, 2007, p. 202.
20 Stafford, *op. cit.*, p. 1.
21 Stafford, *op. cit.*, p. 2.
22 Stafford, *op. cit.*, p. 105.
23 Stafford, *op. cit.*, p. 2.
24 Stafford, *op. cit.*, p. 6.
25 D. Edwards, *Artscience: Creativity in the Post-Google Generation*, Cambridge, MA: Harvard University Press, 2008, pp. 6–7.
26 Edwards, *op. cit.*, Ch. 7.
27 E. Currid, *The Warhol Economy: How Fashion, Art and Music Drive New York City*, Princeton, NJ: Princeton University Press, 2007; and R. Florida, *The Rise of the Creative Classes*, New York, NY: Basic Books, 2002.

28 The term "Bajalta" is an amalgamation of the Spanish words *baja* (meaning lower) and *alta* (or upper), which were used by the Spanish to describe two provinces of New Spain, Baja California and Alta California. See Dear and Leclerc, *Postborder City*, *op. cit.*, pp. 1–3.

29 The quotations in the following paragraphs are taken from the catalog of the Mixed Feelings show: *Mixed Feelings: Art and Culture in the Postborder Metropolis/ Sentimientos Contradictorios: arte y cultura en la metropolis posfronteriza*, co-curated by M. Dear and G. Leclerc at the Fisher Gallery of the University of Southern California, 2002.

Figure 1.3 Colonial Atmosphere. Source: Einar and Jamex de la Torre. Used by permission.

2

Experimental geography
An interview with Trevor Paglen, Oakland, CA, February 17, 2009

Michael Dear

Michael Dear: Can you tell me about the two high-profile projects you were involved with recently that got a lot of publicity: the tracking of the rendition flights, and the shoulder patches of quasi-secret organizations? What were the projects about, and how did you get involved with them?

Trevor Paglen: Those projects spun out of my doctoral work on state secrecy. Over the course of doing my dissertation work, I gathered a bunch of files on what turned out to be the rendition program, although it was unclear at the time what exactly was going on. I was noticing some weird things about airplanes and connections that didn't make a lot of sense, but I started a bunch of files on it and kept adding to them over time. Over the years it became clear that I was looking at the infrastructure behind the CIA's rendition program. I realized that this was a big deal, that I knew a lot about a covert CIA program that few people knew about at the time. I decided to look into it a little more seriously, so I met with A. C. Thompson, a friend of mine who was an investigative journalist, and we wrote an article which turned into the cover story for a number of weeklies across the country. After seeing that article, a publisher approached us about writing a book about the rendition program. We hemmed and hawed for quite a while before deciding to do it. At the end of the day, it came down to the fact that I felt responsible for the research – if the public had put me in a position to do this kind of research, then maybe I owed it to the public to publish it. That's how we decided to do the *Torture Taxi* book.[1] Then I sat down for a year and a half and worked like crazy. It was an intense process because I was doing the field work and the writing at the same time.

The patches project also came out of the fieldwork for my dissertation. Because I have a background in art, I pay attention to metaphors, metonyms, images, and so on. I watch out for the "a-ha" moments in my research, moments where details speak to the whole, becoming something like non-fiction metaphor or allegory. While conducting interviews for my dissertation, I spent a lot of time talking to people working on secret projects. I realized that there was a whole visual culture surrounding secrecy that was generated by people who worked on "black" projects, usually in the form of patches and memorabilia, coins, and souvenirs. I thought they were very interesting

Figure 2.1 Trevor Paglen. The Salt Pit, Northeast of Kabul, Afghanistan (2006). C-Print, 24" x 36." Altman-Siegel Gallery, San Francisco.

as cultural artifacts. Here was a whole visual language developed to speak about things that aren't supposed to be spoken about. It turns out this is a very old theme in art, especially mystical art: "How do you represent something that can't be represented (usually something like a god)?" To me, the symbols these people developed posed a really interesting set of questions about identity, epistemology, and the formation of public cultures. At the same time, of course, these images offered a very rare peek into the culture of classified military and intelligence programs.

The book was entitled: *I Could Tell You But Then You Would Have To Be Destroyed By Me.*[2] To get the thing published I had to practically beg on my hands and knees. The publisher gave me no advance to speak of and did the book as a kind of favor to me. We expected to sell 500 copies, mostly to friends of mine. But it turned out that we got the timing right, and it turned out that lot of people were interested in this topic. It was a moment in American culture where people were getting fed up with secrecy, the Bush administration, and the terror hype. These images were a way to make fun of it and not be afraid of it. I think the book was pretty successful in doing this.

MD: Who was the publisher?

TP: Melville House, it was the same people who did *Torture Taxi* and the more recent book *Experimental Geography*.[3]

MD: Was what you did in the renditions and patches projects "experimental geography"?

TP: Experimental geography is a name that I made up a while ago. Nato Thompson [the curator of an exhibition entitled "Experimental Geography" and editor of the

Figure 2.2 Classified flight test. This was the original version of a patch commemorating a flight test series involving a B-2 "Spirit" stealth bomber. Air Force officials insisted that "Classified Flight Test" not appear on the design; it was replaced by the words "To Serve Man," a reference to an episode of the classic TV series, *The Twilight Zone*. The dog-Latin phrase *Gustatus similis pullus* translates as "Tastes like chicken." Galerie Thomas Zander, Cologne.

accompanying book] and I have a long history of collaborating. He works as a curator at Creative Time in New York and does a lot of public art. Nato and I were roommates as undergraduates at Berkeley, and after we graduated we ended up opening and running several alternative art spaces, then going to graduate school together. We basically grew up together. When we were 18 or 19 years old, we found out about the Situationists and it made a huge impression on our thinking. In a nutshell, experimental geography is a set of guiding ideas about the intersections of spatial practice and cultural production that comes from 15 years of conversations between Nato, myself, and some other friends of ours. It takes the Situationists as a starting point. What both of us were (and are) generally interested in is what happens if, instead of using art theory, or literary theory, or the semiotic tradition to think about art and images – the kind of approaches typical in the humanities – you instead used geographic theory to think about the means of creating art and a radical culture. Geography gives you a different theory of cultural production than the more traditional literary models that most art students learn.

Experimental geography involves thinking about cultural production as the production of space; moreover, it means being self-reflexive about the kind of spaces we produce through our own work as geographers or artists. What exactly are the relations of production that go into our work, and how can we create new spaces through our own cultural practices? Experimental geography is about being hypersensitive to politics, self-reflective about our own practices as geographers or as artists, and about recognizing the fact that the production of culture is the production of space. It's a relatively obvious set of propositions, in my opinion, but they can have a big effect on how we think about our own work if we truly take them seriously.

MD: Besides the Situationists, what are some of the other sources or inspirations for your work?

TP: Oh, punk rock! It was really important. I grew up in the second generation of punk rock, not with the Sex Pistols. I came up around bands like Fugazi and zines like *Maximum Rock N Roll*. When I was a part of the punk scene, the definition of punk had moved away from simply signifying a style of music; it had become much more about the kind of economic model that culture was created under. Let me give you an example. I used to work at a club in Berkeley called 924 Gilman Street. It was a punk rock club that was all volunteer-run by the kids. The idea was that kids would run the club for other kids; that was our first principle. The second thing was that to play there you could not record for a major label. So, if a band had put out a record with one of the "big five" labels like Warner Bros. or EMI, they could not play at the club. The only people who could play there were those who recorded for independent record labels. The band Green Day, which is really famous now, used to play there all the time. They used to record for a label in Berkeley but when they started to record for a major record label overnight they ceased to be a punk rock band for the purposes of our club, even though they played the exact same music! The question of whether something was "punk" or not wasn't about the music being played, but was about the economic relationships or relations of production that went into creating the music. In retrospect, that understanding of the politics of culture had a huge influence on me. It struck me as a really sophisticated way of thinking about culture because it trained me – pretty literally – to see the political economy of culture.

MD: What about the commitment to public discourse in experimental geography?

TP: It all starts with my overriding concern with questions of form, politics, and the spaces of cultural production. When, as scholars and researchers, our research touches on things in the public interest, maybe we should be responsible for communicating what we know to the general public. I think that many physical scientists are quite good at this, for example Carl Sagan. In the department of geography at Berkeley, physical geographers like Kurt Cuffey have stepped up to address global warming in public because he realizes that research doesn't happen in a political vacuum. I think it's a shame that so many of us are actually taught not to engage in the public sphere. This has bad implications for our culture as a whole, because people who know the most about national problems, and have something to contribute to the search for solutions, haven't been encouraged to make it a part of their practice. Even if they have, they haven't received professional credit for it. That's a real disincentive for scholars to develop a public practice side to their work. In the long term, this is not good for the future of society.

Another dimension of engagement is not only about being responsible to the public for the work that you do, but also about the kind of space(s) your work produces and the politics of those spaces. Frankly, we don't spend enough time worrying about opening up spaces of communication. And we have to learn how to communicate beyond professional circles! Of course, some ideas are big and complicated and difficult to explain, but too many times we let our language and jargon get in the way. Some of my physical scientist friends complain about social scientists and cultural geographers speaking a language that doesn't make sense. Some of them are scientists dealing with large, difficult ideas, but they somehow manage to find a way to explain things that the average person can grasp. If I can't explain something to my dad then I feel I'm not trying hard enough!

MD: A lot of people in America don't want to hear the things that academics, scholars, and artists are saying. Isn't there a long history of anti-intellectualism in this country?

TP: That's true, but you have to be prepared to meet people half way. Look at someone like Slavoj Zizek, for example. His ideas are about as complicated as any in the social sciences. Yet he has an enormous audience. There are lines around the block when he gives a lecture! Here's another example: Thomas Friedman, author and columnist for *The New York Times*. A lot of people in our field like to beat up on Tom, and maybe for good reasons. But I think we can do better: If we write academic papers that are only intelligible to specialists, then give them – for free – to multibillion dollar corporations like Wiley whose business model is about restricting access to our work, then we can't complain when Tom Friedman's ideas get a lot more public traction than our own. My point here is that form, content, the production of space, the political economies of ideas, and public discourse are all interrelated. I think scholars take for granted way too many of the formal and economic conventions of academia. A lot of them are completely outdated in my opinion.

MD: Let's go back to the problem of method, which you began talking about earlier in connection with the rendition flights and patches projects. How would you describe the general approach to method in experimental geography?

TP: Well, first of all, I want to point out that I'm not at all interested in defining "experimental geography" too tightly. There's no "school" or consistent approach – it's more about thinking experimentally and creatively about the forms we use in a very broad sense. On the question of method, I've come to really appreciate Allan Pred's wisdom on this, even though it drove me crazy when I was first starting to work with him. Allan's approach was something like "forget methodology; listen to the materials you're working with and they'll suggest their own ways of being researched." He understood in a deep way what we've learned over and over from Nietzsche, Foucault, Bourdieu, and many others: the way we posit questions strongly suggests the answers we come up with. A lot of us understand this theoretically, but putting it into practice is an altogether different set of challenges. It takes a lot of patience and trust to go into a new research project with this kind of attitude, but sometimes it can lead to some really creative, innovative, and insightful work. Shiloh Krupar is particularly good at this.

For me, Allan Pred was such an important geographer because his whole project was a meta-theoretical critique, an embodied critique. He was a remarkably talented theorist, but I was personally more in tune with the sense of practice in his work – which is exactly what we've been talking about. I couldn't even be having this conversation without acknowledging his influence on me.

MD: Who else is working in the field of experimental geography?

TP: A lot of different kinds of people in the arts have intersected with us, such as Lize Mogel or Ashley Hunt, the folks doing radical cartography. There's a whole group of people that I've had these sorts of conversations with; for example, Daniel Tucker from *Area Magazine* in Chicago, and Emily Forman out of New York. Curator Nato Thompson, of course, is my longtime friend, collaborator, and interlocutor when it comes to this material. Closer to home, Shiloh Krupar and I have been good friends and co-conspirators for a long time. She's a genius. There seem to be more and more people experimenting in these ways: Hayden Lorimer, Michael Gallagher, David Pinder, Colin Flint, Stephen Graham, Nick Brown, Jake Kosek. I think a lot of geographers have really started to experiment with their own practices. It's great. I also do a lot of work as a visiting artist in art departments and art schools. There is a tremendous interest in geography in the arts right now – I think a lot of artists see in geography the same thing that got me initially interested in the field: geography provides a far more robust theoretical framework for understanding landscape than more conventional art traditions. As "landscape" in art has moved far outside the frames of painting and photography, a lot of artists are turning towards geography for methodological and analytic inspiration. I really hope that more geography departments become open to the idea of having MFAs (Master of Fine Arts degree) come in to do PhD work. There's just a huge amount of interest in our field from artists, and I think geographic ideas could really become influential in the arts. Obviously, I also think geographers have a lot to learn from artists.

MD: What's next in experimental geography? What are your new projects?

TP: What's next in experimental geography? Who knows? Hopefully something we haven't thought of yet! For myself, I'm doing a project examining the refuse from secret plane crashes and other strange things that have fallen from the sky. The first part of the project is a close look at the materiality of the shoot-down of an Iran-Contra aircraft over Nicaragua in 1986. There are several projects along those lines. I'm also continuing my work tracking and trying to understand classified satellites. That work is really interesting – I feel like I'm just beginning to understand it even though I've been working on it for years. Another project involves working with a team of Russian aerospace engineers to design an "art" satellite. There are more projects than I know what to do with!

MD: How do you see these projects working out for you professionally? Doesn't anyone committed to the experimental risk becoming marginalized from established academic communities and educational institutions?

TP: This is an issue that has always been in the back of my mind and is now getting played out in real-time. In the last year or so, I completed my PhD. Now I'm grappling with whether there's a place for me in existing geography institutions. Of course I think the ideas I'm working with have a place in the discipline, but it's looking unlikely that I'll end up in a geography department. The structure of the university and disciplinary boundaries is still fairly rigid and unavoidable, and I've come to learn that hiring decisions are often compromises among people with very different interests. That's fine and fair, but somehow I don't seem to be the compromise candidate very often, haha! All the things I'm interested in – geography, art, and visual studies – require a kind of transdisciplinary practice that's not easily accommodated in a conventional institution. To me, the future of this kind of work partly depends on trying to figure out ways to support and sustain formally and methodologically experimental scholarly and creative practices. It's a big challenge. I'm confident experimental geography will continue to make inroads and open up new territory – the media landscape we're living in almost demands it.

MD: This is one of those situations where the academy has some catching up to do, because your ideas are gaining a lot of public and national media attention. Experimental geography is certainly catching on, as a collaborative intellectual enterprise that spans art, performance, academics, and politics. It's also fun! Your leadership has been very important in this adventure. Thank you for sharing with us your personal recollections of this movement.

TP: This has been a good conversation for me. I wrote a lot about these issues in the *Experimental Geography* book, but this part of the project is definitely something I haven't previously talked about a lot. I've thought about writing a book about it, but somehow the idea of using a relatively standard form like a university-press book contradicts the whole spirit I'm trying to advocate for!

See www.paglen.com/gallery for more of Trevor Paglen's work.

Notes

1   T. Paglen and A. C.Thompson, *Torture Taxi: On the Trail of the CIA's Rendition Flights*, New York, NY: Melville House, 2006.
2   T. Paglen, *I Could Tell You But Then You Would Have To Be Destroyed By Me: Emblems From the Pentagon's Black World*, New York, NY: Melville House, 2008.
3   N. Thompson, and Independent Curators International, *Experimental Geography; Radical Approaches to Landscape, Cartography and Urbanism*, New York, NY: Melville House, 2008.

3

Drive-by Tijuana

René Peralta

I am sitting at the moment in a middle-class suburb in what is now defined as the geographic center of the city of Tijuana. The center where goods and services are produced and the location of political hegemony. I am relatively close to the border where thousands wait hours to cross to the USA. I am a short drive away from where the Pacific Ocean patiently awaits for a gesture of acknowledgment from the city, and always jealous of Tijuana's gaze toward the north. And to the east a New Tijuana is being built and is home to most of the city's assembly plant workers, an urban edge that grows two hectares daily and is rapidly becoming post-Tijuana. I sit in my home located behind the great Tijuana Racetrack, the infrastructure that gave this city its *raison d'être* in 1915 and its legend as the city of sin, gambling, and prohibited recreation originating from the lust, thirst, and money of San Diego, creating (as Richard Rodriguez once said) a city of world class irony.

Like many cities today, Tijuana is uncertain of its future. The city is being reconfigured and rethought constantly by globalization, economic disparity, and by the struggle of a society to cope with the myth of illegality, which has now become a ruthless reality. It is said that to name something is to understand it, and Tijuana has been subjugated to countless baptisms, such as Tijuana Hybrid, Tijuana Third Nation, Tijuana Cultural Mecca, Tijuana Post Border, and so on. Yet *Tijuana is not Tijuana*, because if all truths are fictitious, or partial constructions, Tijuana evidences that she has thousands of relative truths.[1] The city has a mimetic tendency to present itself, at least for a moment, as similar but not quite the same as Banham's LA, or E. B. White's New York, or any other urban image it sees fit to camouflage itself as. Mimesis as an adaptive behavior that (paradoxically), as Michael Taussig explains, creates difference by making oneself similar to something else. A difference that constructs a world of illusion and an intangible product – in this case, the city of Tijuana itself. Tijuana's search for *otherness* is a type of defense mechanism, a means of survival that allows for the creation of a fictional world. As Rafa Saveddra, writer, DJ, and Tijuanologist, explains: "Build your own idea of the city ... TJ is too real to be a simulacrum, too artificial to be a legitimizing act. Tijuana is the chip and the software to recreate, feign and sell our own voices."[2]

This essay is a rough ride from the coast to the emerging new city-within-a-city to the east, a sampling of histories, geographies, and ideas from one of the most dynamic and volatile border cities of North America. How does this 118-year-old city deal with culture, identity, and future? Tijuana is a city always in search of a past in order to project itself into the future. A city always deceiving, always contrasting, a heterotopia in re-mix. Through this journey, it becomes evident that the city evades all definition. Driving through these ideas, we realize that the city had three major identities: Sin city – Tijuana as Old Mexico, a version of Tijuana as a Mexico imagined by others (USA); Tijuana as a cosmopolitan and modern city that for a short time created its own utopias; then, at the end of the twentieth century, Tijuana became known as the laboratory of postmodernism, a poster child of urban informality and hybridity. Our journey will encompass many places that are important in understanding the urban landscapes, histories, and cultural practices that together form the multiple imaginaries of the city.

Tijuana always gazes northward. Its urban growth has followed two main axes: the Tijuana River that flows east to west, and the US–Mexico international border to the north. The urban fabric of the city thrusts against the corroded metal of the international boundary fence, and many homes have been built right up to it, using it as a backyard fence. The Pacific Ocean had always been a secluded "vacation" space for the citizens of the city, requiring a trek across the deep canyons that separate the Tijuana River Valley from the ocean. It was not until 1959 that a few land owners and the state government began the impulse for urban development toward the west side of the city. During this time a second bullring was built adjacent to the border fence where it now plunges into the ocean. *The Plaza Monumental de Las Playas de Tijuana*, a massive concrete structure, was erected in the most northwestern piece of property in the city, often known as the last corner of Latin America. The bullring brought tourism and finally a new middle-class residential suburb, creating the only coastal community of the city with approximately 100,000 residents. The beach community began to build a coastal *malecon* along the water's edge, with restaurants, shops, residential buildings, and a pedestrian strip that gave immediate access to the beach. Yet in the late 1970s the ocean engulfed and destroyed a complete city block including roads and buildings, creating a sense of uncertainty for developers and government considering rebuilding. Since then the coast has been untamed, and awaits new strategies to re-imagine that brief moment when Tijuana made peace with nature's water edge. Other factors have aided to the incredible idea of rebuilding the *malecon*, especially ecological problems that are a consequence of informal settlements built in the canyons to the east which dump contaminating sediments into the Tijuana Estuary and shorelines of Imperial Beach (in San Diego) and Playas de Tijuana. This ecological crisis may be the only real bi-national issue that could foster a collaborative solution to one of many shared natural geographies of both countries.

As development moved inland the ocean-side community remained economically active and became home to prominent citizens and institutions such as the Universidad Iberoamericana, one of the most prestigious private universities in the country. Shopping centers, movie theaters, and large parks made this coastal a place for well-to-do citizens to enjoy relative calm and peace from the bustling downtown of Tijuana, just a few minutes to the east. *Playas de Tijuana*, as it is known to the inhabitants, for a brief time had exclusive gated communities that were home to entertainment stars like the famous Mexican composer Juan Gabriel, and Tijuana's home-grown rock star Julieta Venegas. Rumor has it that the main lieutenants of the infamous Arellano

Figure 3.1 Playas de Tijuana.
Source: Photograph used by permission of Elias Sanz.

Drug Cartel also made it their home. In 1994 the US government implemented a
fierce anti-immigration program known as Operation Gatekeeper, which fenced off
traditional cross-border migration routes. The new boundary line was built with left-
over materials from the Desert Storm operation in the Persian Gulf. At its end-point,
the metal fence plunged into the Pacific Ocean in a gesture as sublime as land art, and
as terrifying as the Berlin Wall. Playas de Tijuana became home to the symbol of US
policy on immigration, and target of pro-migrant groups. Operation Gatekeeper
pushed illegal immigration eastward to more arid and mountainous areas; it has been
held responsible for a 500 percent increase in migrant deaths. The whole community
of Playas has now been reduced to a single line, *la línea*, a symbol of failed bi-national
immigration reform, and a place for activists, artists, and organizations on both sides
of the debate to promote their views.

To the east of Playas de Tijuana there is a whole different and defiant world within
the city known as the Laureles Canyon. Laureles is part of the 33-canyon system in the
Tijuana River watershed and the archetype of informal settlement on the steep hill-
sides common throughout the city. In contrast to Playas, the Laureles Canyon community is made up of low-income units of illegal origin inhabited by low-wage
workers employed by local manufacturing plants (*maquiladoras*). According to the
Tijuana Municipal Planning Institute there are approximately 80,000 people living
in the canyon. The predominant housing type consists of self-built homes constructed
with a variety of recycled materials that come from the USA. Most of these homes are

built on steep slopes, using old tires as retaining systems, planters, and staircases that together provide access to the many canyon-side homes. Most of the settlers work in formal economies, such as manufacturing plants, so they are entitled to government services. Yet since wages in the *maquiladoras* are very low, many residents participate in informal economies such as home-based child care, food preparation, and many other business that serve the immediate community. The accelerated population growth of Tijuana (6 percent per year) and the demand for almost 70,000 new low-income homes every year have encouraged developments that lack sewage and street paving, built primarily in high-risk natural landscapes. The general lack of government intervention and oversight has brought about a series of housing development types ranging from squatter settlements, private-developer housing, and even government-relocated communities from other high-risk areas of the city.

Laureles Canyon is adjacent to the international border and the Tijuana River Estuary which is located in San Diego County and is a national ecological reserve. The main environmental issues are the excessive sediment deposited from the canyon across the border, and the contaminants that end up in the ocean and cause an almost year-round closure of beaches. Yet in this part of the city there is hope for bi-national cooperation, and a dream about a larger region. It is by jointly re-imagining Laureles Canyon and the Tijuana River Valley with our neighbor to the north that a truly "post-border" future can emerge. This idea was best expressed by Kevin Lynch and Donald Appleyard in their report to the San Diego City Planning Department in 1974:

Figure 3.2 Laureles Canyon.

Source: Photograph used by permission of Andrew Malick.

San Diego/Tijuana could be the center of a large international region, a vital meeting point of two living cultures. The metropolis would share its water, its energy, its landscape, its culture, its economy. The border would be converted into a zone of confluence.[3]

Laureles Canyon and its community represent a new way to look at development in Tijuana as well as the relationship across borders. In planning policies and bi-national cooperation, there must be a re-thinking, changing a region of contested geography to a shared zone of natural diversity and cultural interchange.

West from the Laureles Canyon is the oldest part of Tijuana, El Centro. In 1889 Mexico City engineer Ricardo Orozco was hired to give order to the various ranches that had been established since the Mexican–American war of 1846–48. Orozco was trained at the famous Academy of San Carlos in Mexico City, and his education was influenced by French Positivism and urban ideas of Garden City planning from the USA. His vision for a new border city broke from older colonial plans and negated the generic homogeneous block plans of neighboring San Diego. Orozco created the Zaragosa Plan on the edge of the border and next to the Tijuana River Valley. The plan included a series of diagonal boulevards that connected parks and public spaces, as well as a diverse set of blocks and lots for residential and commercial functions. The positivist ideals in the new plan for Tijuana marked the beginning of the myth of the city, and the birth of an awareness of *otherness*, not totally Mexican but not entirely American either. Antonio Padilla, Tijuana urban historian mentioned:

> one of the prime relations between the map and the ideals of positivism influenced by Agusto Comte, was the rejection at the outset of a return to a historical tradition typified by the Hispano-Colonial model conformed by a grid with a center as the seat of religious and political power. The plan of Tijuana is part of a rational and philosophical order based on man's liberty and not only subjected to rational logic.[4]

The Zaragosa plan was an ideal, a utopian endeavor of Orozco and his predecessors. Yet as soon as the plan was laid out and implemented, it went through radical changes. Topography interrupted the philosophical order and changed its form, while the desire for greater profits became the stimulus for illegality. Violent confrontations arose in places where the street diagonals touched a parcel. Landowners began to transgress the axial paths by building into them in order to obtain a greater amount of land. By 1921, the diagonal boulevards had become a crippled dream of order and control, a failed plan to achieve Cartesian logic.

After this transgression of order, Tijuana reincarnated into a more Americanized urban space due to the influence of two major entertainment infrastructures located in the Tijuana River Valley: the Tijuana Racetrack, and the Agua Caliente Casino. In 1928 American entrepreneurs, trying to make a profit by turning Tijuana into an early Las Vegas, founded the Agua Caliente Casino. They employed Wayne McAllister for the design of the building, an 18-year-old San Diegan draftsman who later became an important designer for the major casinos in Las Vegas and Havana. Casino-pampered

Hollywood celebrities such as Buster Keaton and Rita Hayworth won racetrack jackpots in the thousands of dollars, and encouraged the opening of bars and hotels in downtown Tijuana's infamous Avenida Revolución. The casino was such a success that the US government tried to stop citizens from enjoying themselves by closing the border at 9 p.m. every night, which only helped the downtown hotels as more and more Americans stayed overnight. Even during the Depression the casinos and the commercial strips of downtown Tijuana flourished, but all of this came to an end in 1939, when by Mexican presidential decree gambling was prohibited in Mexico. The Casino was converted into a school.

Even after the Casino moved its operation to Las Vegas, downtown TJ had enough night life to continue flourishing. Its economic base was entertainment. Big bands, piano bars, cocktail lounges, and affordable booze catered 24 hours a day to visitors from the north – and "sin city" was born. Musicians such as Jelly Roll Morton, Jack Johnson, and Nat King Cole, and literary figures such as Ernest Hemingway gave a fresh, jazzy flavor to the city. Tijuana became a hotbed for local and foreign jazz musicians. Many local residents benefited economically from the music and entertainment scene. During this period many west-coast jazz players would come south of the border to party, play, or simply in search of inspiration. The jazz bass player Charles Mingus dedicated an album to the Latin sound of Tijuana with his "Tijuana Moods" LP, inspired by the vibe of the city. Other musicians either came to Tijuana to experience the music scene or were inspired to recreate the "Tijuana sound" in their compositions; such was the case of the 1965 album "Tijuana Jazz" by Gary McFarland and Clark Terry. Legendary musicians such as Art Pepper and the Miles Davis Quintet were sometimes seen enjoying their nights south of the border, taking in the bar scene while local musicians played their night shifts. Between 1962 and 1968, LA trumpet player Herb Alpert won six Grammies with his famous Tijuana Brass, inspired by bullfights, curio shops, and the mariachi bands of Plaza Santa Cecilia (a city square named after the patron saint of musicians). A non-stop kaleidoscope of music, cabaret, and inebriation kept the local economy affluent, the city alive, and its musicians working. Today, downtown Tijuana has lost some of its luster and has become a center of strip clubs and cheap bars catering to tourists. With the decline of work beginning in the late 60s, many bands and orchestras began to disintegrate, and musicians opted to work in piano bars or bands of a different genre in order to make a living.

In truth, El Centro had already begun to suffer after the decline of the postwar tourist crowd and due to large-scale developments in the eastern parts of the city. Like many cities in the USA, the postwar development of the suburbs became the dominant urban model. The original inhabitants of downtown began their journey eastward to residential communities based on the American suburban dream. Later, the federal government would begin construction of a new economic and commercial center in the former Tijuana River basin that would forever change the structure of the El Centro. Today, El Centro is in precarious state because tourism has declined due to new border security after September 11, and because of the never-ending roll-out of residential communities being built by government subsidies in the east. The creation of a new tourism corridor along the coastline, connecting TJ with the cities of Rosarito and Ensenada, has shifted economic development and attracted more affluent visitors interested in purchasing beach front property at discount prices away from the central areas. Along Avenida Revolución the economic downturn is evident in the many "for lease" signs along what used to be the most important commercial strip in the city. Curio shops that would sell leather and velvet paintings can no longer sustain themselves, restaurants are closing, and bars are desolate. Buildings all around are empty and obsolescent. Yet, there remains a sense of optimism that the space where many myths of the city were forged will eventually be re-imagined. Land throughout the city is scarce and private developers are suffering losses due to the increasing costs of bringing infrastructure to new projects as the city expands east. Downtown is still the prime location near the border and the economic heart of the city, as well as being equipped with the infrastructure to sustain the new vertical housing prototypes. This is where the *fronterizo* begins, a realization of a "third consciousness" that aimed to forge an identity based on desires and forbidden dreams. El Centro is where the most authentic cultural production occurred. A place that inspired locals and visitors alike, from Charles Bukowski to Charles Mingus; a place that continues to kindle a desire to understand new ideas. As Mike Davis recalled:

> My first exposure ... to the immense heritage of European Marxism and Critical Theory was at the old El Día bookstore off Revolution Avenue. It was in Tijuana that I first began to appreciate the impact of the Cuban Revolution and was first able to see the U.S. civil rights struggle in larger perspective. Tijuana for us was a little bit of Paris, our personal Left Bank, and my fondness for the city and the cultural freedom it represents has never waned.[5]

Today Tijuana has a new, transplanted urban center and a diametrically distinct peripheral *New Tijuana* to the east. The Zona Río (River Zone) is where the Tijuana River once flowed and the place where the first two important leisure infrastructures, the Racetrack and Casino, were built, initiating an economic boom for Downtown. The river had always been a force to contend with; its natural force once destroyed the first racetrack and many bridges. It became an area where informal settlements, for the first time, developed into a radical social force. Since the 1940s squatters along the river had defied the authorities and government-enforced relocations never worked. Up to the 1970s these settlements were known as *Cartolandia* ('Carton-land') for the material employed to build the small housing shacks. After a severe flood, *Cartolandia* lost the struggle for survival, and its inhabitants were relocated to make way to a new vision for the city. A new urban plan included tree-lined boulevards, a cultural center, and a large concrete channel that would prevent further flooding of the Tijuana River. The plan was drafted in Mexico City by renowned urban planner Pedro Moctezuma following a request from the federal government to *Mexicanize Tijuana*. "Tijuana is not Mexico" Raymond Chandler wrote in *The Long Goodbye*, and to the Mexico City authorities Tijuana lacked a sense of national identity and presented to foreign visitors a strong rupture with what was perceived as "Mexican." A wide boulevard was proposed, similar to the Paseo Reforma in Mexico

City, with roundabouts and large-scale statues of national heroes intended to bring national pride to the border. The concrete channel bisected two of the original communities of the city; Downtown and *Colonia Libertad*, TJ's first neighborhood. The channel erased the river and ruined any chance that the city could relate to its core natural amenity.

It was during this period (the late 1970s) that the first *maquiladoras* (manufacturing assembly plants) begin to appear around Tijuana. The PRI, Mexico's ruling party for 70 years, had gambled on globalization and opened its border to free trade, creating a manufacturing oasis for companies looking for cheap labor and relaxed environmental laws. The economic crisis of the 1960s in industrialized countries also forced companies around the world to look for locations where labor and operating costs could be minimized. Tijuana became one of first maquila zones due to its proximity to major urban centers in the USA. Even though the city did not possess all of the necessary infrastructure, companies still opted to settle near the border to enjoy direct accessibility to the US markets. Tijuana became one of Mexico's fastest-growing cities with very low unemployment rates that attracted hundreds of migrants from southern states to the borderland's manufacturing centers. The new workers settled in canyons and riverbeds in the city. When industrial parks began to appear, they generated informal residential communities around them. At that moment, the paradox between government-sponsored urban projects in the cause of nationalism and the opening of the borders to international industries became the moment for the re-codifying of

Figure 3.3 Cartolandia.

Source: Photograph used by permission of Alfonso Caraveo.

Figure 3.4 Zona Río.
Source: Photograph used by permission of Elias Sanz.

Tijuana. For some, the city had now become cosmopolitan; as Nestor García Canclini describes in his seminal text *Hybrid Cultures*:

> From the beginning of this century until fifteen years ago, Tijuana was known for a casino (abolished during the Cardenas government), cabarets, dance halls and liquor stores where North Americans came to elude their country's prohibition on sex, gambling and alcohol. The recent installation of factories, modern hotels, cultural centers, and access to wide-ranging international information has made it into a modern, contradictory, cosmopolitan city with a strong definition of itself.[6]

Concepts of cultural hybridity resonated with local writers and artists who began creating work that represented the struggles within the two new forces that defined the city at the end of the twentieth century. For instance, works by installation artist Marcos Ramirez "ERRE" showcased the political conflicts between two countries and two cultures that were seeking (at least practically) to become interconnected. At the bi-national art event INSITE 94, Ramirez constructed a small self-built home similar to those found in squatter developments; it was installed in the main plaza of the Tijuana Cultural Center, a component of the Rio Zone urban redevelopment plan of the 1970s and distinguished by a strong modernist and monumental formal architecture.

Figure 3.5 The new TJ in the east.
Source: Photograph by author.

The two structures in juxtaposition represented the discordant realities brought about by modern industry and nationalism, and the constricting effect it had in defining Tijuana's hybridity. An alternative view of a postmodern idea of hybridity came from a young writer/philosopher, Heriberto Yepez, who argued that Tijuana does not define itself through fusion or synthesis, but instead through its contrasts and contradictions. In *Made in Tijuana*, Yepez writes:

> It's the asymmetry, stupid. The asymmetry, get it? The Fusion does not define Tijuana, but its contrast. Process City//of post-synthetic dialectics// antinomian laboratory of the glocalization. Maquilandia+Farmaceuticals+Migration = Polemic Metaphors.[7]

Confronting these debates, sometimes in their interstices, a new generation of Tijuana writers, artists, and musicians have integrated their work into what is known as "Art from the North," a definition that moved from Fronterizo to just being Norteño. The imposed consciousness that was inherent in the government-inspired Mexicanization projects of the 1970s as well as the rhetoric of globalization – promising a kind of Hegelian national synthesis – failed because it was a fundamental misconception of the border by national as well as international investors and media. Tijuana remains true to nothing more than itself.

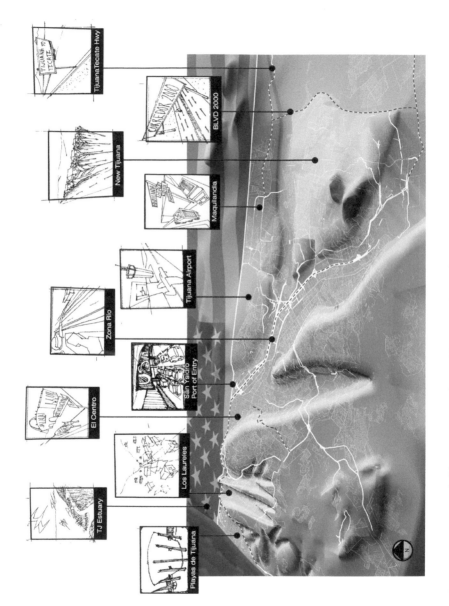

Figure 3.6 Map provided by Arturo Gonzalez Maldonado and Jose Blas Herrera.

Notes

1 F. Montezemolo, "Tijuana is not Tijuana: Fragmented representations at the edge of the border," *World View Cities, Tijuana: The Architectural League of New York*, www.worldviewcities.org/tijuana/fragmented.html, 2005.
2 F. Montezemolo, R. Peralta, and H. Yepez, *Here is Tijuana,* London: Black Dog Publishing, 2006.
3 D. Appleyard and K. Lynch, Report presented to the San Diego City Planning Department on the regional reconnaissance of San Diego, in San Diego, CA, September 15, 1974.
4 A. Padilla Corona, "'El centro histórico de Tijuana': su significado cultural," in *Tijuana Identidades y Nostalgias*, F. M. Acuña Borbolla, M. Ortiz Villacorta Lacave, *et al.* (eds), Tijuana: XVII Ayuntamiento de Tijuana, pp. 121–36.
5 Montezemolo, Peralta, and Yepez, *op. cit.*
6 N. Garcia Canclini, *Hybrid Cultures: Strategies for Entering and Leaving Modernity*, Minneapolis, MN: University of Minneapolis Press, 1995.
7 H. Yepez, *Made in Tijuana*, Mexicali: Instituto de Cultura de Baja California, 2005.

4

[Fake] fake estates

Reconsidering Gordon Matta-Clark's *Fake Estates*

Martin Hogue

One does not impose, but rather exposes, the site.

Robert Smithson[1]

Within architectural thought and process, the site is traditionally thought of as a physical location, a piece of ground that is bound to the earth and subject to its physical laws. Site is also commonly conceived as a location for an intervention; a neutral or unfinished "lot" to be completed by an architectural project. Site and project are often thought to be distinct, one making way for the other.

Work performed in the context of Land and Conceptual Art provides a unique challenge to these assumptions. In these works, the site and the project are understood as interwoven in the production of art. For artists like Robert Smithson, Walter de Maria, and Gordon Matta-Clark, the "site" is integral to the activities of reflection (design) and making (production). The location of the work is established by the artist and the material qualities often emerge from a manipulation of found conditions as much as from new construction. In such projects, the "site" not only invites artistic activity but often constitutes its constructive result.

The *Fake Estates*

Best known for his spatially dynamic extractions of large sections of walls and floors from abandoned buildings, Gordon Matta-Clark in 1975 purchased 13 parcels of residual land in Queens, NY, that had been deemed "gutter space" or "curb property" and put on sale for $25 each. These properties, a 2.33′ × 355′ long strip of land, a 1.83′ × 1.11′ lot, and other similarly unusual lots, were purchased with the goal of highlighting neglected architectural environments that make up the urban and suburban fabric. Many were literally inaccessible and landlocked between buildings or other properties. The artist created an exhibit of his newly acquired "properties" by assembling a photographic inventory of each site, and, with deadpan accuracy, its exact dimensions and location, as well as the deed to the property (Figure 4.1). For Matta-Clark, "the unusability of this land – and the verification of space through the laws of property – is [the] principal object of [his] critique."[2]

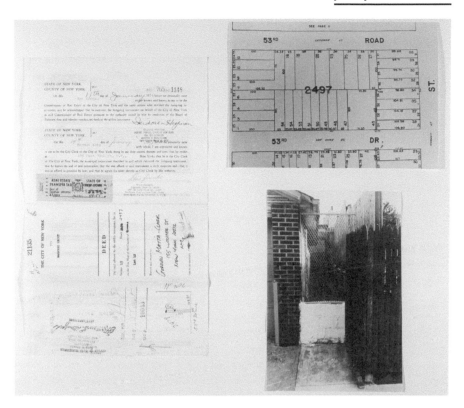

Figure 4.1 Study for Gordon Matta-Clark's Reality Properties: Fake Estates, Little Alley Block 2497, Lot 42, 1974. Collage by the author (2004). Situated between documentation and speculation, this early study is part of a series of collages exploring the 13 sites purchased by Matta-Clark in 1973 and that introduce media and techniques that become more fully developed in later drawings: use of maps as collage material (Sanborn maps like this one, which the artist used as documentary evidence for each *Fake Estate*), visual editing of these documents (erasure, repositioning, and re-scaling of individual map elements), and unfolding of property maps into individual fragments. Several examples of these techniques appear in illustrations throughout this chapter. One of the 13 parcels purchased by Matta-Clark at auction, lot 2497-42 is used as a case study for a number of exploratory collages, comparing the "space" occupied by the iconographic elements that form the map (lines, lot numbers, lot dimensions) to the residual property itself (in solid black).

Source: Sanborn map elements reprinted with permission from the Sanborn Map Company, Inc. © The Sanborn Map Company, Inc. 2004.

As an architect, what has long fascinated me about the *Fake Estates* was the unbuildability of the parcels – that is, their inability to receive a building in the traditional sense. Matta-Clark is suggesting rhetorically that a *site* could be something else than a piece of land to receive a building. Furthermore, I was fascinated with the idea that an act of documentation could constitute an end result in itself. Within architecture, the act of documenting a site is perceived as transitory at best: the *site* eventually makes way for the *project*. Before the period in which the architect is involved intellectually with the site, this location exists merely as a place of unfocused attention – a place that doesn't command any specific meaning attached to architecture or building. In short, the site exists *because* it captures the architect's attention, his or her energies and skills. Architects record impressions and construct representations of the site that will enable them to visualize and conceptualize its attributes while not physically being there, at least not at all times: measurements, photographs of critical features, surrounding context, light orientation, and all other features are noted for later reference. More often than not, the site exists in the mind through these constructed representations. In short, as a product of the architect's own making and decisions, the representations *become* the site. The idea raises some interesting questions: can the site constitute an end product within architecture? Is a constructed site somehow *less than* a building?

The [fake] fake estates

[Fake] *Fake Estates: Reconsidering Gordon Matta-Clark's Fake Estates* emerged as an attempt to address some of these issues. Here, the *Fake Estates* become both a *site* to be critically reinvested as well as the starting point for other, more speculative endeavors. I began this project by spending several months systematically canvassing the entire borough of Queens for residual properties similar to the 13 parcels purchased and documented by Matta-Clark in 1975. Canvassing for these properties occurred in two ways: online browsing through the New York City property database and visual consultation of the 25 Sanborn Map catalogs that make up the borough of Queens.

Browsing through this database occurs in a linear fashion, by moving forward or backward, one property after another.[3] The nearly 1 million properties that constitute the database are organized chronologically: first by borough (Manhattan: 1, Bronx: 2, Brooklyn: 3, Queens: 4, Staten Island: 5), then by block number, and finally by individual lot number within each city block. For example, 5246 70th Street in Queens is also known as lot 4-2497-39 (borough-block-property). In all, nearly 16,500 blocks and 365,000 properties are found in the borough of Queens alone. Properties in this database are assigned individual pages, independent of their size, location, use, and value.

The linear labeling system organizes the 1 million properties chronologically, beginning with the first property in Manhattan (borough 1) and ending with the last property in Staten Island (borough 5). This online browsing offers a peculiar sense of geography in that it does not reveal that consecutive properties in the database (the last lot on one city block and the first on the next) can literally be located miles apart. Through the database, the city is presented as a series of individual fragments, each one click away from the next, undermining the city's organization of streets, fabric, public spaces, and buildings. This single line can, in turn, also be represented in geographical terms by corresponding each point along this line to its physical location on a map of the city (Figures 4.2 and 4.3).

Given the visual rigidity of the online interface and the mechanical nature of the search process – one click at a time, moving from one property to the next – this

Figure 4.2 Linear Browsing – The City as Fragment. Drawing by the author (2006).

Source: Sanborn map elements reprinted with permission from the Sanborn Map Company, Inc. © The Sanborn Map Company, Inc. 2004.

survey took nearly six months to complete. Browsing through more than a few thousand properties at a time proves difficult: each entry must be read through for at least a few seconds (*what is its value? Its size?*) before moving on to the next. While the online search was inevitably quite systematic (my initial search criteria called for all properties under $10,000 to be recorded), the process of browsing through the maps of the Sanborn atlases was less rigorous in nature, for some unique spatial arrangements of properties are often easily recognizable, while others can be difficult to spot with the naked eye. The online database search yielded a number of properties which,

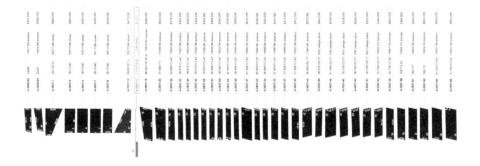

Figure 4.3 Linear Browsing – The City as Fragment. Detail. Drawing by the author (2006).

Source: Sanborn map elements reprinted with permission from the Sanborn Map Company, Inc. © The Sanborn Map Company, Inc. 2004.

by virtue of their minuscule size, are literally invisible in the Sanborn atlases: a 1/8″ × 110′ property (4-8099-145-E), for example would read graphically on a map as no wider than any line marking an edge between two adjacent properties. While these properties share physical characteristics with Matta-Clark's original parcels, these 1,800 properties in this new survey of properties under $10,000 are inauthentic only in the sense that they were not purchased by the artist. If Matta-Clark's purchases were a play on words on the idea of *real* estate, then this new survey could only be identified as comprising *fake* Fake Estates.

The agency of the map

The unexpected discovery of the 1/8″ × 110′ parcel after months of careful data mining constitutes an excellent example of the goals of the project as they were shaped by the difficult online search. This project seeks to visually articulate the agency of maps through a consideration of those moments when conventions for establishing the location and the precise boundaries of a site produce a conceptual "excess of surveying." Maps constitute not simply a way to locate parcels – in a way, one might argue that the *Fake* Estates and the *[Fake] Fake* Estates are by-products of mapping and surveying activities. Indeed, many such lots emerged historically as errors and subsequent adjustments to land surveys, where earlier property demarcations (farms, individual villages, etc...), conflicted with rapid and chaotic development throughout the borough of Queens at the cusp of the twentieth century. While often invisible to the naked eye, the boundaries within a map often split individual blocks or homes, administratively dividing coherent spatial wholes into distinct entities – an interesting reference to Matta-Clark's well known architectural interventions in abandoned buildings (Figure 4.4). Can maps help highlight some of these absurdities?

Suggesting a similarly intense consideration of the city's administrative minutia, the *[Fake] Fake Estates* drawings operate at a variety of scales: the urban scale of New York City as a whole (Figures 4.2 and 4.3), the borough, a city block (Figure 4.4), a typical urban lot – here, cobbled together from various scraps of land (Figure 4.5)– or even the more intimate contact with a single, full-scale plot (Figure 4.6). In the former, an installation of some of these very same properties, inspired by Robert Smithson's *Nonsites* (1967), allows one to gauge the true size of some of these lots by engaging them at full scale on the floor of the gallery. The lot shapes sit on the floor of the gallery and are presented perspectively to the viewer as *planes* while the maps and drawings (Figures 4.2-5), hung on the wall at eye level, appear in orthographic projection as undistorted *plan* images. Thus the same site is presented two ways: one concerned with the experience of sight, the other with an intellectualization or rationalization of the land.

One might be tempted to read far too literally into dimensions as a means to survey the site. Landscape architect and educator James Corner, writing in *Taking Measures Across the American Landscape*, has warned against the strict instrumentality of measures, adding that measures can instead constitute "a form of contemplative survey," an "act of taking stock of a richly constructed inheritance."[4] Like Matta-Clark, the

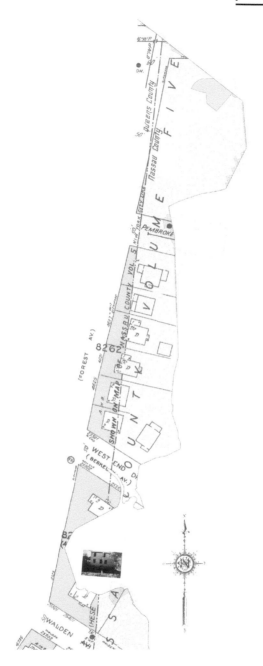

Figure 4.4 *Shared Boundary Between Queens and Nassau Counties.* Drawing by the author (2006).

Source: Sanborn map reprinted with permission from the Sanborn Map Company, Inc. © The Sanborn Map Company, Inc. 2004. Bottom photograph: Gordon Matta-Clark, *Splitting: 4 Corners,* 1974. © 2004 Estate of Gordon Matta-Clark/Artists Rights Society (ARS), New York.

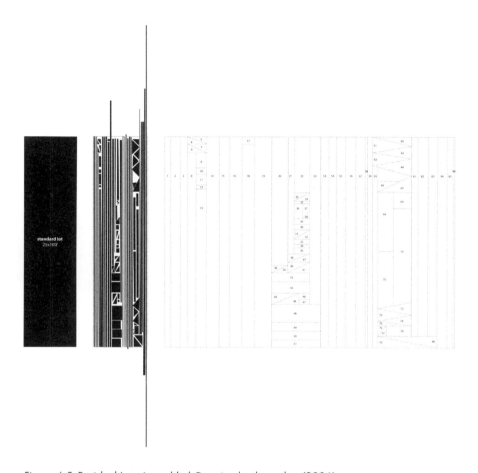

standard lot
25x100'

Figure 4.5 Residual Lots Assembled. Drawing by the author (2006).

goal of this project is not to physically intervene on sites which we assume not to have architectural potential; nor should its intended audience be strictly limited to design professionals. Rather, it is for all those for whom the earthly surface becomes a dynamic territory understood through various processes of physical, cultural, social, and political abstraction. Architects, landscape architects, artists, and geographers alike can take delight in the fact that the city as a whole – literally, every square inch of it – is fully accounted for.

Figure 4.6 Floor Pieces. Full scale rubber cut outs of selected *[Fake] Fake Estates* lots (with property number) by Milo Bonacci, 2006. Installed at Ohio State University, Knowlton School of Architecture, Banvard Gallery, January–February 2007.

Notes

1 Robert Smithson, "Toward the Development of an Air Terminal Site", *Artforum* 6/10 (1967). Reprinted in Nancy Holt (ed.), *The Writings of Robert Smithson,* New York: New York University Press, 1979, p. 47.

2 Pamela Lee, *Object to Be Destroyed-The Work of Gordon Matta-Clark*, Cambridge: MIT Press, 2000, p. 104.

3 City of New York, Department of Finance, http://nycserv.nyc.gov/nycproperty/ nynav/jsp/selectbbl.jsp

4 James Corner and Alex MacLean, *Taking Measures Across the American Landscape,* New Haven, CT: Yale University Press, 2000, p. xvi.

5

The City Formerly Known as Cambridge

A useless map by the Institute for Infinitely Small Things

kanarinka

About the map

The City Formerly Known as Cambridge is a hypothetical (but entirely possible!) map of Cambridge, Massachusetts. During 2006–7, the Institute for Infinitely Small Things invited residents and visitors to the city to rename any public place in Cambridge. This was a big experiment to see what the city would look like if the people that live and work in Cambridge renamed it, right now. We collected over 330 new names along with reasons that ranged from vanity to politics to silliness to forgotten histories to the contested present. This map is a collection of the public's names and stories about Cambridge, its diverse inhabitants and visitors, its traditions and ever-changing attitudes.

This project makes the case that names matter. Who gets to name things? Whose stories get remembered? Whose history is consecrated and whose is forgotten? Most Cambridge history books, for example, begin in 1636 with the founding of Newtowne, though there were Native Americans living here long before that time with their own names for its geography. On this map, we have mixed in some real renamed places from the city's history to recall those geographic places that have been contested, disputed, and transformed over time. A city's names, like its communities, are a living, breathing organism.

How it worked

All of these names were collected through face-to-face conversations with people in Cambridge. The Institute for Infinitely Small Things set up its Renaming Booth in commercial centers, at farmers markets, at community festivals, and even at a four-square tournament to collect names. Renaming was open to everyone and the Institute did not play any editorial role other than correcting spelling mistakes. Everyone who contributed a new name to the map received a free copy mailed to them.

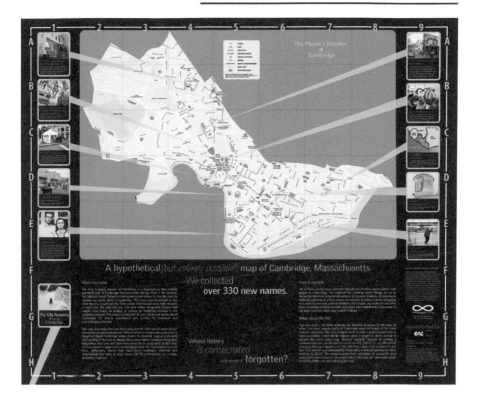

Figure 5.1 The City Formerly Known as Cambridge, produced by The Institute for Infinity Small Things, 2007.

What about the $$?

You may notice the dollar amounts ($) attached to names on the map. In effect, the most popular parts of Cambridge were auctioned off for small amounts of money. This was designed both to make a political point and to solve a problem: what should happen when two people or more wished to rename the same place (say, Harvard Square)? Instead of making a curatorial decision, we decided to mirror real life, which is to say: the person who pays the most money wins. The first time a place was renamed it was free. Each time it was subsequently renamed by a different person, the price went up by $0.25. The money earned from auctioning off names (in total about $20) went towards the production of this map and represents about 0.0025 percent of the cost of producing the map.

The new names

People renamed streets for many reasons ranging from vanity to politics to silliness to the commemoration of local histories. For example, Long's Funeral Home was renamed to Sam Azzam Plywood Palace by a leader of the neighborhood council who had protested Sam Azzam's construction for many years. Massachusetts Avenue was renamed to Prince Hall Blvd to commemorate the famous African-American Mason. And a gradeschooler renamed the King/Amigos School to the Farming Cows School because she thought it would be fun to have cows at school.

So, we spent two years creating a useless map with public parks named after pet goldfish that have long since passed away. Why? Because names are really important! Names of public places reflect who is included (and who is excluded) from a city's history. Names of public places also create a shared sense of community and belonging. For example, despite its astronomical property prices, Cambridge is a very diverse city with significant Brazilian, Asian, and Ethiopian populations. Speaking strictly demographically, one neighborhood is 30 percent African-American and another is 40 percent Asian/Pacific Islander. But you would not guess these facts from biking around Cantabrigian neighborhoods with their Anglophone street names.

But renaming public spaces is a long, contentious process. As the head of a Brazilian community organization said, "Well, after several years we did finally get a street named after a Brazilian religious figure but then they spelled his name wrong on the sign." Rather than undertaking actual renaming, the Institute wanted to symbolically open the renaming process to the public and allow anyone to rename a public place that meant something to them. In the process, of course, we ended up with a bit of silliness and some really great stories. Hopefully we also piqued public interest in the place names around them and planted the seed that things could look a lot differently than they do right now.

Figure 5.2 A detail from the map, The City Formerly Known as Cambridge.

Where can I get a useless map of Cambridge?

You can purchase the map for $5 at www.infinitelysmallthings.net.

About the Institute for Infinitely Small Things

The Institute for Infinitely Small Things conducts creative, participatory research that aims to temporarily transform public spaces dominated by corporate and political agendas. Using performance and conversation, we investigate social and political "tiny things." These have included corporate ads, street names, and post-9/11 security terminology. The Institute advocates for public engagement through its research reports in the form of maps, books, and videos. This interdisciplinary group has a varied and open membership which includes artists, filmmakers, computer programmers, historians, and hula hoopers. For this project, the Institute was Catherine D'Ignazio, Dave Raymond, Heather Ring, Heloisa Escudero, Jaimes Mayhew, James Manning, Katharine Urbati, Matilda Sabal, Max Sabal, Nicole Siggins, Rob Sabal, Savic Rasovic, Shannon Coyle, and Toby Kim Lee.

Acknowledgments

Cartography by Nat Case and Hedberg Maps. Map design by Maegan Rzasa, Ryan Torres, and Tarek Awad. This project was made possible with the support of the Cambridge Arts Council, the LEF Foundation and iKatun. Special thanks to Aliza Shapiro and Fotini Lazaridou-Hatzigoga (modular table design), James Manning (construction and documentation), Rob Coshow (map photography), Jane Beal and Elizabeth White at the Cambridge Arts Council, the Cambridge Historical Commission, and Rick Rawlins and the Senior Design Studio (InFlux) at the Art Institute of Boston.

6

Undisciplined geography

Notes from the field of contemporary art

Emily Eliza Scott

This essay explores strategies that well-disciplined but restless, even rebellious, geographers might learn from contemporary artists who *enact* geography and, in so doing, creatively model interdisciplinary and/or extra-disciplinary inquiry and production.[1] It is intended as a toolbox, or field kit, for practicing geography in unexpected ways and places, in decidedly unruly fashion. No doubt, there has been remarkable activity at the crossroads of art and geography in recent years, reflected in the myriad exhibitions and publications devoted to the intersection of the two fields.[2] As Lize Mogel discusses in Chapter 20, artists of late have taken great interest in maps as well as the politics and technologies inherent to mapmaking.[3] Artists who practice what they describe as "radical" or "tactical" cartography recognize the generative power of maps: that "the map has always been a political agent,"[4] that maps "don't merely represent space, they shape arguments,"[5] that maps hold the potential "to produce reality."[6] Often, their work exposes the social inequities and conflicts that, in the words of art historian Rosalyn Deutsche, "produce and maintain all spaces," the goal being the further "democratization" of space itself.[7] The three projects discussed here – the Center for Land Use Interpretation's online Land Use Database; Heather Frazar's digital video, *Core Matters (or GISP2 Chronologies)*; and *Malibu Public Beaches* "safaris" led by the Los Angeles Urban Rangers (an interdisciplinary collective I founded in 2004) – have a more ambiguous relationship to the political and are *not* primarily concerned with mapping, but rather with facilitating critical and open-ended investigations of place, or, in the case of Frazar, elucidating the role that specific places and materials play in the production of geographical knowledge. These projects, I will argue, not only help to carve out workspace between disciplines and/or between the academy and the world beyond, but also stretch the field of geography via experimental research methods, formats, and modes of distribution.[8]

It should be said from the outset that I am not an official geographer, but rather a long-time interloper in the field and, on the books, a historian of contemporary art and a practicing site-specific artist.[9] My work – within and outside of the academy – has been more deeply influenced by geography than any other discipline, however, and especially by the nuanced theoretical and practical tools it lends to the study of cultural landscapes.[10] Like others, I am attracted to the built-in breadth of a field that embraces perspectives from the humanities, the social, and the physical sciences.

Figure 6.1 Video still from *Core Matters.*
Source: Used by permission of Heather Frazar.

Without question, there is often a palpable and institutionally reinforced rift between the social and physical camps of geography, a divide Frazar addresses in her work. That said, coming from the field of art history, with its persistent commitments to the intricate analysis of individual artists and discrete objects, geography has always seemed to me refreshingly expansive and *grounded.*

It is not surprising that artists are among those at the helm of experimental inter-/ extradisciplinary inquiry and production, due in part to the fact that they are in a position, via their unusually malleable professional lives, to probe unfamiliar terrain, cobbling and crafting working methods as they go. Artist and writer Claire Pentecost has gone so far as to say that artists have a responsibility to exercise this relative freedom by assuming the role of "public amateurs," serving as conduits "between specialized knowledge fields and other members of the public sphere."[11] Artists, in this sense, operate as expert communicators, mediating complex information and questions across traditional divides (e.g., from one discipline to another, from academic to non-academic arenas). Effective relay depends upon the thoughtful correlation of content and form, or an active interrogation of the format(s) most appropriate for representing the subject at hand, as well as careful attention to how and with whom interaction is intended. Nato Thompson, curator of "Experimental Geography: Radical Approaches to Landscape, Cartography, and Urbanism," suggests that artists are especially good at "discovering new forms for conveying ideas or impulses" as well as exploiting the potential of "ambiguity" as a "productive intellectual tool."[12] He clarifies the distinction between straightforward geography and its less orthodox manifestations:

> The task of the geographer is to alert us to what is directly in front of us, while the task of the experimental geographer – an amalgam of scientist, artist, and explorer – is to do so in a manner that deploys aesthetics, ambiguity, poetry, and a dash of empiricism.[13]

The artists discussed in this essay employ ambiguity not to muddle matters; on the contrary, it is used as a *means* to activate their public(s) and to provoke new ways of seeing and actively participating in the world.[14]

I begin by turning to the Center for Land Use Interpretation, an entity already familiar to many in both the art world and the field of cultural geography, and its experimental research methods. One way to describe what this non-profit organization produces – which has always been tricky to pin down – is a *framework* for geographical research. Specifically, the Center creates interpretive exhibitions, exhibition venues, tours, lectures, publications, archives, a residency program, and an online database, all devoted to investigating "the language of land use and teaching it to others,"[15] with the ultimate goal of "improving the collective understanding of the human/land dialectic."[16] Critic Ralph Rugoff identifies the Center's "real subject" as "how we look at, and conceptualize, the world."[17] The Center is simultaneously informational and enigmatic; one of its achievements is the mastery of how to be both at the same time. Maintaining the (anti-)aesthetic veneer of a government agency or other administrative body, its core participants, most notably director Matthew Coolidge, have generally eschewed the question of whether or not their work is "art," yet the Center has received ample funding from and largely circulated within art and architecture contexts since its inception in 1994. For the purposes of this essay, I am interested in how the Center conducts and perpetuates non-resolute, or open-ended, geographical research, implicitly asking each of us to begin with the question: "What form of inquiry does my subject call for?"

The Center's projects focus on places that are both everyday and overlooked, places that constitute the simultaneously strange and banal landscape of contemporary American life. Their recently published *Overlook: Exploring the Internal Fringes of America with the Center for Land Use Interpretation*, for instance, surveys such wide-ranging sites as show caves, towns underwater as a result of large-scale engineering endeavors like dams, mock cities designed for police-force drills, miniature models of the Earth's surface, and nuclear proving grounds. Recent tours in the Los Angeles area have visited landfills and other highlights of the city's "landscape of waste;" Terminal Island, whose ownership is split between the Ports of Los Angeles and Long Beach, which together comprise the fifth busiest port complex in the world; and exurban gravel pits where the foundational materials of the urban environment are extracted. In its own words, the Center targets those places "left unexplored by scholars, scientists, and other specialists," seeking to give due attention to "these corridors and vistas and to trip over all the protruding artifacts of the present on the way to explaining the extraordinary conditions we all find ourselves in all the time."[18]

Much of their research naturally takes place in the field, where messiness and surprise are inevitable, where "paradox is a continuous predicament, and is recognized as another form of truth."[19] Carried out by numerous individuals – some affiliated with the Center over the long-haul, others in passing via participation in a particular project – research often arrives at the Center in the form of motley arrays of hand-written observations, photographs, maps, newspaper clippings, and other found materials, to eventually be "sorted and stored in electronic and physical files – or discarded."[20] While adherence to a pre-determined field regimen is forgone, the Center has produced various tools for on-the-ground investigation, including a "Field Researcher Site-Characterization Form." Upon first glance, this document resembles something used for official purposes; closer inspection, however, undermines this impression. Its "Land Use Classification List," for instance, presents a curious set of checkbox categories (e.g., "abandoned," "analog environment," "art," "bottled water source," "pedestrian," "toll booth," "utopia"), leaving unclear whether these non-hierarchical classifications are meant to be comprehensive, arbitrary, literal, meta-phorical, earnest, ironic, or somewhere in between. Utilized onsite, the list catalyzes a free-play of associations, propelling both field and field observer into motion, ulti-mately drawing attention back to the limitless process of interpretation itself (as well as to the dialectical relationship between habits of observation and classification).

The Center's Land Use Database – with over a thousand "unusual and exemplary" site entries, written in plain language, and available to the public via the Center's website – is the destination for much of its ever-accumulating and collaboratively col-lected data. Considered "the informational bedrock"[21] of the organization, the data-base provides source material for the Center's programming, exhibits, and publications, and enables others "to explore remotely, to search obliquely, and to make creative col-lisions and juxtapositions that render new meanings and explanations of America – and of the many ways of looking at it."[22] The database – a manifestation of the Center's vast research efforts to date, in facilitating others' investigations and modeling an inventive process for geographical inquiry itself – moreover reflects its subject: the dynamic and uncontainable nature of the American built landscape. While arguably specialists in a unique brand of creative cultural production, the Center, as Rugoff notes, specializes "in non-specialization," employing a "non-disciplinary" approach to trace "an underlying logic that connects disparate fields and perspectives, linking them to the common ground of land use and its interpretation."[23] As such, it offers an anti-authoritarian antidote to much academic production, in which specialized knowledge is disseminated, via exclusive language, from one expert to his or her peers and/or in a hierarchical flow from teacher to student.

Heather Frazar's digital video work, *Core Matters (or GISP2 Chronologies)*, crafted alongside her MA thesis in Cultural Geography at the University of California, Los Angeles,[24] illustrates the potential of engaging multiple formats to explore different facets of one's object of study.[25] Both Frazar's video and its written counterpart evolved from the same body of research, and investigate the "material life" of a two-mile long scientific ice core, namely, the Greenland Ice Sheet Project Two (GISP2) core, extracted from central Greenland and since archived at the National Ice Core Laboratory (NICL) in Denver, Colorado. Incorporating found and self-produced moving images, her video traces the ice core's entanglement with these places, its material and aesthetic specificities, as well as the relay of encounters it passes through as an object of scientific inquiry. I focus upon *Core Matters* here because it points to the value of questioning the limits of any given format (e.g., academic journal article, dissertation, conference paper) and in seeking *alignment* between the content of one's research and the form it will eventually take. We might well ask ourselves, in the throes of any project: "What form(at), or medium, most fittingly represents my subject? What will particular formats allow, or not allow, me to convey?"

For Frazar, the medium of video offered not only an opportunity to inhabit her subject differently than through writing alone, but also a means to communicate its visceral qualities, "while at the same time allowing room for viewers to make their own observations."[26] By way of close-up views (e.g., a colorful mosaic of ice crystals under magnification) and slowed images (e.g., the ice core's murky, banded column streaming vertically behind closing credits), Frazar emphasizes the ice's objecthood. She cites her own visual pleasure as one driving force behind the work:

> I am unabashedly drawn to the stratified shades of white, blue, and gray; the minimalisms of snow; the stark repetitions of a massive frozen archive. I find that the visual stoicisms of science produce inherently meditative spaces, and video was a medium I could use to "spend time" in these spaces without having to ... carve [them] out with words.[27]

As viewers, we ponder the specificity of this mute and crystalline substance as well as how information is extracted from it by procedures both brute (e.g., a table saw shaving its surface) and precisely controlled (e.g., a computer measuring its isotope values). Further, we witness its geographical origins – the bleak, white landscape from which it emerged, with subterranean corridors excavated from frozen ground, the sleek and serial space of the NICL archive where the core indefinitely resides (Figures 6.1 and 6.2). As Frazar rightly perceives, "It turns out that watching an ice-core technician's beard grow icicles in the course of an interview does a much better job at communicating what a −36°C degree freezer is like, than does writing about [it]."[28]

While resembling a straightforward documentary in its assemblage of raw footage dispersed with interviews, *Core Matters* departs from this genre by way of subtle informational gaps and blurred distinctions (e.g., which footage is Frazar's own?). Clarity is eclipsed by unresolved questions, most pressingly: What exactly is being documented? The aesthetic value of the ice or of the spaces where ice core research is produced? The translation of the three-dimensional world into two-dimensional information? The stereotypic contrast between field scientists, here macho and youthful with Dire Straits playing in the background, and laboratory scientists, most vividly personified by the nasal-voiced expert she interviews in his book-filled office? Are we to glean a message about the role of ice core research in the ever more urgent study of climate change? Frazer comments on one liberty afforded to her by producing an artwork versus a scholarly piece, namely, the possibility of leaving these sorts of questions "intentionally unanswered ... questions that as a non-academic I am (happily) not obligated to spell out in exacting detail."[29]

Frazar's work is especially relevant in the context of this volume because it effectively bridges cultural and physical geography, pointing to their shared roots in a discipline that has long struggled to co-embrace and integrate them. Her interactions with physical geographers – in the research stage and at later screenings of her video – speak to the potential fruitfulness of interdisciplinary (or more accurately, intradisciplinary) exchange:

> I have encountered a number of "hard" scientists who are closet appreciators of the social vagaries and aesthetics of their discipline, but have no outlets for expressing their observations. In such cases, quantitative scientists are more than happy (sometimes relieved, even) to talk at length about the very things that interest social scientists and artists.[30]

The making of *Core Matters*, in other words, produced a space for mutually satisfying dialogue and collaboration between different kinds of geographers. Not only does the video provide a platform for ice core scientists to share stories and impressions typically excluded from their own research accounts, and in so doing reflect their subject back to them in a new light, but also it greatly enriched Frazar's and her audience's insights into her object of study.

Some artists work to expand dialogic space "out in the world," or beyond the academy, with actual sites acting as a medium for participatory geographical exploration. In the case of *Malibu Public Beaches* (2007–2010) by the Los Angeles Urban Rangers, the journalistic research of one group member – about long-standing territorial battles over the distinction between public versus private property along Malibu's coastline – was collaboratively transformed into an interactive public tour, or "safari."

Figure 6.2 Video still from *Core Matters*.
Source: Used by permission of Heather Frazar.

Together, the tour format and its onsite enactment, model a form of spatialized civic education. Specifically, the safaris teach participants how to access and utilize public-private beaches, which are often remarkably well concealed, along 20 of the 27 total miles of Malibu's beachfront lined with high-dollar private development. Like other "distributed art projects," these safaris treat "the city street and the urban grid, along with all of the social, economic, and political relationships present there … not as the background to an imaginary story or a theatrical play but as the very substance with which one performs."[31] The Urban Rangers additionally designed a user-friendly *Malibu Public Beaches Guide*, distributed during events and on their website (and eventually on public buses throughout the city), with a map locating all 18 public access ways. My discussion of *Malibu Public Beaches* centers on its simultaneous activation of site and user by means of a directed encounter, and its attentiveness to *where* geographical work happens. This project compels us to consider the destination of our labors, by asking: "Where will my work most effectively operate, especially in terms of the communities/publics I intend to engage?"

The Los Angeles Urban Rangers comprises artists, architects, geographers, urban designers, and others "who aim, with both wit and a healthy dose of sincerity, to facilitate creative, critical, head-on, oblique, and crisscrossed investigations into our sprawling metropolis."[32] Since its inception in 2004, the group has produced campfire talks, guided hikes, maps, field kits, and other devices to creatively (re)interpret Los Angeles' everyday environments (e.g., empty downtown lots, Hollywood Boulevard, freeway landscapes). In the guise of park rangers – complete with green and grey uniforms branded with their own palm tree and freeway cloverleaf logo – the Urban Rangers use accessible and democratic language, low-tech props, and down-to-earth hospitality, to spark curiosity about places that are experienced day to day, but are often taken for granted. One goal is to extend the same sense of wonder typically reserved for extraordinary places "out there" (e.g., national parks), to ordinary places on the urban home front. Another is to impart critical interpretive skills, or know-how, to urban citizens navigating their habitual surroundings.

Malibu Public Beaches safaris are geared toward practicing public space "survival skills," and specifically, "how to find, park, walk, picnic, and sunbathe on a Malibu beach legally and safely."[33] As it turns out, this isn't as simple as one might expect. Although all beaches in California are public below the mean high tide line, and the state constitution forbids development that excludes "the right of way to such water whenever it is required for any public purpose,"[34] as stated before, the vast majority of public tidelands are blocked by multi-million dollar homes (with the lawns, potted plants, periodic valet parking, and occasional privately hired and publicly menacing security guards that go along with them). In many cases, individual homeowners have ceded control of partial property between their homes and the water in exchange for permission to build or renovate, resulting in dry sand "public easements." While virtually all beaches in Malibu have enough dry sand easements to support a satisfying day at the beach, however, they remain unmarked, making it nearly impossible for beach users to delineate public easements from private land.

The Urban Rangers' safaris hinge upon interactive exercises that make use of participants' embodied presence onsite to negotiate and clarify such legally complex terrain. After watching Urban Rangers demonstrate how to identify, measure, and stake out a public easement, safari-goers break into groups to "trail-blaze the public-private boundary" on specific easements. Later, safari-goers embark on a "no-kill access way hunt," competing to find all the public access ways along a given stretch of coastline, many of which are camouflaged by imposing "no entry" signs, driveways, and various other forms of defensive architecture. In a final activity, safari-goers use magnets to add alternative signage, or "magnetic public commentary," to an un-permitted "No Parking" sign installed at the request of local homeowners near a public coastal access way.

Malibu Public Beaches, and the work of the Los Angeles Urban Rangers more generally, seeks to stimulate public discourse about public spaces, and to model a form of participatory "place-making" on the ground. The Malibu project has spurred conten-tious debate, and tangible resistance on the part of certain homeowners. Alongside ample press coverage about the safaris in local media, there have been forced interrup-tions of safaris by irate homeowners, letters-to-the-editor accusing the Urban Rangers of inciting unwanted activism, and a lawsuit threat on the part of a local homeowners property association (falsely) accusing Rangers and safari-goers of trespassing on private property. In effect, the Malibu homeowners have become an unexpected participant, or actor, in the project. This, in turn, has forced the Urban Rangers to reconsider the relationship between their art practice and straightforward activism (as well as the strategic value in the ranger's "neutral" persona, which has been useful in navigating controversial waters). One distinction, they maintain, is that their work addresses the layered and open-ended meanings of place, encouraging others to actively engage and (re-)imagine the world around them.

To be sure, each of the three projects discussed in this essay is based upon substan-tial and earnest research, reflecting more than a dash of empiricism, and possesses concrete pedagogical value: the Center for Land Use Interpretation contributes to understandings of the contemporary built landscape, Frazar's *Core Matters* illuminates the ways in which scientific knowledge is shaped by specific materials, places, and actions, and the Urban Rangers impart practical skills for identifying and using urban public space. At the same time, in their divergence from more straightforward informational modes and structures (i.e. governmental land management agency, documentary film genre, uniformed authority figure), or precise deployment of ambi-guity, they suggest that geographical interpretation is necessarily, and richly, unfixed and ongoing. Each project has built within it room for participants/viewers to make their own discoveries, with a larger aim of provoking further critical, imaginative, and grass roots engagements with place. As such, they encourage a "democracy of space," though in a somewhat different sense than articulated by Deutsche or implied by the artworks Mogel describes, which have overt activist agendas. Ultimately, the projects discussed here speak to the vitality and relevance of specific geographies, or why place matters, seeking to compellingly communicate this most basic of geographical principles to diverse audiences "on the ground." But of what relevance is this to me, might ask the geographer content to remain within the edges of her discipline and expected professional role? This art, I believe, pries open opportunities for academic geographers as well: to creatively re-imagine their processes of research and representation, for example, by experimenting with new formats that are more content-responsive, or allowing one's subject matter to direct the project, its methods, and the form it eventually takes; to extend their reach and effect; and to inspire students to create as inventive and meaningful work as possible.

Notes

1 By "enacting geography" I mean actively performing and putting into action the work of geography, taken as the investigation and description of the Earth's surface, including its physical features, cultural formations, and the interrelation of the two.

2 As pointed out in the *AAG Newsletter* from October 2007, the convergence between art and geography is evident in numerous recent exhibitions. These include *An Atlas* (curators: Alexis Bhagat and Lize Mogel; various venues; 2007 to present), *Experimental Geography: Radical Approaches to Landscape, Cartography, and Urbanism* (curator: Nato Thompson with Independent Curators' International; various venues; 2008 to present), *Geography and the Politics of Mobility* (curator: Ursula Biemann; Generali Foundation, Vienna; 2003), *Global Navigation System* (curator: Nicolas Bourriaud; Palais de Tokyo, Paris; 2003), *Just Space(s)* (curators: Ava Bromberg and Nicholas Brown; Los Angeles Contemporary Exhibitions, Los Angeles, CA; 2007), *Mapquest* (curator: Elena Sorokina; P.S. 122, New York City, NY; 2006), and *Weather Report: Climate Change* (curator: Lucy Lippard; Boulder Museum of Contemporary Art, Boulder, CO; 2007). Many of these exhibitions were accompanied by exhibition catalogs including critical essays about contemporary art and geography.

3 Mogel attributes the "current mapping impulse" in contemporary art as well as in everyday life to "a number of shifts in the way we think about representation and space. ... The way we visualize spatial and personal relationships has ... radically changed—the absolute centrality of the Internet to metropolitan citizens, saturation of electronic communication, and increased mobility have taught us to understand information as embodied in map/networked form." See her essay in this volume (Disorientation Guide, Chapter 20).

4 Lize Mogel interview by Trevor Paglen: www.artwurl.org/interviews/INT057. html.

5 Institute for Applied Autonomy, "Tactical Cartographies," in *An Atlas of Radical Cartography*, eds Alexis Bhag at and Lize Mogel (Los Angeles: Journal of Aesthetics and Protest, 2007), 35.

6 Kanarinka, "Art-Machines, Body-Ovens and Map-Recipes: Entries for a Psychogeographic Dictionary," *Cartographic Perspectives* 53, 2006, 25.

7 R. Deutsche, *Evictions: Art and Spatial Politics*, Cambridge, MA: MIT Press, 1996, p. xi.

8 As artist kanarinka, from the Institute for Infinitely Small Things, has remarked, "Artists have always borrowed from other fields of study and activity. Perhaps it is only the pace that has accelerated." What is very rare, she argues, is work that

"not only borrows techniques from disciplines," but that also "creates a transdisciplinary space" that supports "ongoing dialogue" and "[allows] work 'outside' the disciplines to flow back 'inside' and contribute to shaping and changing the borders of these disciplines themselves." kanarinka, *op. cit.*, p. 38.

9 I imagine the term "site-specific" carries different connotations among geographers than among art historians. Within contemporary art history and criticism, however, its usage has often been haphazard. For more on the historical and conceptual evolution of "site-specific" art, see Miwon Kwon, *One Place After Another: Site-Specific Art and Locational Identity*, Cambridge, MA: MIT Press, 2002.

10 I am indebted, beyond words, to the mentorship of the late Professor Denis E. Cosgrove, with whom I minored in geography at UCLA, participated in his long-running UCLA Cultural Geography Methods Workshop, and co-organized a year-long interdisciplinary study group which culminated in a two-day symposium at the UCLA Hammer Museum, *Field Works: Art/Geography* (May 2005). It is with tremendous sadness over his death and gratitude for the privilege to have worked with him, the most generous and gracious of teachers, that I dedicate this essay to him.

11 Claire Pentecost website: www.clairepentecost.org/.

12 "Interview with Nato Thompson," by Lauren Cornell: http://rhizome.org/. Thompson borrows the term "experimental geography" from Trevor Paglen, one of the artist-geographers in the exhibition, who explains the term in his catalog essay: "Experimental Geography: from Cultural Production to the Production of Space," in *Experimental Geography: Radical Approaches to Landscape, Cartography, and Urbanism*, Nato Thompson (ed.), New York: Independent Curators' International, 2008, pp. 26–33.

13 Independent Curators International website: www.ici-exhibitions.org/exhibitions/experimental/experimental.htm.

14 Paglen is wary of experimental methods or new technologies used for novelty's sake, and reminds us that they are "only helpful when [they suggest] new kinds of analysis or different ways of seeing." "Mapping Ghosts: Visible Collective Talks to Trevor Paglen," in A. Bhagat and L. Mogel (eds), *An Atlas of Radical Cartography*, Los Angeles: Journal of Aesthetics and Protest, 2007, pp. 42–3.

15 M. Coolidge and S. Simons (eds), *Overlook: Exploring the Internal Fringes of America with the Center for Land Use Interpretation*, New York: Metropolis Books, 2006, p. 16.

16 CLUI website: www.clui.org/.

17 R. Rugoff, "Circling the Center," in M. Coolidge and S. Simons (eds), *Overlook: Exploring the Internal Fringes of America with the Center for Land Use Interpretation*, New York: Metropolis Books, 2006, p. 35.

18 Coolidge and Simons, *op. cit.*, pp. 16–17.

19 "Interview with Matthew Coolidge," *Badlands: New Horizons in Landscape*, ed. Denise Markonish (North Adams, MA: MASS MoCA, 2008), p. 140.

20 Coolidge and Simons, *op. cit.*, p. 17.

21 *Ibid.*

22 Coolidge and Simons, *op. cit.*, p. 25.

23 Rugoff, *op. cit.*, p. 39.

24 H. Frazar, "Core Matters: Greenland, Denver, Los Angeles and the GISP2 Ice Core," MA thesis (UCLA, 2005). She has subsequently published "Icy Demands: Coring, Curating and Researching the GISP2 Ice Core," in K. Yusoff (ed.) *Bipolar*, London: The Arts Catalyst, 2008, pp. 40–1; and "Core Matters: Greenland, Denver and the GISP2 Ice Core," in D. Cosgrove and V. della Dora (eds), *High Places: Cultural Geographies of Mountains, Ice and Science*, London: I. B. Tauris, 2008, pp. 64–83.

25 Originally commissioned by the UCLA Center for Modern and Contemporary Studies for a symposium called *Field Works: Art/Geography*, which addressed "emerging relations between geographical science and artistic production," Frazar's work has since been screened in both academic geography and art venues.

26 Heather Frazar, e-mail message to the author, August 22, 2008.

27 *Ibid.*

28 *Ibid.*

29 *Ibid.*

30 Heather Frazar, e-mail message to the author, Spring, 2007.

31 Kanarinka, *op. cit.*, p. 33.

32 Los Angeles Urban Rangers website: www.laurbanrangers.org.

33 *Ibid.*

34 Constitution of the State of California, Article X, Section 4.

7

Codex profundo

Gustavo Leclerc

I began my first sketch book in 1991, five years after coming to the USA from Mexico. The books were vehicles to help me establish a new home. I see them as very connected to being an immigrant, to leaving one home and trying to find another. This search has been an internal process and while they were not diaries, the fact that I dated them shows they served a similar purpose. They became mediations on cultural shock, allowing me to process all the new things I was bombarded with. My personal experience as an immigrant was not your typical one. It was so full of contradictions; I arrived in Bel Air to live here without knowing what Bel Air was. I didn't have any money or a car and the bus stop was miles away.

In my initial situation in the USA, I felt under a cultural microscope, with people who had so many pre-conceptions, myths, and expectations about me as a Mexican. They were intrigued and full of a strange curiosity. I looked like the gardener but lived inside the house. These books became a safe place to explore ideas about this experience – a refuge. I didn't let anybody see them for many years.

As I learned about American culture, I started to discover many new and different things but I didn't know how to react – sometimes I'd fully embrace them – hamburgers and donuts for example. TV became a big attraction in the first years, watching shows that were so American at heart, *I Love Lucy* for example. I was also mesmerized by *Three's Company*.

Being here I became rapidly aware of race dynamics, which I had never thought of much in Mexico before. This awareness was at first just on an intuitive level, related to how I was observed, addressed, interacted with, everything soft. Later people's attitudes became more obvious. My immature political consciousness had been previously only of class – rich versus poor. Maybe this naiveté came out of the experience with my own immediate family, where everyone was already so different – my grandmother was Indian, my grandfather was white, my brother and sister are both light skinned, and I was the darkest one of all.

From this framework each book goes in a different direction.

In *The Book about Buildings* influences relate to ideas of architecture and imagination such as Italo Calvino's *Invisible Cities* and *Empire of Signs* by Roland Barthes. I was also interested in pre-Colombian myths and forms, the idea of a complex, large urban city, popular culture, folk art, writing passages, songs, poems, and so on that would relate to notions of space and place. In this book, I would imagine possible buildings, certain spaces, where function was not a consideration, but instead only forms responding to the themes above in different combinations and recombinations.

Often I would have a perspective of distance, looking at things from far away, checking out the moves, so to speak, in architectural trends. At the time, postmodernism and neoclassical styles were popular, also high-tech modernism, which included buildings that looked like machines. These were approaches that were very rational, approaches that favored either history or science. My approach, on the other hand, was looking to the whimsical, magical, and not following logic. I love the process of drawing with my hands.

The Book of Apparitions started with events in the news. La Virgin de Guadalupe had appeared on a tortilla in Texas, on a billboard on the US/Mexico border, and in a tree in a front yard of East LA. It gave me the idea of looking for places La Virgin de Guadalupe might appear as a way to explore my relationship to Catholicism, which had been uneasy – from going to Catholic school and thinking of becoming a priest to being an atheist. At first the book was about religion but then it switched to being more about gender power relationships and my own ambivalence about gender definitions.

The book *Celibacy* continues the themes of *The Book of Apparitions* but intensifies/ amplifies the issues of gender and sexuality. These issues are also complicated by issues of class and race. I felt that under all the dynamics of race and culture, at the very basic level, everything is all about sex, this basic type of interaction. In this book I was also interested in exploring how my enchantment and initial fascination with North American culture had become overridden with the notion that consumer society is a form of oppression, a myth creator. That's why I have a coca-cola bottle and Mickey Mouse in the book, but in negative ways, as "super myths." The book is about the pursuit of happiness in relation to the American dream for immigrants, complicated by relationships related to sex, sexual attraction, repulsion, fear, guilt, and inaccessibility.

The book *Conejo* is about a transborder character named Conejo who has ambivalent feelings about his Mexican/American identity. I gave that character a Chilango (Mexico City) identity of survival/adaptation. It has a very strong ego. This means he'll do anything, accept anything, disregard anything that will help him survive, whether use La Virgin de Guadalupe, the American flag, or Mickey Mouse. His only concern is to make it.

Cuicuilco

How do geometric forms encourage new possibilities? The settlement of Cuicuilco (located in present-day Mexico city) dates back to 1400 BCE when the Lago de Texcoco in the Valle de México nurtured many farming communities. Its circular pyramid is unusual. Cuicuilco was later buried under a lava flow and the Valle's populations combined to create Teotihuacan, the first great urban civilization of pre-Columbian

Figure 7.1 Cuicuilco by Gustavo Leclerc.

Mexico. The Valle landscape has been intensively overwritten with the layers and rhythms of the past.

Nezahualcóyotl was poet-king of Tetzcoco in the mid-fifteenth century. His poem, *Tlamatinime,* reflects Aztec obsessions with predestiny: cycles of birth, death, and re-birth. It foreshadowed doom just at the time when Cortés the conquistador was arriving from Spain.

La carpa

La carpa (literally, big top, or circus tent) is a tradition of occasional performance in the public spaces of Mexican towns and villages. Music, humor, and biting political commentary were part of the performance. Children were forbidden to attend because things could get a bit raunchy, but we usually found a way to get in. Architectural forms – the pyramid, caracol, wall – create opportunities for performance. Ruins are especially important sites for la carpa because there we can chase memories into shadowy spaces.

Figure 7.2 La Carpa by Gustavo Leclerc.

My grandmother's favorite recipe for a drink made from the rind (cáscara) of the pineapple is written in Totonac, a lyrical language of one of Mexico's many pre-Columbian populations. The Totonac still exist in large numbers in Veracruz state. The recipe is recycled in a Spanish-language version. Today, I can taste these memories in Los Angeles, far/not so far from Veracruz.

Xa Taqlhqma Akaxica (Tepache de Piña)

Xqotxqa akaxka (cascara de piña)
Saqsi (panela)
Chuchut (agua)

Kachaqa chu kakalakchuku Xqotxpa akaxka, alistalh
Kamuju nak aqtum tlamank, kamakapini saqsi chu
Kalaqxkutanqalh la pala aqtum u aqti kilhtamaku
Chu tlantiya liqotnankan.

La casa de ramas rojas (house of red branches)

Red is an important color; it speaks about life, death, danger, and sin. I was reading a history of color and pigments, and simultaneously re-imagining architecture's basic forms such as the "box." Red branches remind me of the Floating Shrine on the island of Miyajima in Japan. Gold is a more neutral color than red, even though gold itself inspires the basest human actions. For me, gold is a connecting tissue or sheen. It links my images, and provides a continuity reminiscent of the unfolding harmonies in a pre-Columbian codex.

Figure 7.3 La Casa de Ramas Rojas / House of Red Branches by Gustavo Leclerc.

Huastecan popular songs typically employ violins to express strong emotion. While the musicians themselves remain stone-faced, the passion is all in the music. In contrast, jarocho music from the neighboring south of Veracruz is joyful and noisy, played by musicians who are totally engaged and joyful. Huastec music is nowadays enjoying a renaissance through fusion with rock and other contemporary forms. More layers.

Exotic fish

The first time I attended a lecture by Rem Koolhaas, the famous Dutch architect, I desperately felt the need to wrench his shapes into something more exotic and tropical. Red waters drive the up-ended forms.

Nahuatl is the best known of the Aztec languages. The Aztecs revered artists, particularly the codex painters. Aztec cosmology found expression in many spatial forms, such as the famous feathered serpent known as Quetzalcoatl. This moment of tribute to art and creativity relates to my concern with the "will to form," that is, how belief systems find concrete expression in art, architecture, and geography. Architecture thus becomes a form of speculation to explore new identities, places, texts, bodies, fears, and desires.

El Día de los Muertos (The Day of the Dead) is a set of rituals that combine to slow time and open spaces for contact with the departed ones. As well as formal performances, there are installations (altars) and processions (along a walk decorated with streams of yellow flowers). The celebrations unfold just like a codex, just like my

Figure 7.4 Exotic Fish by Gustavo Leclerc.

sketch books; the images are variously connected by gold, flowers, ritual, and narrative. My sketch books reach back to the most ancient times, to a Mexico profundo, and then connect forward to the present. They are my witness to our becoming.

The Artist

Disciple, abundant, multiple, restless.
The true artist, capable, practicing, skillful,
Maintains dialogue with his heart, meet things with his mind.
The true artist draws out all from his heart:
  works with delight; makes things with calm, with sagacity;
  works like a true Toltec; composes his objects; works
  dexterously; invents;
  arranges materials; adorns them; makes the adjust.

Nahuatl Text

2 SPATIAL LITERACIES
Geotexts

Sarah Luria

An interest in the role of place in the life of literature, and the role of literature in the production of place is one of the exciting points of convergence between geography and the humanities. The collection of essays and vignettes in this section argue the benefits of a deep collaboration between the two fields: geography can help us understand literature and literature can help us understand geography. For example, literature scholar Janelle Jenstad reports that she has her students read the Renaissance literature of London not only with the *Oxford English Dictionary* but with a London map. One needs to know London, Jenstad argues, in order to get the jokes of the era's "city comedies," to appreciate, for instance, that a chaste, virginal heroine lives in "the main market street where all things were for sale." Geographer Tim Mennel explains why he decided to write his dissertation on Robert Moses and New York City as a novel. "Fiction," he argues, "is an essential tool in clarifying both how we think about Robert Moses, and, more broadly, how we think about geographical discourse."

Such decisions are evidence of the fascination in the humanities with the earth sciences that has come to be known as the "spatial turn." This critical shift has been wide ranging, informed in part by Henri Lefebvre's Marxist critique of the production of place, Michel Foucault's "archaeologies," which spatialized historical knowledge, and postcolonial and global criticism.[1] The essays and vignettes here testify to the accelerated critical energy being focused on a geography of literature and literary geography: they chart new intersections between poetry and mapped space, cities and novels, time and place; they show how the fluid terrain of literary geography can provide us with room to expand our sense of ground truth; and they showcase the importance of such familiar literary elements as narrative form, point of view, and metaphor in how we see, understand, and construct place.

Perhaps the most obvious way literature and geography can expand each others' horizons is to challenge their shared faith in, as literary scholar Barbara Eckstein puts it, the "artistic recreation of the empirically observable and nameable world." Such verbal recreation is evident in both literary realism and geographic description. Eckstein draws upon Lefebvre to question literary realism's confidence that any space is "thoroughly accessible" and knowable. Geographer Tim Mennel is similarly dissatisfied by geographical "narratives" that "have a substantiating and totemizing power," "positing themselves as definitional rather than as interpretive, discursive, and metaphoric." In Eckstein's view, what is needed is a way of reporting place that "engages more than the best efforts of individual human consciousness moving alone at ground level through the spatial scales discernible to the naked human eye and ear and hand ... – in other words, realism."

Both Eckstein and Mennel find a solution to the limits of realist representation in the possibilities of experimental narrative form. For his geography dissertation, Tim Mennel wrote a "synoptic novel" to challenge the story Robert Caro told in his highly influential book *The Power Broker* about Robert Moses and the making of modern New York. In Caro's tale, Mennel argues, Moses figures as the villain – the omnipotent master-planner whose "urban renewal" eviscerated the soul of New York. Mennel's "synoptic" form widens its portrayal: in addition to Moses, the novel follows three fictional characters associated with him – a chauffeur, a saxophonist, and a labor organizer – who introduce other views of the city and critique Moses' work, and so demonstrate "the contextual nature of what we know about him." New York's cultural geography – its "jazz and performance-art subcultures"– plays an important role in building a context for Moses' work as well. The result, evidenced in the excerpt from Mennel's novel included here, is a dynamic assessment of "individual action and urban complexity," a revealing look at Moses as a part and product of his world.

Eckstein analyzes the place-shaping role of narrative form in her comparison of two novels about New Orleans. Frederick Turner's realist novel *Redemption* reinscribes the conventional view of New Orleans looked for by tourists and exploits the fascination with Storyville, the city's notorious neighborhood where jazz was born. Turner's method and form can't help but leave out crucial stories and troubled areas of the city – such as the white racist riots that occurred when the novel takes place– that are less known but all too real. Eckstein finds a better approach in Michael Ondaatje's *Coming Through Slaughter*, which attempts to imagine the New Orleans of early jazz great Buddy Bolden, who died all but forgotten in a state mental hospital. As Eckstein argues, "Ondaatje drives us through Bolden's geography – another concept or abstraction: not just a cruise through the neighborhood of his birth but a surreal trip through a world of poverty, race, corruption and genius." Ondaatje's experimental form complicates our sense of place: "The pastiche of the novel repeatedly pulls us into and out of the story so that we both care about Bolden and know we cannot know him. What we can struggle to know instead is the space and time he occupied and what forces produced that geography and history." The more fluid space of experimental narrative form, not bound by fictions of realism, can thus help widen our perspective on place and make us more responsive to it. One has only to recall, as Eckstein does, our inability to see what was happening to the residents of New Orleans' Ward 9 during Hurricane Katrina to conclude that we need a broader sense of the city and welcome the role literature might play in helping to supply it.

Both Eckstein and Mennel remind us that point of view is everything to a creative writer just as it is everything to making place. Whether a story is told from the first person, limited third person, or omniscient point of view determines what can be seen and told and known in a story. This is dramatically illustrated in geography, where the point from which a place is viewed – from the ground or the air, and the infinite numbers of angles within and between those modes – frames the space that is seen. Every view both shows a scene and locates the subject who viewed it, whether omniscient (aerial) or first person (ground, limited to one character). Invoking multiple points of view has grown increasingly popular in modern and postmodern works.

Several essays here reflect upon the profoundly different sense of place we get from aerial and ground level views and demonstrate how literary geography can integrate these opposed perspectives in revealing ways. I show how Thoreau plays past, present, and objective and subjective views off each other, as he practices several ways of seeing – as a property surveyor, a naturalist, a poet, and a Concord native in his remarkable survey of the Concord River. Howard Horowitz melds aerial views and ground truth in his "wordmaps" that show the shape of a landmass artfully composed by words that tell of its history and topography. His verbal depiction of the Oregon coast thus names all "major capes and coastal landmarks ... at their correct locations," while it constructs a poetry out of location, a narrative of place that fits the contours of the land.

In contrast to the aerial/ground views of Horowitz and Thoreau, Robbert Flick gives us an overview at street level as he stops his car in auto-centered Los Angeles to take photographic portraits, or, more precisely, tableaus, of street life. In Flick's series of photographs, we move slowly down the street. The elements of the built environment change slightly but remain connected, recognizable pieces of the same block: the lights, the signs, the railings, the luggage piled at the entrance of the store for quick sale. The characters on this stage also change slightly as people come and go: a man wearing a white t-shirt just emerging from the shadows in one is still there emerging in the next, but out of frame in the last. These related views lead you to dissect the scene – what's the same, what has changed?– and to pay attention to all the colorful elements of street life made artistic here by this way of seeing. Flick's walk down Los Angeles's Broadway reminds one of Walt Whitman's verbal snapshots walking down New York's Broadway in the 1850s, or Eadweard Muybridge's late-nineteenth-century motion studies using time-lapse photography. Flick's street views are like frames in a film, but they are not continuous: people come in and out of frame, but we don't get to see them every step of the way. There are gaps just as in life; we can only see so much. But what Flick's images do let us see is how the most mundane activities – a car making a turn in an intersection, a group of people entering a store – are compelling. They reveal, as Jane Jacobs might have put it, the drama of the street.

The power of literary techniques to enrich and alter our sense of geography is concentrated in what is perhaps literature's most creative site: metaphor. Peta Mitchell argues that metaphor has been a central engine for the spatial turn – Foucault's metaphor of "archaeology," and later Deleuze and Guattari's "geology," helped to bring the humanities to the earth sciences by resolving the split between geography and history introduced by Kant. Metaphor, the poet Donald Hall reminds us, "compares the seemingly incomparable" and "wins us with [the] energy of resolved contrast."[2] Mitchell finds the most powerful metaphors of the spatial turn in the fiction and literary non-fiction of writers such as Anne Michaels and Tim Robinson. The limestone terrain at the center of Michaels's *Fugitive Pieces* and Robinson's study of the Aran Islands brings history and geography together so fully that one wonders they could ever be considered apart: the porous rock is so responsive to the touch of time it is "a uniquely tender and memorious ground." It is with Mitchell's helpful review of the critical shift in the spatial turn and how it has been catalyzed by metaphor that this section begins.

In her book the *Language of Landscape*, landscape architect and historian Anne Whiston Spirn identifies several essential components of place: "territory, boundary, path, gateway, meeting place, prospect, refuge, source, and sign." Composed of such elements, human habitats, Spirn notes, are "performance spaces" "generated by active processes," and are not "simply formal and fixed."[3] What might literary studies, with its focus on the narratives played out in our spaces, and with its essential components of metaphor, setting, point of view, narrative form, and style, add to the study of such language? These essays provide some answers. There is what the land tells us but then there are the words we use to describe it, and the changes that are made by that very act.

Notes

1 See for instance H. Lefebvre, *The Production of Space*, tr. D. Nicholson-Smith, Malden: Blackwell Publishing, 1991; M. Foucault, *The Archaeology of Knowledge*, tr. A. M. Sheridan Smith, London: Tavistock, 1972; and E. W. Soja, *Postmodern Geographies*, London: Verso, 1989.

2 D. Hall, *To Read a Poem,* 2nd edition, Boston: Heinle & Heinle, 1992, p. 35.

3 A. Whiston Spirn, *The Language of Landscape,* New Haven: Yale University Press, 1998, p. 121.

8

"The stratified record upon which we set our feet"

The spatial turn and the multilayering of history, geography, and geology

Peta Mitchell

In Thomas Mann's tetralogy of the 1930s and 1940s, *Joseph and His Brothers*, the narrator declares history is not only "that which has happened and that which goes on happening in time," but it is also "the stratified record upon which we set our feet, the ground beneath us."[1] By opening up history to its spatial, geographical, and geological dimensions Mann both predicts and encapsulates the twentieth century's "spatial turn,"[2] a critical shift that divested geography of its largely passive role as history's "stage" and brought to the fore intersections between the humanities and the earth sciences.

In this essay, I wish to draw out particularly the relationships between history, narrative, geography, and geology revealed by this spatial turn and the questions these pose for the disciplinary relationship between geography and the humanities. As Mann's statement exemplifies, the spatial turn has often been captured most strikingly in fiction, nowhere more so than in Graham Swift's *Waterland* (1983)[3] and Anne Michaels's *Fugitive Pieces* (1996).[4] Both novels present space, place, and landscape as having a palpable influence on history and memory, and the geographical/geological line that runs through them continues through Tim Robinson's non-fictional, two-volume "topographical" history *Stones of Aran*.[5] Robinson's work – which is not history, geography, or literature, and yet is all three – constructs an imaginative geography that renders inseparable geography, geology, history, memory, and the act of writing.

The interdisciplinarity at the core of the "spatial turn" is, perhaps, most immediately apparent in metaphor. Mann's geological metaphor of history as a "stratified record" is echoed by Gilles Deleuze, nearly half a century later, when he argues that "the world is made up of superimposed surfaces, archives or strata. The world is thus knowledge."[6] More recently, in her study of John McPhee's acclaimed geological survey of North America, *Annals of the Former World* (1998), Norma Tilden takes up the geological term "stratigraphies," noting that it "suggests a parallel between written histories and the earth's own life story."[7] As Tilden points out, narrative can also be stratigraphic since it too can "respon[d] to the ebb and flow of the land, and its constant refusal to stand still long enough for us to pin it down – to stake our claim to it."[8] This cross-disciplinary metaphor of stratigraphy is particularly germane to the narratives of Swift, Michaels, and Robinson – narratives that spatialize, that sedimentarize, the notion of history; narratives that blend history with folklore, with geology, geography, landscape, memory; narratives that both inscribe and describe the stratified ground they tread. In this essay I want to draw out the stratigraphic line that runs through these texts – all are based on a striated ground that collapses and problematizes history and geography, time and space. All three are also explicitly chronotopic in the Bakhtinian sense[9]–at both the thematic and narrative level, all three foreground an irreducible space-time compression.

Before turning to these stratigraphic narratives, I wish to make a historical detour through the Enlightenment, for, I argue, the interdisciplinarity at the heart of the spatial turn is a response to, and an ongoing dialogue with, the eighteenth-century emergence of geography, geology, and history as separate disciplines. The eighteenth century was a time in which, as John Gascoigne explains, "the old all-encompassing categories of the knowledge of Nature – natural philosophy and natural history – began to be broken down into the embryonic scientific disciplines," such as geography and geology.[10] Immanuel Kant occupies a noteworthy position in the history of geographical thought, standing as he does at the intersection between the disciplines of philosophy and geography and on the threshold of the age of scientificity and discipline-formation. In the 1750s, Kant introduced physical geography as a discipline in his lectures at the University of Königsberg. In these lectures, which were first published as *Physische Geographie* in 1802, Kant rigorously defined geography in contrast to history: geography related to description and space; history to narration and time.[11] According to Richard Hartshorne, Kant was the first to argue for geography's status as a separate scientific discipline, preceding both Alexander von Humboldt and Alfred Hettner's claims for geography's disciplinary difference.[12] Moreover, just as Kant was describing the boundaries of geography, geology too was shoring up its disciplinary borders. Only two decades after Kant began his physical geography lectures at the University of Königsberg, the term "geology" was used for the first time in its contemporary scientific sense.[13]

While Kant's *Physische Geographie* had little direct influence on the discipline of geography until the twentieth century, it arguably had a far greater influence on nineteenth-century philosophy.[14] Michel Foucault – a critical theorist who, along with Henri Lefebvre, is almost synonymous with the spatial turn – claims that "[s]ince Kant what is to be thought by the philosopher is time."[15] For Foucault, Kant's separation of geography and history effectively separated space and time. It also accorded history the role of active "becoming," while the descriptive science of geography was relegated to the passive role of "being."[16] As a result, Foucault argues, nineteenth- and early twentieth-century philosophy largely ignored geography and space and concerned itself with studying history as "becoming" or advancement over time.

In his 1969 work *The Archaeology of Knowledge*, Foucault suggests an alternative, more spatially aware, method of historical and philosophical inquiry. Foucault's "archaeological" method – which he later revised as "genealogy," following Nietzsche's *On the Genealogy of Knowledge* – mobilizes spatial metaphors of cartography and archaeology to bring geography and geology back into the historical/philosophical frame. By employing this "archaeological" method, the historian does not describe or fix the origins or limits of disciplines.[17] Rather, the historian of ideas seeks to map "discursive practices in so far as they give rise to a corpus of knowledge, in so far as they assume the status and role of a science" and to reveal the ways in which disciplines or discursive practices, figured as strata, "articulate their own historicities onto one another."[18]

For Gilles Deleuze, Foucault's conception of historical formations as strata creates a new understanding of knowledge and its relation to power. Foucault shows us, Deleuze says, that where knowledge is "stratified, archivized, and endowed with a relatively rigid segmentarity," power is "diagrammatic: it mobilizes non-stratified matter and functions, and unfolds with a very flexible segmentarity."[19] In its compound form, power – knowledge is a shifting terrain shaped by stratifying and non-stratifying forces, and Foucault presents his archaeological-genealogical method as a means by which these forces might be excavated and mapped. Indeed, there are marked resonances between Deleuze's analysis of Foucault's archaeology-genealogy and his own work with Félix Guattari, which is similarly characterized by geographical and geological metaphors such as stratification, cartography, territorialization, and deterritorialization. In effect, as a number of critics have pointed out, Deleuze and Guattari "transform [Foucault's] genealogy into geology."[20] This strategic linguistic shift from genealogy to geology makes explicit the shift from layered time to layered space – time implicit in the spatial turn. According to Claire Colebrook, where genealogy "is an attempt to think of time … as *effective* history," Deleuze and Guattari's geology is "an attempt at a grammar of space: different series, plains, territories, paths and maps."[21]

As a new "grammar of space," Deleuze and Guattari's revisioning of genealogy as geology requires also a rethinking of geography. Underpinning their framework for a geological approach to epistemology is Deleuze and Guattari's preoccupation with what they term "geophilosophy."[22] What geophilosophy offers philosophy, they argue, is a release from the restrictive subject-object dualism: "[t]hinking," they argue, "is neither a line drawn between subject and object nor a revolving of one around the other. Rather thinking takes place in the relationship of territory and the earth."[23] Moreover, geophilosophy, as a phenomenon of the spatial turn, is founded upon an awareness of geography as active and strategic rather than descriptive. In a move reminiscent of Foucault's critique of Kant, Deleuze and Guattari claim:

> [g]eography is not confined to providing historical form with a substance and variable places. It is not merely physical and human but mental, like the landscape. Geography wrests history from the cult of necessity in order to stress the irreducibility of contingency. It wrests it from the cult of origins in order to affirm the power of a 'milieu' … 'Becoming' does not belong to history.[24]

For Deleuze and Guattari, and Foucault, change, or "becoming," is not the province of history alone in the same way that the spatial science of geography does not simply equate to passive, descriptive "being." The history of ideas requires also a geography and a geology of ideas: time and space are the two axes that enable its critique. Moreover, the language these theorists use is intimately bound to their critique. By employing spatial geographical and geological metaphors, they lay bare the spatio-temporal striations of emergence, divergence, and convergence that run through the histories of these disciplines. Thus, underpinning the twentieth century's spatial turn is a historical-geographical-geological nexus that both raises and speaks to the question of disciplinarity and interdisciplinarity. This shift in thinking about the relationships between knowledge, history, geography, and geology further resonates not only in such self-confessed "spatial histories" as Paul Carter's *Road to Botany Bay* (1987)[25] and Manuel De Landa's "geological" history *A Thousand Years of Nonlinear History* (1997),[26] but also – returning to my early theme of stratigraphic narrative – in Graham Swift's *Waterland*, Anne Michaels's *Fugitive Pieces*, and Tim Robinson's *Stones of Aran*. These texts attest to a particular form of postmodern chronotope, in which, as Paul Smethurst argues, "space is not merely in the service of time, but has a poetics of its own, which reveals itself through a geographical or topological imagination rather than a historical one."[27]

History as siltation: Graham Swift's *Waterland*

Indeed, Swift's *Waterland* appears almost perfect in its commingling of history and geography. The waterland of the title of Graham Swift's 1983 novel is the silt-clogged, eel-ridden wetlands and peat bogs of the Fens in eastern England, whose constant geographical flux between earth and water suggests the very nature of history. *Waterland*'s narrator is Tom Crick, a history teacher living, tellingly, in Greenwich. At zero degrees longitude – the geographical site of the zero hour – the prime meridian is perhaps the exemplary chronotope. Greenwich is, as Smethurst notes, "the zero space-time, from which the rest of the world was marked out."[28] A family and, by extension, career crisis has led Crick to begin to remember, to recount to his history students the story of his life, a life dominated by the landscape of the Fens in which he grew up.

The Fens is also a landscape whose history is intertwined with the history of his ancestors, and is a landscape whose form suggests an alternative vision of history. History, Tom Crick tells his students, departing from the curriculum, "goes in two directions at once. It goes backwards as it goes forwards. It loops. It takes detours. Do not fall into the illusion that history is a well-disciplined and unflagging column marching unswervingly into the future" (135). His own "humble model for progress," Crick tells them, "is the reclamation of land," and he enjoins his students to "forget ... your revolutions, your turning-points, your grand metamorphoses of history. Consider, instead, the slow and arduous process, the interminable and ambiguous process – the process of human siltation – of land reclamation" (10). In this waterland – this hybrid space, which "of all landscapes, most approximates to Nothing" (13) – human progress must pit itself against the most ambivalent, the most equivocal of sedimentary forces: silt, "which shapes and undermines continents; which demolishes as it builds; which is simultaneous accretion and erosion; neither progress nor decay" (8–9).

According to William Howarth, humans have throughout history treated wetlands in an ambivalent way, both in a physical and a representational sense. Until relatively recently, he argues, wetlands were maligned and systematically destroyed because they impeded human progress in the landscape. However, he states, "[b]y the mid-nineteenth century, shifting attitudes changed wetlands from economic liability into cultural asset. Writers began to read places not as reflected power or virtue, but as states of emotion and perception." The ambiguous position of wetlands in the cultural imaginary reflects the very ambiguity of wetland itself. As Howarth argues:

> We cannot essentialize wetlands, because they are hybrid and multivalent: neither land nor water alone, they are water-land; a continuum between terra and aqua. In rhetorical terms they are not syntax but parataxis, phrases placed side by side without apparent connection, a term Joseph Frank used to describe spatial forms that evoke a great variety of response. In their wildness, wetlands dispossess readers of old codes and lead toward new syntax, where phrases may begin to reassemble.[29]

Swift's wetlands – the Fens – can neither be essentialized nor reduced. In Pamela Cooper's reading, they are at once "densely literal but embedded in the multiple transformations of metaphor; obtrusively material yet (like history) always already mediated by prior inscription." As a result, "[a]t once a geography and a topography, *Waterland*'s marshes become effectively a phantasmagoria, more hallucination than fairy-tale; and here the contradictory physical properties of the Fens acquire other dimensions of significance."[30] The physical flatness of the Fens, its tendency to collapse geography and topography and space and time, and the horizontality of the process of siltation suggest again a paratactic rhetoric of spatiality and temporality. Siltation is a geological process, but, as Smethurst suggests, the "horizontal nature of the water – land chronotope" does not suggest a deep, vertical geological time.[31] Rather, Swift's novel tends to invoke the geology of Deleuze or the genealogy-archaeology of Foucault. In effect, *Waterland* provides a metanarrative on the interrelationship between history and geography by way of a microcosmic, fictionalized spatial history of the Fens, which has itself been sheltered from many of the global historical events going on beyond its watery borders.

"Lyric geology": history, geology, and metaphor in Anne Michaels's *Fugitive Pieces*

Where *Waterland* is geographically relatively insular and focused on the ways in which individual and local history speak to grand historical narratives, Anne Michael's 1996 novel of the Holocaust, *Fugitive Pieces*, is necessarily more global. *Fugitive Pieces* also opens in a liminal space between land and water – a peat bog near the Polish archaeological site of Biskupin, in which the seven-year-old Jewish boy Jakob Beer hides – buried like a bog body – after his family is murdered by German soldiers. Jakob is saved by a Greek geologist, Athos, whom he describes as being "dedicated to a private trinity of peat, limestone, and archaeological wood (19)." Of the three, Athos has a "special affection for limestone – that crushed reef of memory, that living stone, organic history squeezed into massive mountain tombs" (32). In *Fugitive Pieces*, limestone bears with it witness to the suffering of the Jews set to work in the Golleschau quarry who "carried their lives in their hands" as they "were forced to haul huge blocks of limestone endlessly" (53). Also, like a geological layer within the text, limestone links the three geographical "sites" of the novel: Poland, Greece, and Canada.[32]

Jakob flees with Athos to the Ionian island of Zakynthos with its limestone cliffs, where he remains hidden in a room for four years until the Germans leave the island in 1944. Soon after, Athos takes up a position in the new geography department at the University of Toronto to teach a course entitled The History of Geographical Thought, and Jakob eventually enrols as well, taking courses in literature, history, and geography (94, 108). As a boy, Jakob is "transfixed by the way time buckled, met itself in pleats and folds" (30), by "Athos's tales of geologists and explorers, cartographers and navigators" (54). Jakob mulls upon the Catalan Atlas, the "most definitive mappamondo of its time," which in its quest for truth and fact left *terrae incognitae* blank, unlike "maps of history," which he sees as being "less honest" (136–37). While he is fascinated by these romanticized objects and narratives that stand testament to human endeavor and scientific progress, Jakob is also constantly attuned to the watermark that stains them. He is similarly attuned to his own watermark, his genetic and historical burden of remembering the dead, and to the way in which these stories, maps, and rocks lay bare to him that memory cannot be separated from landscape; history from geography. "It's no metaphor," Jakob states,

> to feel the influence of the dead in the world, just as it's no metaphor to hear the radiocarbon chronometer, the Geiger counter amplifying the faint breathing of rock, fifty thousand years old. (Like the faint thump from behind the womb wall.) It is no metaphor to witness the astonishing fidelity of minerals magnetized, even after hundreds of millions of years, pointing to the magnetic pole, minerals that have never forgotten magma whose cooling off has left them forever desirous. We long for place; but place itself longs. Human memory is encoded in air currents and river sediment. Eskers of ash wait to be scooped up, lives reconstituted. (53)

However, the only means by which Jakob (and Michaels) can convey this complex interpenetration of history, geography, memory, and landscape is through metaphor, a mode "annihilated" by the German language, which turned "humans into objects" (143). In his attempt to untangle history, and inspired by Athos's "lyric geology" (209), Jakob becomes a poet. His anthology, aptly titled *Groundwork*, documents his childhood experiences and "recounts the geology of the mass graves" (209) – in one move contradicting Theodor Adorno's injunction that to write poetry after Auschwitz is barbaric[33] and providing a corrective to Simon Schama's observation that "In our mind's eye we are accustomed to thinking of the Holocaust as having no landscape."[34] Indeed, Jakob's choice to become a poet suggests how vital poetry is to his task of reconnecting history and geography. Poetic language, particularly metaphor, allows for a kind of border-crossing that exposes disciplinary and linguistic striations even as it traverses them.

Méira Cook has examined the ways in which *Fugitive Pieces* itself responds to Adorno's dictum, which she states:

> is not merely an indictment against lyric poetry as a genre but against all writing that in the wake of the Holocaust it must find new ways to represent the elisions and failures of grief when it is used as a system of discourse.[35]

Michaels's response to Adorno, Cook argues, lies in her use of metaphor – by speaking in a "foreign language," Michaels's aim is "to bring to the prose of the traumatic narrative the unruly compulsions of poetry, and in so doing to restore to language what Adorno once mourned as necessarily lost for ever."[36] Meredith Criglington also remarks upon Michaels's use of "chronotopic metaphors drawn from fields such as archaeology, geology, and physics," arguing that Michaels employs these metaphors strategically "in order to explore the relative, shaping perspective of the person who witnesses, remembers, or researches the events of the past."[37] Cook, however, is more ambivalent about Michaels's intensely metaphorical language, describing it as at certain times "over-lush," "contrived," and even "clichéd," but at other times as "unparalleled."[38] Nonetheless, she maintains that Michaels's use of metaphor as a "device of memory" is central to the way in which she investigates the problematics of witnessing and memory in relation to trauma.[39]

Certainly Michaels's figurative language is strategic and active, rather than ornamental as in the Classical theory of metaphor. It represents at both the textual and thematic levels the impossibility of literality, of perfect witnessing or perfect memory. As Howarth notes, metaphor itself is both constructive and erosive; metaphor "alters the meanings of words, undermining their stability until land and sea intertwine."[40] Just as it is suffused with metaphors of geography, geology, and archaeology, *Fugitive Pieces* demands a reader who is at once an archaeologist, geologist, and geographer, a reader who, like its character Athos, is at all times attentive to the stratification of history, memory, language, and landscape and who can read obliquely through their layers.

Hybridity, disciplinarity, and the linguistic politics of the spatial turn: Tim Robinson's *Stones Of Aran*

Before I turn to the question of polymathy or interdisciplinarity inherent in the spatial turn, I will look at one final chronotopic text, Tim Robinson's non-fiction *Stones of Aran*. *Stones of Aran* is a two-volume "topographical" or spatial history of Árainn, the largest of the Aran Islands, which lie off the west coast of Ireland. The first book, *Pilgrimage* (1986), "makes a circuit" of the island's coastline, while the companion volume, *Labyrinth* (1995), explores the island's interior. Robinson's tours are driven by his quest for a theory of the "adequate step" – to him, "walking is a way of expressing, acting out, a relationship to the physical world," an "intense cognitive and physical involvement with the terrain."[41] Like Swift's and Michaels's novels, *Stones of Aran* is a stratigraphic narrative, but through its non-fictional yet still literary blending of geography, geology, and history, Robinson's work plays out in interesting ways both the poetics and politics of the spatial turn.

Robinson himself is, like Michaels's character Athos, a polymath fascinated by karst limestone topography. A former visual artist (under the name Drever) and student of physics and mathematics, Robinson and his wife moved to the Aran Islands in the early 1970s. Captivated by the entirely soil-less topography and the geological and mythic history of the islands, Robinson recasts himself as a cartographer and begins to draw and publish maps of the islands. But as he does so, he also begins to write the islands as well. Robinson relates the humorous anecdote of his and his wife Máiréad's first encounter with Aran. "On the day of our arrival," he writes, "we met an old man who explained the basic geography: 'The ocean,' he told us, 'goes all around the island.'" As Robinson and his wife allow the man's remark to guide their "rambles," he discovers that the absurd obviousness of the statement belies a more profound truth about the island's geography: "indeed the ocean encircles Aran like the rim of a magnifying glass, focusing attention to the point of obsession."[42] This fascination with the island would, within the space of a few months, lead Robinson and his wife to decide to leave London for Aran.

Arriving with Robinson on Aran's limestone terrain, we have moved from the water–land chronotope of Swift's *Waterland*, through the water–land–rock trinity of *Fugitive Pieces*, to this almost alien landscape that we might think of as a water–rock chronotope. In *Pilgrimage*, Robinson describes the "bare, soluble limestone" of Aran as "a uniquely tender and memorious ground":

> Every shower sends rivulets wandering across its surface, deepening the ways of their predecessors and gradually engraving their initial caprices as law into the stone. This recording of the weather of the ages also revivifies much more ancient fossils, which are precisely etched by the rain's delicate acids, so that now when a rising or setting sun shadows them forth, prehistory is as urgent underfoot as last night's graffiti in city streets. ... Further, this land has provided its inhabitants ... with one material only, stone, which may fall, but still endures.[43]

Not only does Robinson's lyrical description of the limestone topography recall Athos's "crushed reefs of memory," but also the narrative voice of W. H. Auden's 1948 poem "In Praise of Limestone." Auden's limestone, with its peculiar ability to dissolve in water, is innately different from the "immoderate soils"– the "granite wastes," the plains of clay and gravel – colonized by humans where "there is room for armies to drill" and where "rivers / wait to be tamed." Limestone might metamorphose into marble – that symbol of monumental history, of civilization, and classical art – but in Auden's poem the stone's truer form is its soluble one: "an older colder voice, the oceanic whisper." No tempered granite or marble for Auden; limestone is reserved his highest praise: "when I try to imagine a faultless love / Or the life to come, what I hear is the murmur / Of underground streams, what I see is a limestone landscape."[44]

Likewise for Robinson, Aran's limestone topography presents itself as the ideal landscape upon which to practice a history and cartography that does not neglect the stratified record beneath his feet. Robinson hints that this kind of chronotopic spatial history might be impossible, or at the very least vertiginous in scale, if practiced in a landscape that had experienced more human development – for instance, on Auden's colonized and civilized granite wastes or plains of clay and gravel. Robinson writes of:

> our craggy, boggy, overgrown and overbuilt terrain, on which every step carries us across geologies, biologies, myths, histories, politics, etcetera, and trips us with the trailing *Rosa spinosissima* of personal associations. To forget these dimensions of the step is to forgo our honour as human beings, but an awareness of them equal to the involuted complexities under foot at any given moment would be a crushing backload to have to carry.[45]

Unlike these heavily overwritten landscapes, Robinson continues:

> Aran, of the world's countless facets one of the most finely carved by nature, closely structured by labour and minutely commented by tradition, is the exemplary terrain upon which to dream of that work, the guide-book to the adequate step.[46]

Aran, though, is not a simple landscape – it is at once rugged and enduring and fragile and protean, both geologically and sociologically. In *Pilgrimage*, when Robinson reaches the narrowest point of the island – Blind Sound, or An Sunda Caoch – he writes that "[h]ere it is that the sea will some day break through and divide the island – the belief recurs in Aran folklore and, I have been told, in Aran people's dreams."[47] In his description of Aran's turloughs, we see Robinson's attentiveness to etymology and to geology as he explains the island's unpredictable geography:

> The genitive form in this toponym, Róidín an Turlaigh, shows that the final syllable of *turlach* is not, as the OED states and as the anglicization 'turlough' implies, the word *loch* (genitive *locha*), a lake or lough, but is in fact a mere postfix of place; thus *turlach*, from *tur*, dry, could be explicated as 'a place that dries up.' Turloughs … form in enclosed depressions and are filled and emptied not by streams but from below, through openings in their beds which act alternately as springs and swallow-holes, as the general level of ground-water held in the fissures of the limestone rises or falls. In Aran the joints in the limestone have been opened by solution only down to a depth of twenty or thirty feet, below which they are tight, with the result that the water contained in them is draped like a mantle over the island's core of unfissured rock. The thickness of this mantle varies in average with the seasons, and fluctuates with every shower of rain, and wherever and whenever it rises above ground-level in the bottom of a hollow, a marsh or pond or lake appears. In An Turlach Mór, this hydrology works like a dream: one day you see cattle grazing in a meadow; the next, when you pass, water lies there like a drawn blade.[48]

This passage, which I have quoted at length, also provides a characteristic example of Robinson's use of language. Throughout *Stones of Aran*, he consistently, almost seamlessly, modulates between unembellished, quasi-scientific language and lyrical, metaphorical language.

While Robinson might readily recognize Aran as the "exemplary terrain" for his spatial history, the question of the language in which this history might best be conveyed is always at stake. Written language is both an escape and a trap for Robinson – writing is, he maintains:

> my way out of this labyrinth. But I am no abstract, deep-sea philosopher; if I raise a metaphor as a sail to catch the winds of thought, I am soon overturned by shoals, or fly to the horizon and lie becalmed there.[49]

In his collection of essays *My Time in Space*, Robinson speaks of his desire to discover a language that will express this complex relationship to geography – a language that will not simply anthropomorphize, sexualize, or spiritualize the landscape.[50] Indeed, in *Labyrinth* he describes his practice of writing as being inextricable from his practice of walking in the landscape:

> If I cannot lay my hand on the phrase I am searching for in my room, I stroll out, scramble over the back wall and go rooting for words among the crevices of the rock. The crag is my testing-ground for the aerodynamics of sentences, a rebounding-place to prance upon when a chapter comes to its own conclusions and sets me free.[51]

Robinson's intensely beautiful and detailed works are virtually unclassifiable – subtly and deftly he overlays geographical, social, and mythic histories of the island one with the other, constructing an imaginative geography of the island. These histories are themselves like the sedimentary layers of the limestone topography of the islands; they are distinct but inseparable, and open to erosion. Yet, while the extraordinary hybridity of Robinson's work has largely been praised by critics, at least one, whose review Robinson quotes in *Labyrinth*, is equally sceptical of this blending of disciplines: "Striding roughshod," the critic writes, "over the bounds of specialisms and genres, ... and in trying to be 'not just' a historian or geologist or botanist or even a poet, Robinson ends up being nothing in particular."[52] Robinson's bad review might simply be that – a bad review – but I would argue it raises a central problematic for any interdisciplinary study of the spatial turn: the stratification of academic/scientific disciplines. What I have attempted to outline in this essay is the ways in which "postmodern" fictional and non-fictional spatial and stratigraphic theories and narratives blur disciplinary borders, privileging hybridity. To use Swift's metaphor of land-reclamation, these postmodern texts could be considered the water that threatens to obscure the borders of those disciplinary fields fought for and marked out by the Enlightenment tradition.

Metaphor, as I have argued, is central to these questions of hybridity and interdisciplinarity, and over the past twenty years it has become commonplace for scholars in geography and in the humanities to comment upon the prevalence of spatial and geographic metaphors in later twentieth-century theory and fiction. Tracing metaphors of geology and geography through fictional and theoretical texts of the later twentieth century opens them up to a stratigraphic inquiry that does not separate history from geography or geography from language but that also does not simply elide disciplinary borders. In effect, I argue, what the spatial turn requires of scholars who investigate the dynamic intersection where the disciplinary strata of geography and the humanities abut and blur is an attentiveness to the stratified record of those disciplines and its relationship to language. And the question for a project that conjoins geography and the humanities is the question of how we negotiate between this desire for convergence and a recognition of the inescapable stratification of disciplinary ground.

Notes

1 T. Mann, *Joseph and His Brothers*, New York: Knopf, 1983, p. 121.
2 For the earliest articulations of the "spatial turn," see E. W. Soja, *Postmodern Geographies*, London: Verso, 1989, p. 16; and F. Jameson, *Postmodernism, Or, The Cultural Logic of Late Capitalism*, Durham: Duke University Press, 1991, p. 154.
3 G. Swift, *Waterland*, London: Picador, 1992; hereafter cited in text.
4 A. Michaels, *Fugitive Pieces*, London: Bloomsbury, 1998; hereafter cited in text.
5 Tim Robinson's first volume in the *Stones of Aran* duology was published in 1986 as *Stones of Aran: Pilgrimage*, London: Penguin, 1990. The second volume was published in 1995 as *Stones of Aran: Labyrinth*, Dublin: Lilliput, 1995.
6 G. Deleuze, *Foucault*, tr. S. Hand, London: Continuum, 1999, p. 98.
7 N. Tilden, "Stratigraphies: writing a suspect terrain," *Biography* 25, no. 1, 2002, 26.
8 *Ibid.*
9 In his 1937–38 essay outlining his concept of the "chronotope" Mikhail Bakhtin states:

> [w]e will give the name *chronotope* (literally, 'time space') to the intrinsic connectedness of temporal and spatial relationships that are artistically expressed in literature. ... In the literary artistic chronotope, spatial and temporal indicators are fused into one carefully thought-out, concrete whole. Time, as it were, thickens, takes on flesh, becomes artistically visible; likewise, space becomes charged and responsive to the movements of time, plot and history.

> M. M. Bakhtin, "Forms of time and of the chronotope in the novel: notes toward a historical poetics," in M. Holquist (ed.), *The Dialogic Imagination*, Austin: University of Texas Press, 1981, p. 250.

10 J. Gascoigne, "The eighteenth-century scientific community: a prosopographical study," *Social Studies of Science* 25, 1995, 575.

11 I. Kant, "Introduction to 'Physische geographie,'" in J. A. May, *Kant's Concept of Geography and its Relation to Recent Geographical Thought*, Toronto: University of Toronto Press, 1970, p. 258.

12 See R. Hartshorne, "The concept of geography as a science of space, from Kant and Humboldt to Hettner," *Annals of the Association of American Geographers* 48, no. 2, 1958, 99–107.

13 According to Gabriel Gohau, two Genevan naturalists – Jean-André Deluc (in 1778) and Horace-Bénédict de Saussure (in 1779) – were the first to use the term *geology* in the disciplinary sense. G. Gohau, *A History of Geology*, rev. and tr. A. V. Carozzi and M. Carozzi, New Brunswick: Rutgers University Press, 1991, pp. 2–3.

14 R. Hartshorne, *op. cit.*, p. 107.

15 M. Foucault, "The eye of power: a conversation with Jean-Pierre Barou and Michelle Perrot," in C. Gordon (ed.), *Power/Knowledge: Selected Interviews and Other Writings*, New York: Pantheon, 1980, p. 149.

16 P. Mitchell, *Cartographic Strategies of Postmodernity: The Figure of the Map in Contemporary Theory and Fiction*, New York: Routledge, 2007, p. 52.

17 M. Foucault, *The Archaeology of Knowledge*, tr. A. M. Sheridan Smith, London: Tavistock, 1972, p. 179.

18 M. Foucault, *op. cit.*, pp. 190, 191.

19 G. Deleuze, *op. cit.*, p. 61.

20 S. D. Ross, *The Gift of Touch: Embodying the Good*, Albany: State University of New York Press, 1998, p. 296. See also "plateau," number three of G. Deleuze and F. Guattari, *A Thousand Plateaus: Capitalism and Schizophrenia*, tr. B. Massumi, Minneapolis: University of Minnesota Press, 1987, pp. 39–74; C. Colebrook, *Gilles Deleuze*, London: Routledge, 2002, p. 58; and C. Colebrook, "A grammar of becoming: strategy, subjectivism, and style," in E. Grosz (ed.), *Becomings: Explorations in Time, Memory, and Futures*, Ithaca: Cornell University Press, 1999, pp. 117–40.

21 C. Colebrook, *op. cit.*, p. 132.

22 For a fuller account of Deleuze and Guattari's "geophilosophy," see M. Bonta and J. Protevi, *Deleuze and Geophilosophy*, Edinburgh: Edinburgh University Press, 2004.

23 G. Deleuze and F. Guattari, *What is Philosophy?* tr. H. Tomlinson and G. Burchell, New York: Columbia University Press, 1994, p. 85.

24 G. Deleuze and F. Guattari, *op. cit.*, p. 96.

25 P. Carter, *The Road to Botany Bay: An Essay in Spatial History*, London: Faber and Faber, 1987.

26 M. De Landa, *A Thousand Years of Nonlinear History*, New York: Zone, 1997.

27 P. Smethurst, *The Postmodern Chronotope: Reading Space and Time in Contemporary Fiction*, Amsterdam: Rodopi, 2000, p. 15.

28 P. Smethurst, *op. cit.*, p. 166.

29 W. Howarth, "Imagined territory: the writing of wetlands," *New Literary History* 30, no. 3, 1999, 525; 520.

30 P. Cooper, "Imperial topographies: the spaces of history in *Waterland*," *Modern Fiction Studies* 42, no. 2, 1996, 376.

31 P. Smethurst, *op. cit.*, p. 162.

32 Jakob and Athos's first encounter with Toronto is the "great limestone hall" of Union Station (90).

33 T. W. Adorno, "Cultural criticism and society," in *Prisms*, tr. S. Weber and S. Weber, Cambridge, MA: MIT Press, 1981, p. 34. A number of critical essays on *Fugitive Pieces* have linked the novel with Adorno's famous statement: see, for example, M. Cook, "At the membrane of language and silence: metaphor and memory in *Fugitive Pieces*," *Canadian Literature* 164, 2000, 12–33; S. Gubar, "Empathetic identification in Anne Michaels's *Fugitive Pieces*: masculinity and poetry after Auschwitz," *Signs* 28, no. 1, 2002, 249–75; and D. Coffey, "Blood and soil in Anne Michaels's *Fugitive Pieces*: the pastoral in Holocaust literature," *Modern Fiction Studies* 53, no.1, 2007, 27–49.

34 S. Schama, *Landscape and Memory*, New York: Vintage, 1996, p. 26. See also D. Coffey, *op. cit.*, p. 40, who examines the "unsettling affinities" between *Fugitive Pieces* and Schama's analysis of Nazi pastoral fantasies in *Landscape and Memory*.

35 M. Cook, *op. cit.*, p. 12.

36 M. Cook, *op. cit.*, p. 29.

37 M. Criglington, "The city as a site of counter-memory in Anne Michaels's *Fugitive Pieces* and Michael Ondaatje's *In the Skin of a Lion*," *Essays on Canadian Writing* 81, 2004, 141.

38 M. Cook, *op. cit.*, pp. 17–19.

39 M. Cook, *op. cit.*, p. 26.

40 W. Howarth, *op. cit.*, p. 526.

41 T. Robinson, *My Time in Space*, Dublin: Lilliput, 2001, p. 103.

42 T. Robinson, *Pilgrimage, op. cit.*, p. 10.

43 T. Robinson, *Pilgrimage, op. cit.*, p. 4.

44 W. H. Auden, "In praise of limestone," in J. Parini (ed.), *The Wadsworth Anthology of Poetry*, Boston: Wadsworth, 2005, pp. 439–41.

45 T. Robinson, *Pilgrimage, op. cit.*, p. 12.

46 T. Robinson, *Pilgrimage, op. cit.*, p. 13.

47 T. Robinson, *Pilgrimage, op. cit.*, p. 65.

48 T. Robinson, *Labyrinth, op. cit.*, p. 91.

49 T. Robinson, *Labyrinth, op. cit.*, p. 455.

50 T. Robinson, *My Time in Space, op. cit.*, p. 103.

51 T. Robinson, *Labyrinth, op. cit.*, p. 294.

52 T. Robinson, *Labyrinth*, op. cit., p. 307.

9

Monument of myth

Finding Robert Moses through geographic fiction

Timothy Mennel

> If we cannot understand the landscape of modern New York City without refer-
> ence to its changing economic base, the rise of an industrial economy and
> architecture, and the impact of the Progressive Movement, so too will we not
> understand either the making or the meaning of that landscape without reference
> to the brothers Roebling, Louis Sullivan, ... the Rockefellers and Harrimans, and
> especially Robert Moses.[1]

My 2007 dissertation in geography, *Everything Must Go: A Novel of Robert Moses's
New York*, was a substantially fictional depiction of the life and influence of the long-
time master builder. Usually, when a scholar writes fiction, it is either a hobby or an
act of malfeasance; here, it is the opposite. I used this medium precisely because I
believe that fiction is an essential tool in clarifying both how we think about Robert
Moses and, more broadly, how we think about geographical discourse.

Journalist Robert Caro's biography of Moses has long been the standard text on the
scope and nature of Moses's power.[2] Moses had been the driving force behind the
physical transformation of New York from the 1930s onward, yet he was not much of
a figure in the public imagination prior to Caro's work, which constructed a narrative
framework that rendered sensible New York's jumble of highways, redevelopments,
institutions, and structures. Caro's book is the source of what nearly "everyone knows"
about Moses, and the tales in it are by turns shocking, inspiring, and memorable. It is
a magnificent accomplishment of rhetoric, if weakly edited. But as many scholars have
noted, it is not an entirely accurate or fair assessment of either Moses or New York,
and it lacks psychological nuance.[3] Caro's empathy for those harmed by Moses's work
is profound, but he does not distinguish well between structural and contingent
forces, he takes little account of geography, and deeper questions of intent and action
are often shoved aside. To his credit, Caro was able to elicit interviews and commentary from a large number of well-placed sources, but he is not a careful judge of emotion and bias. Caro was consistently willing to retail an anti-Moses story from an
anonymous source, even when the record indicates that matters were more complex.[4]

In short, Robert Caro wrote a fantastic novel – something in evidence early on,
when he told Moses that:

the function of biographer or historian includes that of literary technician. I would feel I was doing a lousy job if I wrote, "One day, Moses thought of making a park system on Long Island." But by asking where you thought of it, how you thought of it, etc., I can give the reader a sense of place, of activity, of reality.[5]

This triumph of narrativity – abetted by a generational shift in the conception of the political relationship between the individual and the city – helps explain the book's enduring popularity and the force with which Caro was able to define Moses in the public mind. People remember and believe Caro's stories about Moses: the one about the Long Island bridges (built low to keep buses of blacks from Long Island beaches); the one about the swimming pools (kept cold to keep blacks out); the one about the Cross-Bronx Expressway (rammed through a lovely neighborhood for no reason). These narratives are routinely invoked as the epitome of evil Modernist city planning and are used as rhetorical cudgels against large-scale development.

The problem is that these stories aren't true, and acting as if they are diminishes our understanding of the world and our politics.[6] At the very least, our narratives change if we acknowledge that Robert Moses was not a planner, not a Modernist, and usually not evil.

Such narrative reductionism is unfortunately also characteristic of too much contemporary geographic discourse. Some geography fetishizes its status as (social) science; other geography characterizes itself as political interventions over questions of hegemony, resistance, and identity. And as with Caro's stories about Moses, geography's disciplinary narratives have a substantiating and totemizing power, positing themselves as definitional rather than as interpretive, discursive, and metaphoric.

Eight years after first reading *The Power Broker* and slowly beginning to doubt its depictions, I saw a novel/dissertation as a way to redefine both Caro's portrait and – more important – some of the standard methods of writing about place.

It is less notable that my dissertation was the first overtly fictional one in geography than that one had not been undertaken sooner. Earlier generations of geographers – including Ralph Hall Brown, J. Wreford Watson, Douglas C. D. Pocock, Yi-Fu Tuan, and Donald Meinig – viewed the harmonious integration of literary sensibilities into their field as a desideratum, much as historians have.[7]

Brown, especially, took considerable care to establish not simply the physical realities of a location but also the evolving understanding of its significance and structure. He was alert to the perils posed by facile or unreliable sources yet used fiction as a tool toward truth – a clarifying lens through which to assess the state of specific geographical knowledge.[8] His *Mirror for Americans* (1943) purports to be the analysis of a fictional colonial-era protagonist, whom Brown uses as a filter to humanize colonial conceptions of geography and to stress their contingent and contextual nature.[9]

Much as images of place can shape actual places, fictional writings exert influence on our perception of places. Following historiographer Hayden White, literature *is* the nexus between what happens and how we begin to make sense of it.[10] Fiction is thus not merely reportage indexical of society's ills, nor is it only a tool of social resistance.[11] Rather, literature is a non-teleological testing ground for experience-based understandings of the world.[12] That is, literature organizes perception and experience alike, while also generating a product that can be critiqued. A good novel articulates contrasting modes of perception and overlapping understandings of the world – not merely conflicts in interpretation but irreducible interweavings of individual subjectivities, physical realities, and cultural norms. For this reason, *Everything Must Go* is its own statement of geographic principle.

Unlike most academic discourse, novels do not have abstracts or introductions that give away everything to come; they often lack clear chains of causality. Novelistic language alludes, echoes, and defers closure. It is descriptive and tends to close in on meaning slowly. It can encourage the development in the reader of what John Keats called negative capability – the ability to suspend judgment and analysis of contradictory material and impulses, resisting the drive toward explanation and resolution.[13] There is not much point to using literature to say obvious things or to say things that can be said in more explicit ways.

Moreover, whereas much non-fiction strives for a voice that denies its own reliance on metaphor and abstraction, fiction is premised on a writer's and readers' shared awareness of the work's constructed nature. In a work that focuses on the material, actions, and underlying mentalities of daily existence as the constructs that they are, this recursive structuring abets a complex engagement with not always commensurable understandings of the world.

Fiction might thus be a better vehicle for contesting narratives like Caro's than non-fiction, fighting story with story. Certainly, academic critiques of Caro have had little impact on public discourse. To re-envision Moses's actions and experience as mortal and contingent requires the self-reflective, self-structuring mechanisms of fiction. My Moses is a man torn by knowledge and ability, sexuality and community, action and resistance. He is also the father of a black saxophonist – a depiction that is no less ridiculous (yet arguably more consistent with the record) than some of what Caro put forth. Most important, my Moses is placed within a web of other New Yorkers, real and imagined – most with less political power than he, but all wrestling similar influences.

These enduring tensions are entangled with the subjectivities of characters. This interpellation of subjective and objective states brings geography and fiction together, since narratives and understandings of place entail both perspectives. This does not mean that a novelist or geographer need only take account of how people "feel" about a place; rather, it is a matter of interrogating how, when, and why individual perception matters – or doesn't. The career of Robert Moses was full of events that engaged this conflict directly.

Everything Must Go is what Blanche Gelfant defined as a synoptic novel, in that it attempts to depict the complexity of the city, rather than the perspective of an urban individual or individuals alone. Such works are characterized by "contrasting and contiguous social worlds … multifarious scenes … [a] tenuous system of social relationships, meetings, and separations, and its total impact as a place and atmosphere upon the modern sensibility."[14] Accordingly, there is a great deal of material in it that does not address Moses directly. Stories about three fictional men – Moses's early chauffeur (Riley Pageler), a deranged mid-century saxophonist (Ray Pageler), and an embittered labor organizer (Miles Pageler)–are used to illuminate different periods and aspects of Moses's life and to draw attention to the contextual nature of what we know about him. There is also much that is not overtly geographic – such as the lengthy descriptions of the mid-century jazz and performance-art subcultures.

This material does, however, develop a thesis about Moses and New York's physical and cultural geography by self-consciously addressing the ways that people and their actions become representative (or non-representative) of a mode of thought. Ray Pageler's attempts to cultivate power and influence contrast with those of Moses, who is typically reviled as the embodiment of blind, heartless Modernism, imposing rationalistic solutions on vibrant, organic communities. While there is a degree of truth to this portrait, it sidesteps the issue of how Moses accrued such disproportionate force, both as a shaper of the physical environment and as a symbol. The contemporaneous discourse of jazz – with its arcane hierarchies, rarefied discourse, and valorization of individual physicality – can be seen as the twisted shadow of Modernist rationality. Moreover, Ray Pageler's worldview embodies what Peter Sloterdijk termed "kynicism," a philosophical recognition of the need for the individual to believe in his or her own efficacy despite the preponderance of evidence to the contrary for most.[15]

These overlapping and intersecting narratives take inspiration from John Dos Passos's work, especially *Manhattan Transfer*, which embraces the fragmentation of the modern city as a (non-)organizing principle.[16] The point of articulating various perspectives and attitudes is not to resolve their complexity but to heighten it by focusing not on the "likeability" of various characters but on their social and historical contingency. That people with and without power can be ugly, prejudiced, untrustworthy, violent, and ill-educated while also being cultured, thoughtful, funny, and attractive is central to the case. No one perspective dominates any contested landscape – something novelists at least since Gustave Flaubert have recognized to be true of urban form and novelistic structure alike.[17]

In looking to the nexus of individual action and urban complexity, we can see in both real landscapes and novelistic depictions the influences of both authorship and social constraint.[18] The nature and characteristics of certain places and depictions are, indeed, attributable largely to the will and machinations of single individuals; many others arise from clusters of influences, from structural forces, or as secondary products of other developments. But neither a responsible geographer nor a compelling novelist can ascribe all the good or evil in the world to individual or structural agendas alone.

Everything Must Go is not a lecture on Robert Moses. It is not merely a reinterpretation of Caro's take on him. It is an embodiment of a metaphoric and intrinsically incomplete worldview that is shared by the structures of fiction and geographical perception alike. For all of its faults, backwardness, and laughable introspection, the discipline of geography has always been at the fore in asking, "How do we comprehend our physical environment?" The answers have been both descriptive and prescriptive yet often disappointing. But a recurring theme in the literature has been the insufficiency of existing structures to fully account for the complexity of the world. There are methods, discourses, perspectives, and topics we have yet to incorporate, and until we do geography's conclusions will remain parochial and unsatisfying. It may fall beyond the power of fictional discourse to remedy these disciplinary shortcomings, but deploying it can only help our attempts to grasp our world in less partial and more truthful ways.

These two excerpts from *Everything Must Go* illuminate both the technocratic optimism that underlay urban renewal and the lived geography of it.

In the first, set in July 1949, Robert Moses gives his perspective on the planned Cross-Bronx Expressway to Frank Lloyd Wright as they are being chauffeured near the Washington Bridge over the Harlem River. The years of correspondence between the cousins – who called each other Mole and Skylark – reveal a host of commonalities and a friendly antagonism.[19]

> "We're beginning the Cross-Bronx limited-access road here," said Moses. "Here and at the other end, in Soundview, in the marsh. Most of it is just lines on a map still, but we can strengthen this bridge, begin to build the connections we'll need. This bridge has a lot of unused capacity.[20] We're adding lanes, removing some sad plants – where is a root structure to go on a bridge? I ask you. But this is an easy preliminary. There's no one to displace here, only traffic, and that stream always comes back stronger for being diverted. Look at it, Frank."
>
> The tiny old man craned his head out of the puttering car to see the enormous double-arched underside of the bridge across the Harlem.[21] It wasn't hard to imagine endless trickles of cars poking across it. Wright had no mental picture of the rest of the projected journey across the Bronx, however, and assumed Moses would find a way to make it much like the Harlem River Drive and Henry Hudson Parkway: smooth, wooded, and at one with the environs.
>
> "I thought I heard you say some time ago that the cost of a road across the Bronx would be prohibitive. Has the Mole become inconsistent as he ages?"[22]
>
> "Not a bit. The road will be blindingly costly.[23] But *not* building it will be the death of this city. Anyway, personal preference doesn't enter into my responsibilities. Do you think I *want* to build such a road? Far easier to route traffic southward from here and then onto an improved Bruckner Boulevard.[24] Those roads together make a V around the Bronx, using low-lying water spaces and industrial lands, instead of a hashmark across it. We're trying the same thing in Brooklyn – going around it on the Belt rather than hacking across. But that's no longer enough – your own Broadacre requires monstrous mobility, and that means more traffic everywhere, including here."[25]
>
> "Is that why I hear of your plans for expressways across Manhattan?[26] Well, anything to tear out some of the senseless piles of money-turned-brick, but surely those don't qualify as easy either."
>
> "No, no, but any fool can see how crucial they are. To fly through the Hudson tunnels or across the East River bridges only to be met with the very congestion and cantankering traffickism that those projects were meant to deny is an insult to man's ability to surmount even the worst problems that his own ingenuity can cook up. Some simple elevated runs on Broome and Thirty-first and eventually 125th will speed things up so well that people will wonder how they could have lived without them."

At this, Wright shook his head the smallest bit. "Can it be that the city's prisoners would venerate a highway? No true church, that. Horrible to envision."

"We have to make the space back, of course. We're *saving* the city with traffic, Frank, not emptying it out."[27]

In this excerpt, set years later, Moses's fictional son – troubled veteran and saxophonist Ray Pageler – surveys the urban-renewal panorama visible from the Cumberland Street Hospital in Brooklyn, where he is a resident in the detox ward.[28]

Ray Pageler goes back to looking out the window.

It was a while before I stopped taking buildings for granted. Look at them as process. *Everything is becoming.*

Those projects on all the demapped blocks between Myrtle and Park haven't been there forever. They're still new to someone. Walked past them a hundred times before I asked myself what was there before – who had lived there and what it had been like. Look at Saint Michael's down there to the left. The turrets sticking up above the public housing. Nurses here call it Saint Edward's still.[29] You can see scraps of the old mixed right in, and that makes the new shine even newer. By the church there, look, you got that cute old library, and that behemoth school, too.[30] That was community before, and I guess it's community now. The support system outlasted the residents it was built to serve. Nice wide sidewalks now.[31] Filling the rest of the view are those six-story brick double-crosses, one next to the other, punctuated by the perfect grids of eight-panel windows. There's something cold there, something inexorable. These things don't go up for no reason. Someone wants them. Someone's making money on them. Someone likes the view.[32]

Where I guess the last of the tenements had been, now they got a community center that looks like a concrete octopus. From up here you can see that it's like the houses themselves: plopped in the middle of a cleared lot, tentacles sprouted this way and that. Dunno what's in there. Bulletin boards, most likely. A sewing circle? Gym class? Whatever it is, I suppose it used to be different, like pretty much everything else. Before the war, there wasn't any public housing out here or anywhere else.[33]

This whole area is under military occupation. Has been for a long time. Put your forts by the rivers and you control the land. Even those idiot Pilgrims knew that. And the Dutch, the Dutch were real smart: they got a city that comes to a point right where it's most vulnerable, so they slap a garrison there. Put another on the big, flat island in the harbor. Presto: the fort of New York.

Now, this part of the archipelago is kinda sheltered. Look how that inlet off the East River protects what's happening at the Yard. Look at everything you'd have to pass upstream to get in there – past the narrows, past the harbor forts, under the Heights, under the bridges angling like a compass from Tillary and Flatbush.[34] Safe place. Most the jobs here come from that Navy Yard, directly or indirectly, and the jobs are what keep people from rising up and screaming to the sky. Buncha places already closed on account of the Civic Center, but not over here yet.[35]

So by the time of the big war, the Navy was already the all-and-all here, which meant that it could get whatever it wanted double-quick. Robert Moses dreams he has that kind of power. Moses didn't build all these houses, but they were a test run, and he was watching. They threw up those cheap-ass Wallabouts first, and then the rest of them, knocking down block upon acre not because they cared about the poor folks in the wood-and-plaster shacks from Civil War and beyond. Hell no.[36]

The war just provided an excuse for all kinds of hurry-up. Gotta build things quick, get ships in the water, planes in the air, boots on the ground, and houses on the ground for war workers to sleep in, can't ask people to live in tents. But at the same time everything was getting speeded up, they were also getting slowed down. All the good metal, all the rubber, basically anything that got made industrially, went to the war. So that housing had to come from somewhere else – no one was going to be dropping a nice solid saltbox on Jerry after all. So make those rooms out of cardboard, wood chips, and smelly glue. Throw some mattresses and hot plates in and let's go! Solve the problem and get on to the next one! Get things done! Don't you know there's a war on?[37]

Now this place is surrounded. City just put a new name on those Fort Greene Houses. Fact, they put two.[38] Didn't change much of nothing about the place. They're the same brick, the same patchy grass, the same poor people. You have to look beyond the name on the thing. Maybe I finally know that. The way it is is the way it is.

So in the war we learned to build things fast fast fast, and when we won we figured that must have been why: we threw a lot of shit together in a hurry, hard after a single big idea. But no reason to stop with Europe – we turned the speed on ourselves and brought the wreckage home. No Red's got to drop a bomb on Sands Street or the Lower East Side or Harlem 'cause we blasted flat the terrain all by ourselves.[39] There's more to it than military convenience, but that's one way to look at why Brooklyn's changing so much.

In France, guys told me about Paris, with the streets wide enough to drive tanks down, a clear field of fire, space as a weapon.[40] We're not doing that here. Here, we're building the road *on top of* the people and places that might object to it, or at the very least between them, separating like from like. Now why do we do that? Far as I can see, it's more a question of where roads can't go than where they do. Once we eliminate the impossible, the political seems more probable. "Can't" is a funny word; run it through that political-economic-social-racial urban wringer and sometimes it winds up meaning "will," depending on what hand holds the pens and paperwork, what voice intones the truth. All you have to do is look toward Brooklyn Heights to see that.[41]

Acknowledgments

My work would not exist without the support of John S. Adams, Judith A. Martin, Katherine Solomonson, David Treuer, and especially the late Roger P. Miller, of the University of Minnesota. I thank Sarah Luria for her editorial guidance and patience.

Notes

1 M. S. Samuels, "The Biography of Landscape: Cause and Culpability," in D. W. Meinig (ed.), *The Interpretation of Ordinary Landscapes: Geographical Essays*, New York: Oxford University Press, 1979, pp. 65–66.

2 R. Caro, *The Power Broker: Robert Moses and the Fall of New York,* New York: Alfred A. Knopf, 1974.

3 See, among others, E. N. Saveth, "The Moses Model," *Reviews in American History* 4.3, 1976, 451–57; and J. H. Kay, "The Master Builder and His Works," *The Nation*, 28 September 1974, 277. The most thorough professional critiques are in J. P. Krieg (ed.), *Robert Moses: Single-Minded Genius,* Interlaken: Heart of the Lakes, 1989.

4 See K. E. Markoe, "Robert Caro and His Critics," in J. P. Krieg, *op. cit.*, pp. 47–54.

5 Robert A. Caro to Robert Moses, September 14, 1967, in folder C, Box 53, Robert Moses Papers, Manuscripts and Archives Division, New York Public Library, Astor, Lenox and Tilden Foundations (RMP). More recently, Caro said, "I have always thought ... that in nonfiction, the level of the writing has to be as good as any novel if it is going to endure." J. Darman, "The Marathon Man," *Newsweek*, February 16, 2009.

6 For other debunkings, see R. Bromley, "Not So Simple! Caro, Moses, and the Impact of the Cross-Bronx Expressway," *Bronx County Historical Society Journal* 35.1, 1998, 4–29; and most of the essays in H. Ballon and K. T. Jackson, (eds), *Robert Moses and the Modern City,* New York: W. W. Norton, 2007; as well as the ambivalence in M. Berman, *All That Is Solid Melts into Air: The Experience of Modernity*, New York: Simon and Schuster, 1982.

7 On historians, see L. Gossman, "History and Literature: Reproduction or Signification," and H. White, "The Historical Text as Literary Artifact," both in R. H. Canary and H. Kozicki (eds), *The Writing of History: Literary Form and Historical Understanding*, Madison: University of Wisconsin Press, 1978, pp. 3–39, 41–62.

8 For example, R. H. Brown, "Over-Imaginative Travelers," *Hobbies*, September 1942, 109, 124.

9 R. H. Brown, *Mirror for Americans*, New York: American Geographical Society, 1943.

10 H. White, "The Value of Narrativity in the Representation of Reality," *Critical Inquiry* 7.1, 1980, 5–27.

11 Contemporary human geographers tend to take this perspective. See S. Smith, "Qualitative Methods," in R. J. Johnston *et al.* (eds), *The Dictionary of Human Geography*, 4th edn, Malden: Blackwell, 2000, p. 661.

12 See D. C. D. Pocock (ed.), *Humanistic Geography and Literature: Essays on the Experience of Place,* Totowa: Barnes and Noble, 1981, p. 12. On the importance of literature's "nondirectedness," see Y.-F. Tuan, "Literature and Geography: Implications for Geographical Research," in D. Ley and M. S. Samuels (eds), *Humanistic Geography: Prospects and Problems*, Chicago: Maaroufa, 1978, pp. 197–98.

13 See D. Walsh, *Literature and Knowledge*, Middletown: Wesleyan University Press, 1969, p. 70.

14 B. H. Gelfant, *The American City Novel*, Norman: University of Oklahoma Press, 1954, pp. 14, 235; see also P. Keating, "The Metropolis in Literature," in A. Sutcliffe (ed.), *Metropolis 1890–1940*, Chicago: University of Chicago Press, 1984, pp. 129–45.

15 P. Sloterdijk, *Critique of Cynical Reason*, tr. M. Eldred, Minneapolis: University of Minnesota Press, 1987.

16 P. Keating, *op. cit.*, p. 143.

17 See R. Alter, *Imagined Cities: Urban Experience and the Language of the Novel*, New Haven: Yale University Press, 2005; and Berman, *op. cit.*, pp. 255–75.

18 M. S. Samuels, "Biography of Landscape," *op. cit.*, p. 64.

19 Their correspondence is in Special Collections and Visual Resources, Getty Research Institute Research Library, Los Angeles.

20 A good summary of this bridge's history is at www.nycroads.com/crossings/washington-heights.

21 Detailed accounts of Moses's relationship with Wright are in C. Rodgers, *Robert Moses: Builder for Democracy*, New York: Henry Holt, 1952, pp. 249–53 – which establishes the potential date of this meeting – and R. Moses, *Public Works: A Dangerous Trade*, New York: McGraw-Hill, 1970, pp. 855–72. Wright may have been in New York to promote his book *Genius and the Mobocracy*.

22 Wright's rhetoric is derived from *Frank Lloyd Wright: The Mike Wallace Interviews*, VHS, New York: Archetype, 1994 [1957].

23 "The cost of slashing a major artery crossing the Bronx easterly from the George Washington Bridge would be prohibitive," R. Moses to F. LaGuardia, November 11, 1940, in Triborough Bridge Authority, *Vital Gaps in New York Metropolitan Arteries*, New York: TBA, 1940, p. 5. See also R. Moses, memo to the City Planning Commission, December 10, 1940, folder 006, roll 6, microfilm records of NYC Department of Parks, Office of the Commissioner (Robert Moses), Municipal Archives, New York (MANY). Moses may have been spiting those who proposed such a road (R. L. Duffus, *Mastering a Metropolis*, New York: Harper and Bros., 1930, p. 158).

24 See Triborough Bridge and Tunnel Authority, *The Traffic Improvement of Bruckner Boulevard from the Major Deegan Expressway and the Triborough Bridge to the Bronx River Expressway near Longfellow Avenue, Borough of the Bronx*, New York: TBTA, 1950. The route of the Cross-Bronx was fixed no later than January 1944. R. Moses, "Memorandum to the Mayor on the Cross Bronx Expressway," April 19, 1946, folder 022, roll 31, MANY.

25 See F. L. Wright, "Broadacre City: A New Community Plan," *Architectural Record* 77, April 1935, 243–54.

26 For Moses's projects around this time, see R. Moses to the Mayor, July 21, 1947, folder Robert Moses Correspondence (Miscellaneous), Box 90, RMP.

27 C. Rodgers, *op. cit.*, p. 300.

28 The hospital opened in the early 1920s and was expanded in 1953.

29 See J. K. Sharp, *History of the Diocese of Brooklyn, 1853–1953: The Catholic Church on Long Island*, New York: Fordham University Press, 1954.

30 See Fiorello H. LaGuardia Community College, La Guardia and Wagner Archives Photo Exhibit, *New York Transformed, 1939–1967, Photos from the New York City Housing Authority Collection*, "Spared the Wreckers Ball," at: www.laguardiawagnerarchive.lagcc.cuny.edu/PhotosVirtualExhibit/ShowPhotosDetails.asp?photo=02.003.26874&ShowPage=9&PhotoID=816. On New York housing see R. K. Plunz, *A History of Housing in New York City: Dwelling Type and Social Change in the American Metropolis*, New York: Columbia University Press, 1990.

31 On the relation of sidewalk width to livability, see J. Jacobs, *The Death and Life of Great American Cities*, New York: Random House, 1961, Chs. 2–4; and A. Duany et al., *Suburban Nation: The Rise of Sprawl and the Decline of the American Dream*, New York: North Point, 2000, pp. 12–17.

32 On the overhaul of the Brooklyn waterfront districts, see J. Schwartz, *The New York Approach: Robert Moses, Urban Liberals, and Redevelopment of the Inner City*, Columbus: Ohio State University Press, 1993, pp. 229–47.

33 This assertion is incorrect. See R. K. Plunz, *op. cit.*, Ch. 7, and C. Mele, *Selling the Lower East Side: Culture, Real Estate, and Resistance in New York City*, Minneapolis: University of Minnesota Press, 2000, Ch. 3. The Fort Greene Houses were, however, the first state-financed project of their kind and larger in scale than their predecessors.

34 Pageler is here echoing Moses's observations about Navy Yard access during the Brooklyn-Battery Bridge controversy of 1939. See R. Moses, *Public Works, op. cit.*, pp. 202–7.

35 See City Planning Commission, *Master Plan of the Brooklyn Civic Center & Downtown Area*, New York: City of New York, 1945. On the elimination of industry, see J. Schwartz, *op. cit.*, Ch. 9.

36 P. Marcuse, "Housing Policy and the Myth of the Benevolent State" (1978), in R. G. Bratt, C. Hartman, and A. Meyerson (eds), *Critical Perspectives on Housing*, Philadelphia: Temple University Press, 1986, pp. 248–63, holds that urban renewal was simply an economic shell game.

37 On the shoddiness of military wartime housing, see, for example, "300,000 Temporary War Houses May Find These New Uses," *The Architectural Forum* 8, 1945; on wartime materials restrictions, see H. C. Mansfield and Associates, *A Short History of OPA*, Office of Temporary Controls, Office of Price Administration, general publication no. 15, Washington, D.C.: GPO, 1948. On this specific project, see R. A. M. Stern, T. Mellins, and D. Fishman, *New York 1960: Architecture and Urbanism Between the Second World War and the Bicentennial*, New York: Monacelli, 1995, p. 901.

38 Administratively, the Fort Greene Houses were split into Walt Whitman Houses and Raymond Ingersoll Houses in 1957–58. R. A. M. Stern et al., *op. cit.*, p. 901.

39 R. A. M. Stern has suggested that America's urban-renewal plan was in part motivated by guilt over wartime firebombings (R. Burns (dir.), "The City and the World," part 7 of *New York: A Documentary Film*, 2001). Fear of nuclear attack on

urban centers was profound at this time, which influenced urban policies. See, for example, *Building in the Atomic Age*, proceedings of a conference sponsored by the Department of Civil and Sanitary Engineering, MIT, Cambridge, MA, 16–17 June 1952, and R. K. Plunz, *op. cit.*, pp. 277–79. On reconstruction on Sands Street, see "City Planning Commission Speeds Civic Center Project for Brooklyn," *The New York Times*, March 22, 1945, 25.

40 Georges-Eugène Haussmann's reconstruction of Paris is often characterized as a facilitation of police power (e.g., D. Harvey, "Money, Time, Space, and the City," in *The Urban Experience*, Baltimore: Johns Hopkins University Press, 1989, p. 193). J. M. Chapman and B. Chapman, in *The Life and Times of Baron Haussmann: Paris in the Second Empire*, London: Weidenfeld and Nicolson, 1957, pp. 184–86, note, however, that Haussmann, like Moses, was obsessed with issues of health and congestion. Moses's take on Haussmann is R. Moses, "What Happened to Haussmann," *The Architectural Forum*, July 1942, 57–66.

41 Pageler is referring to the fact that the Brooklyn-Queens Expressway is widely thought to have been rerouted to avoid gutting a largely white and politically connected community (R. A. M. Stern *et al., op. cit.*, pp. 896–98). The tale behind this is more complex than is conventionally understood.

10

Fate and redemption in New Orleans
Or, why geographers should care about narrative form

Barbara Eckstein

To answer the question announced in the title, let me begin with two specific narratives: Michael Ondaatje's *Coming Through Slaughter* and Frederick Turner's *Redemption*. On the first pages of Ondaatje's 1976 novel *Coming Through Slaughter*, he places two images. One is a blurry reproduction of a photograph taken at the turn of the last century. It is a picture of a New Orleans band typical for the period. This one, though, contains the only known image of Buddy Bolden, the legendary cornet player, whose virtuosity has been much admired in the memories of early New Orleans jazz musicians but never recorded. Because Bolden allegedly had a psychotic break, authorities sent him to the state asylum in Slaughter, Louisiana – where he died years later – just as the technology of recording was coming into its own. Thus, the picture introduces a paradox: a photograph of music. The other image emphasizes this synaesthetic, that is, cross-sensory paradox: it is a string of three sonographs, a scientific image of dolphins' squawks, clicks, and whistles. Both images are underwritten by captions that act like ekphrasis, verbal descriptions of images that we are seeing. After these two images, Part 1 of the novel commences with the curious words "His geography." These words are followed first by the creation of a scene of reading for us, the reader – "Float by in a car today and see the corner shops. ..." Then Ondaatje introduces the early twentieth-century tropes that have defined – still define – New Orleans as a place. These tropes, that is, common representations, are, for example, the prostitutes classified by race, the piano players in the brothels, and the patron of New Orleans underworld in 1900, Tom Anderson. The multifaceted opening of the form of Ondaatje's novel implies that *how* we read these tropes defines "his [Bolden's] geography."

Even more than Bolden, Frederick Turner's central character Francis, Fast-Mail, Muldoon, is one dependent upon the patronage of the notorious Tom Anderson. Turner's 2006 novel *Redemption*, like Ondaatje's novel 30 years before and many others besides, deploys the platial tropes of New Orleans: pre-preservation buildings modest and grand; prostitutes classified by race and class; the photographs of them by E. J. Bellocq, the musicians who entertained their johns; and Tom Anderson, the "mayor of Storyville," whose power exceeded any held in official hands. Both Ondaatje and Turner reproduce the space of the city of New Orleans as that which lies within or radiates from the neighborhood Storyville – that small prostitution district extant between 1897 and 1917 and legendarily controlled by Tom Anderson.

But the narrative forms of the two novels differ in important ways, and these different forms steer our reading in separate directions, including our reading of the novels' and New Orleans's geography. Ondaatje starts his narrative not with words but with two images – one historical and the other contemporary and technological – that cross-reference the workings of our senses. Before any story unfolds, these images and the accompanying text raise questions about how knowable Bolden, his music, or the time and place of his life can now be. They, in fact, raise questions about the reliability of knowledge derived from human senses, the very basis of what we humans think of as empirical observation. Then Ondaatje places us, the contemporary readers, in a car traveling, in the present, through the poor downtown neighborhood once home to Buddy Bolden, a neighborhood apart from Storyville. We enter the story, cruising the present streets in search of the past. Different voices, sometimes Bolden's, sometimes others, try to tell Bolden's story, backing over and around the same history, returning often to the geography through which Bolden traveled to get from New Orleans to Slaughter: "Passing wet chicory that lies in the fields like the sky."[1]

By contrast, Turner, who begins his historical novel in the same downtown neighborhood, relies instead on mimesis – artistic recreation of the empirically observable and nameable world – to open and sustain his realist tale. In this realist form, the place and the past are both delivered by a third-person narrator to the reader as though thoroughly accessible. Sensory details affirm the narrator's authority to speak for this downtown neighborhood at the turn of the last century. Turner writes, for example:

> Up and down Washington Street lamplight began to seep from between the sun-blistered shutters of other cottages catching edges of the furious decorations of their facades – the carved fans and furbelows of the carpenter's jigsaw, the scrollwork of the lintels, the gingerbread icing of the eaves. Beyond the high, darkening bulk of the Double X Brewery down on Tchoupitoulas he could just make out portions of the steamers from Liverpool and Antwerp, Bordeaux and Genoa, moored at the Stuyvesant Docks whose broad wharves lay below street level.[2]

In Turner's novel, the redemption emphasized by his title comes, in this realist world, only for the individual, central character, Fast-Mail, who finally separates himself from the incorrigible Anderson and his minions. A character of limited education and insight, Fast-Mail lags behind the narrator and the reader in understanding the corrupted world that he inhabits. Dramatic irony separates his limited knowledge from the greater awareness shared between the narrator and the reader. When his redemption finally comes, as in the novel's final pages he confronts and then walks away from Anderson, that redemption is Fast-Mail's alone. There is no redemption for Anderson's violent minions or for the city itself. There is no reimagining of space, questioning of its susceptibility to time, or doubting of the author's and reader's sensory acuity that represents that space. Instead, the novel defines New Orleans through familiar representations of sexual exploitation and graft and by the accommodating art forms – mostly jazz – practiced to entertain those who embrace the exploitation and to distract those who ignore it. The city is static space. As is conventional for realist fiction, only the consciousness of human character moves, only one character at the center of the tale.

Rather than shelve Turner's novel as ably written yet unremarkable realism or as simply not my cup of tea, I suggest it provides an opportunity to think about two important and related points. First, a novel about New Orleans entitled *Redemption* and published post-Katrina in 2006 deserves special attention from all those who want to understand the role of story in the production of space. Furthermore, the narrative form of any story and how perceptively we read that form affects the kind of role it plays in the production of the places it represents.

Sociospatial theorist Henri Lefebvre claimed in the 1970s that realist writers, more than his sometime friends the surrealists or the situationists, could not disrupt the production of space by the forces of capitalism and uneven development.[3] Instead realist narrative only held a mirror up to space, representing it as inevitably or invisibly developed by capitalist assumptions – assumptions such as economic growth in a given territory benefits everyone in the territory or pollution and poverty are necessary bedfellows. Similarly, scientific rhetorician Donna Haraway argued in the same post-60s era that we have to look deep within the microscope to find metaphors of space and organization that will disturb what she speaks of as our heteronormative, bourgeois practices of reality and realism.[4] Anyone who has ever curled up on a rainy night with a well-wrought, realist novel – and there are many, of course – will object mightily to Lefebvre's and Haraway's presumption. Finely tuned psychological insight, humor, cathartic loss, descriptions of a city's underbelly and a nation's ideological flaws, all are among the rewards of being lost in a vivid, realist novel. One of my recent favorites is Ethan Canin's *America, America*.[5] As the gentle narrator unravels this tale about personal and political innocence lost, the title turns from triumph to keening. The beauty and pleasures of such art surely deserve our attention, we might argue, and yet, Lefebvre and Haraway reply, ours is a guilty pleasure if we fail to recognize in such work the reification of spatial and social arrangements as human history, human senses, and dominant economic systems offer them to us. US urban and suburban spatial arrangements, class structure, circumscribed mobility, monoracial vision – all are features of Canin's novel. We question these assumptions only to the extent the novel's form raises the questions or we the readers resist its form. By contrast, the novelistic form that questions human sensory constructs of space (and time), like the biological study that reveals surprising social and spatial arrangements of other life forms, provides designs and metaphors for alternative arrangements of human space and society.

I chose Turner's novel to re-examine these judgments of literary realism not only because the site of his novel is New Orleans, but more emphatically because as a folklorist of American Indian tales, a critic of Manifest Destiny, a biographer of John Muir, and a contributor of platial prose to art exhibits and books, he has had many opportunities to study the role of story and literature in the production of space.[6]

That is, it seems particularly fair to expect Turner to be sensitive to the production of space. Taken as a whole, his work expresses a commitment to the environmental and cultural value of places, especially wilderness places, and still, the human sensory and temporal conventions of realist prose, such as that exhibited in *Redemption,* inhibit any attempt to question spatial practices predicated on human mastery, especially human mastery by the world's materially privileged.

Because Ondaatje and Turner are preoccupied by the same territory and much of the same history, yet the forms of their novels differ, the two together can help us understand Lefebvre's claim and Haraway's. Turner's narrator delivers information about early twentieth-century New Orleans and its wildest side – quotidian violence against prostitutes, ruthless containment of Negroes – without betraying any judgment. Only Fast-Mail has an overtly reflective consciousness and his is limited. The twenty-first-century reader is left to observe the horrors described by the narrator and the moral limitations displayed by the central character and consider what choices Fast-Mail has, what choices we might make in his place. We are pulled into the story and on these terms. By contrast, Ondaatje pulls us into history – with the old photograph, for starters – and then yanks us out of the story – with the sonograph, for example. We are pulled between stories of the past and concepts in the present, making the reader, rather than the character, the one struggling to understand. Ondaatje drives us through Bolden's geography – another concept or abstraction: not just a cruise through the neighborhood of his birth but a disorienting trip through a world of poverty, race, corruption, and genius to a state of being and a place called s/ Slaughter. In that other corrupt place, we eventually realize, a founding father of New Orleans jazz went forever silent. There he apparently lost self-assertion and became, instead, an amalgam of those who presume to put words to his character, story to his photograph, or sounds to his music – those like Ondaatje and the reader. The pastiche of the novel repeatedly pulls us into and out of the story so that we both care about Bolden and know we cannot know him. What we can struggle to know instead is the space and time he occupied and what forces produced that geography and history.

By contrast, Turner takes us inside the mind of a crippled former policeman, now Anderson lacky, who lost his identity when an injury stole his talent as a runner and gains a moral center when he mistakenly attempts to redeem a young woman within Anderson's power. Ondaatje creates a geography underneath the radar of history that has always focused on Storyville. "History was slow [in Bolden's neighborhood]. It was elsewhere in town, in the brothel district of Storyville, that one made and lost money."[7] In Bolden's geography that Ondaatje imagines, Bolden's downtown barbershop and then, in the end, the off-stage asylum at Slaughter circulate through the better-known streets and the music of New Orleans. Turner creates a geographic scale conventional for realism, a scale focused most on one human body, one central character and his psyche. This individual character learns in the corrupted bars and streets of New Orleans that there is no innocence to rescue but his own.

If the stakes of this investigation of literary form seem confined to a narrow aesthetic realm, I have not yet made myself clear. The fate, in this case of New Orleans, hangs in the balance. Fate, says the *Oxford English Dictionary,* is simply a derivative of the Latin verb *fari,* to speak. *Fatum* is that which has been spoken. Inflected by its

Greek ancestor, linguistic *fatum* becomes a person's lot or portion – a person's space, you might say. What has been spoken about a person's space – *how* those words have been formed – produces that space, perhaps even becomes the destiny of that space.

And yet that destiny might, with difficulty, be disrupted. Rather than space being a static product of what has been spoken about it, Lefebvre declares that fate is a dynamic trial by space. That is, it may be possible to change what is spoken about a space, how it is represented, the tropes, metaphors, and scales that govern human perception. As a result it may be possible to change the production of space itself. Although transformational trials may also be traumas, in them lies hope that humans can recognize how their spatial habits, especially habits under capitalism, reproduce socioeconomic inequity and environmental catastrophe. To put it in Lefebvre's terms, it is possible, though difficult, to change abstract space into differential space.

In *La production de l'espace*, he writes – as translated by Donald Nicholson-Smith:

> The [spatial] transition here considered is characterized first of all by its contradictions: contradictions between (economic) growth and (social) development, between the social and the political, between power and knowledge (*connaissance*— [i.e. what is known from experience rather than received knowledge—*savoir*]) and between abstract and differential space.[8]

These are contradictions familiar to planners and anyone who grapples with concepts and practices of sustainability. Economic development, wedded to real estate development, is traditionally the governing force in US city planning, despite a host of social, environmental, historic, and aesthetic interests often present to resist this presumption of economic and spatial growth or sprawl. Supporters of sustainability assert that the presumed oppositions among economic growth, environmental health, and social equity are instead a web of mutual dependence.

Lefebvre's interpretation of the contradictions is, of course, distinctly Marxist in its orientation (although, not, in the end, in its views on history). In Lefebvre's keywords, abstract space is that space created by capitalism, a kind of soup of commodity culture in which users passively float thinking they have real choices. Even those at the short end of uneven development are not necessarily awake to the true choices obscured and suppressed in and by abstract space. In *Redemption*, the representation of the pervasive underworld of Storyville limits characters' choices, but this constraint would not, I think, be the most worrying to Lefebvre because those limits are so apparent to the reader. In fact, the plot depends upon Fast-Mail experiencing redemption through recognition that there may exist alternative moral spaces that the reader already knows. The more dangerous and abstract space is, in fact, that presumed but not created by the novel. It is the unexplored space outside Storyville, where many individuals enable, even demand, a Storyville, where no one interrogates the intersection of "tourist trade" and sexual exploitation, for example. Abstract space is that which pervasive economic, social, political, and cultural forces condition us to see as economically, morally, and environmentally inevitable and even innocent.

Differential space, by contrast, is that which some users – various users, some of them literary artists – have been able to appropriate for alternate purposes. We the reader may be invited to begin Ondaatje's novel as tourists gliding through Bolden's old neighborhood with our car windows rolled up, but the opening cross-sensory images have already warned us that something unsettling is up. By the novel's end, we are instead positioned as second liners, those that follow alongside New Orleans parades. Through the body of a young female fan, the reader joins the second line lured by Bolden's cornet and machismo to follow him down Canal Street. At the climax of charisma and sensuality, when Bolden collapses, the reader is in the crowd on the street to experience a kind of instantaneous transubstantiation of Bolden and of the sociospatial dynamics of the street itself. That is, the form of the novel fractures (again), unable to know the details of what might have happened following Bolden's collapse in favor of versions of that story that all somehow lead to the trip through the fields of chicory to the asylum at Slaughter. Returning repeatedly to the transitional space between New Orleans and Slaughter, "passing wet chicory," the reader begins to understand "his geography." Without the power of his music or his body, a black man of Bolden's class and time disappears through s/Slaughter into a state institution for the insane, and jazz history goes on down Canal Street with whatever version of him it chooses to create. In the trial by space that is fate – whether Bolden's or ours – contradictions, even paradoxes, are present. To keep the discussion within the world of Ondaatje's novel, we might say that the *connaissance* of local knowledge visible in the single, silent, crumpled photograph of Bolden exists in contradiction to the power of recording that made New Orleans jazz an international phenomenon, yet both are features of New Orleans geography through history. From the time of Bolden's life, to the time of Ondaatje's creation, to the time of our reading, that geography has been repeatedly on trial as racialized, class-inflected, ecologically challenged, economically uneven space.

A transition recreating space and place anew will not occur without great human effort and sacrifice – an effort and sacrifice beyond a faith in human senses and a representation of those findings in realist narrative (and realist mapping as well). A transition from the production of things in space, as occurs under capitalism and controlled by the state, and the re-production of space itself is just plain hard, as struggles to rebuild New Orleans demonstrate. The desired differential space writ large is, claimed Lefebvfre in the 1970s, "a goal and a significance that are still far distant ... – and that will never be realized without risking catastrophe, or without bittersweet leave-takings of everything once valued, everything once triumphant." "Sooner or later," he predicted,

> the cultivated elites find themselves in the same situation as peoples dispossessed through conquest and colonization ... because nothing and no one can avoid *trial by space* – an ordeal which is the modern world's answer to the judgment of God or the classical conception of fate.[9]

If we are willing to imagine this spatial apocalypse – perhaps via global warming we are already living it – then, I think, we also must imagine redemption of a communal sort, and a means to that redemption that engages more than the best efforts of individual human consciousness moving alone at ground level through the spatial scales discernible to the naked human eye and ear and hand and rendered in prose via rhythms of human clock time and logic – in other words, realism.

As Haraway encourages us to do, we might look to geographic scales other than those immediately discernible to the human eye or conventional in human social and political arrangement. We might instead imagine the scalar realm of the observant human loner interacting with other scalar alternatives of a larger or smaller and more sensorially unconventional sort. For example, John Mutter, Director of the Earth Institute at Columbia University and an environmental scientist, interprets the scope of the post-Katrina ecological disaster on the Gulf Coast and the death toll in New Orleans on a reconfigured scale of the tropics encircling the Earth. It is in this region – not the South but rather the globe's belt – including, for example, Darfur and Sri Lanka, where some of the world's greatest so-called natural disasters create the highest death tolls. In these regions where water is too present or not present enough, the degree of human poverty is higher. Those who can, live in more temperate climates or are the sheltered or mobile few in the tropics. The existence of extensive poverty, Mutter easily demonstrates, creates disasters, that is, it turns serious environmental challenges – hurricanes, tsunamis, earthquakes, droughts – into wholesale catastrophes. Like their climatic compatriots in Sri Lanka and Darfur, those many citizens in New Orleans without the educational, economic, or physical mobility to flee natural disasters or those who can flee only as refugees dependent upon the state are left stranded and in physical, economic, and ecological danger.[10]

Consider another alternative. Cultural geographer Achille Mbembe offers a fluidity to scalar imagining, or, more broadly, to mapping, when he writes of itinerate territoriality in Africa, that shifting terrain of empowerment and disempowerment existing outside or across nation-state boundaries. In these places refugees dwell longer than temporarily, and roving violence is practiced not only by extra-legal organizations but also by the policing forces themselves.[11] Just as it is revealing to see New Orleans in the context of the tropics arid and wet, it is useful to think of New Orleans, past and present, as part of an archipelago of itinerate territoriality. The persistence of post-Katrina refugee camps, outside the city and even outside Louisiana, has been for some New Orleans evacuees an obvious condition of this itinerate territoriality. But this geographic instability existed pre-Katrina as well, produced by shifting ecological, economic, social, political, and cultural terrain as much as by the geopolitical structures of parish, state, and nation. In fact, the current controversy over the race and class of its refugee population, for example, is one that has been endemic to the itinerate territoriality inside the metro-region for centuries, long before post-Katrina flooding set large numbers of New Orleanians flowing outside the metro-region. From the movement of slaves out of Africa through the Caribbean and on through the markets of New Orleans's to the movement of postbellum ex-slaves into the city to the dislocations of New Orleans poorest citizens in the face of urban renewal and then post-renewal gentrification, New Orleans's poorest citizens have been, over time, participants in a production of space as itinerate territoriality with the attendant history of official and unofficial violence. Mutter's and Mbembe's spatial constructions are but two examples of the alternate scaling and mapping that can disrupt abstract space and human confidence in our naked senses by enabling us to recognize different spatial dynamics – and thus different social and environmental dynamics – than those given by conventional boundaries and jurisdictions. As creative cartographer Denis Wood has been telling us for over a decade, these alternative spatial conceptions that exist outside the expected geopolitical categories teach us that conventional spatial arrangements (e.g., city limits, territory defined by zoning ordinances, interstate highway systems, national borders) have themselves been conceived by particular individuals for particular reasons (at, I would add, particular moments in history). They are not natural.[12] The conventions of realist narrative, like the conventions of realist mapping, are not simply the products of careful human sensory perception. Instead both realist narrative and realist mapping narrowly reproduce a place as a static object of empirical observation, framed by conventionally – often, hegemonically – defined expectations.

That said, alternative scaling and mapping is necessary but not sufficient to meet our collective trial by space. We must return to the role of the reader. We can add to Mutter's and Mbembe's geographic strategies, postcolonial scholar Qadri Ismail's articulation of a postempirical reading practice. Through this postempirical reading strategy he tries to find a way to deauthorize the kinds of naturalized nationalist histories, nationalist imaginaries, that, for example, have driven the territorial claims and the violence of both majority Sinhalese and minority Tamils in Ismail's native Sri Lanka. When empirical observation covers for nationalist histories – as it so often does – something called postempirical reading might effectively expose this nationalist scalar presumption, that is to say, the presumption that a certain nationalist tale deserves a certain nationally defined territory on the ground.

For Ismail, postempirical reading is first of all literary reading that understands history as a "literary activity."[13] Histories, like maps, are made not found. Interpretation of them requires attention to their narrative forms and the making of those forms. I am adding historical novels to Ismail's category of historical storytelling, a storytelling activity that benefits from a postempirical reading practice.

> The postempiricist reader, [Ismail claims] will insist on identifying the narratological level as being at work here [in a Tamil nationalist history, say]: slipping in interpretation, connecting events that could be seen otherwise, producing a continuous and continuing narrative, producing commonality and comparability (in a word, analogy) from events that could also be understood as discrete and different, effectively creating, [for example] a story of broken Sinhalese promises.[14]

Explicitly, postempirical interpretive work defies tidy narratives about spaces and places and their inhabitants, recognizing how the "narratological level," that is, the narrative conventions and forms, is producing the meaning of history and space. Implicitly, postempirical interpretive work must persist even in the presence of revered platial beliefs born of remembered trauma (think, for example, Israel and Palestine) – perhaps especially in the face of such memories. The common crisis of our trial by space has produced very high stakes. If we are to meet those challenges, then we as interpreters cannot too easily give away our sympathies or our common commitment to the land, water, and air that all humans and all life must share.

In Turner's novel his main character Fast-Mail carries his own memory of his trauma: specifically, crippling and cowardice acquired under the gun of Robert Charles, a figure from New Orleans history. Charles was a Negro who, with a gun, defied unwarranted arrest and then even police siege, to the horror of white New Orleanians in 1900. Negroes in New Orleans in 1900 did not often have guns. If, by some oddity of war or country life they did have guns, they did not use them to defend themselves against white figures of authority. Turner imagines his central character as one of those physically brave but ill-trained police officers attempting to dislodge Charles from a house in a warren of densely built, old frame houses and alleyways. This historical event motivates the fictional plot of former police officer Fast-Mail's passivity, physical injury, descent into Storyville, and moral recovery. The novel's memory of this historical event, however, also turns away from the horror for Africanist New Orleanians recorded in 1900 by famous, anti-lynching activist, Ida Wells Barnett. While Turner's character lost his former identity as a powerful runner – thus, Fast-Mail – in the siege of the house from which Charles was firing, Wells Barnett remembers the white mob rule throughout the city that attended the stand-off with Charles.[15] Not only did authorities at that time, according to local newspapers of the day, invite white citizens to pummel the body – especially the face – of the dead Robert Charles until it was unrecognizable raw flesh, but in virtually every corner of the city police were unable to contain white violence against random Africanist citizens whose bodies, especially faces, similarly lost their individual identities. Turner's story provides the fictional Fast-Mail an identity and that identity produces the space of New Orleans out of the gangsterismo of Anderson and his rivals, a corruption that persists as a foil for Fast-Mail's redemption. In the novel, the historical Africanist identity theft implicit in the white violence and in the stories told by the white newspapers in 1900 remains unaddressed, lost in the space of naturalized racism described objectively by the narrator.

The narrator does briefly describe the drums of racial unrest going silent after white mob rule terrorized the city's Negro citizens, but the narrative voice emphasizes the white press coverage that spoke of the "craven cowardice of the police," for this, through the character of Fast-Mail, is the story *Redemption* tells.[16] After Fast-Mail is shot, following orders to rescue a fellow officer, the larger story of legendary Charles and white mob rule are over for him.

> That was the end of the Robert Charles riot for him and the beginning of the rest of his life, the slowly, steadily lengthening stretch of years in which he efficiently functioned as Tom Anderson's "man about town," while his every other step proclaimed his dishonor.[17]

Inside Fast-Mail's story and the abstract space created by the tropes of corrupt New Orleans, the reader has little means to perceive differential space or to imagine different scales of interpretation smaller or larger than the individual human psyche.

These limitations of realism in this 2006 narrative might well feed contemporary Africanist memories of racial attacks and of pointed neglect. A more detailed postempirical reading of *Redemption* than I have attempted here needs to describe how the novel does this selective narratological work of historical absence as well as presence. That is, it needs to explore both the novel's turning away from the white mob rule that accompanied the Robert Charles incident and the novel's annunciation of corruption in Storyville orchestrated by a ruthless Tom Anderson. A postempirical reading of the novel could argue that emphasis on the extraordinary criminality of Anderson practiced in Storyville is familiar and easy for any reader to understand and, therefore, obscures the more insidious racial dynamics once evidenced in mob rule and more often embedded in the capitalist geographic phenomenon Lefebvre calls "abstract space."

Ismail's model of postempirical reading implies that the interpreter, not the literary artist – or historian or cartographer – bears the burden of identifying the narratological level at work in a text. I like Ismail's insistent interpreter, but I suspect that the storyteller's instruction in reading, seeing, and hearing, built into the form of the story (or map) is far more influential than a hectoring critic like me.

The storyteller, in the very telling of the story, can appropriate the site of reading, announcing where the reader resides in the story. Remember Ondaatje's "float by in a car today and see the corner shops," a sentence that positions the reader in the text. In the altered state that the narrator can create for the reader or listener, one could, like one of the western storytellers of Barry Lopez, for example, see "the realm of life that could not be sensed until one overcame the damage done to perception by a long exposure to inescapable noise" in modern life.[18] In Ondaatje's story of Buddy Bolden, for example, the reader might be said to enter the world of the deaf – remember that opening photograph of Bolden's band and that sonograph of sound – in order to understand the music of a man who had come through slaughter and Slaughter. In this silent world, New Orleans's familiar jazz rhythms and cadences, and with it Storyville and notorious prostitution, are repositioned on the map, away from the Mississippi crescent upriver to an obscure site where a pioneering jazz voice went silent. Perhaps that voice defined New Orleans jazz, as the aural memories of early jazz musicians claim. We can take the photograph of the band and the stories of those who heard Bolden and place them in the trajectory of jazz history as we know it in words and music. Bolden can then become a redeemed hero of an extraordinary American art form like Sidney Bechet or Louis Armstrong. Or perhaps instead the silence of Bolden's voice offstage in an asylum for the indigent questions all that we understand jazz and its origins in New Orleans space to have been in the past and to be now. In that silence is social inequity; uneven economic development; and contested New Orleans spaces thrown into disarray again in the floods of 2005. Through Ondaatje's narrative form – postmodern, fractured, call it what you will – his readers become aware of Bolden's silence and the need to think about it conceptually, not just narratively. Ondaatje creates with his story of jazz a metaphor of silence and offers us words that provide a pastiche of images and events to fill the silence and raise questions about its meaning.

This novel, I am arguing, has a postempirical role to play in the fate of New Orleans as one site in the archipelago of tropical catastrophes. As such it is one example of the lessons literary form teaches about geography. *Coming through Slaughter* can play this role effectively because its form disrupts the naturalized expectations of abstract space – be they the conventional tropes about the territory of Storyville or the jazz tourist's search for the sites where jazz was born. Through its collage and disruption of time and space, it instead moves outside empirical observation to represent New Orleans as differential space. In this represented differential space, the economic, social, and inevitable environmental trial by space besetting the city is apprehensible, palpable. With this apprehension comes the recognition, I am arguing, that other alternate means of mapping New Orleans's trial by space, in fact, of mapping our collective, global trial by space have been imagined and can be imagined. Mutter's mapping on the scale of the water-beset tropics or Mbembe's itinerate territoriality are just two of these imaginings. Human sensory observation and its claims to be the source of reality, as represented in such forms as literary realism, can only return to us what we have given to it – abstract space. It can remind us of what we have forgotten; it can add details to a picture we have seen before. It cannot teach us how to conceive this world differently. It cannot prepare us for our trial by space.

Notes

1 M. Ondaatje, *Coming through Slaughter*, New York: Penguin Books, 1976, p. 85 and elsewhere.

2 F. Turner, *Redemption*, New York: Harcourt, Inc., 2006, p. 2.

3 H. Lefebvre, *The Production of Space*, tr. D. Nicholson-Smith, Malden: Blackwell Publishing, 1991, pp. 29–30 and elsewhere. For an analysis of Lefebvre's relationship with the surrealists and the situationists see, for example, K. Goonewardena, S. Kipfer, R. Milgram, C. Schmid (eds), *Space, Difference, Everyday Life: Reading Henri Lefebvre*, London: Routledge, 2008.

4 D. Haraway, "Otherworldly Conversations; Terran Topics; Local Terms," *The Haraway Reader*, New York: Routledge, 2004, pp. 125–50.

5 E. Canin, *America, America*, New York: Random House, 2008.

6 See, for example, Turner's *In the Land of Temple Caves: Notes on Art and the Human Spirit,* New York: Counterpoint, 2004; *Spirit of Place: The Making of an American Literary Landscape*, San Francisco: Sierra Club Books, 1989; *Rediscovering America: John Muir in His Time and Ours*, New York: Viking Press, 1985; *Beyond Geography: The Western Spirit Against the Wilderness*, New Brunswick: Rutgers University Press, 1983; and his edited volume, *The Portable North American Indian Reader*, New York: Viking Press, 1974; as well as *1929: A Novel of the Jazz Age*, New York: Counterpoint, 2003; and *Remembering Song: Encounters with the New Orleans Jazz Tradition*, New York: Viking Press, 1982.

7 M. Ondaatje, *op. cit.*, p. 9.

8 H. Lefebvre, *op. cit.*, p. 408.

9 H. Lefebvre, *op. cit.*, pp. 409, 416, emphasis his.

10 J. Mutter, lecture, Iowa City, IA, 2006.

11 A. Mbembe, "At the Edge of the World: Boundaries, Territoriality, and Sovereignty in Africa," in A. Appadurai (ed.), *Globalization*, Durham: Duke University Press, 2003, pp. 22–51.

12 D. Wood with J. Fels, *The Power of Maps*, New York: Guilford Press, 1992. See also D. Wood and J. Fels, *The Nature of Maps: Cartographic Constructions of the Natural World*, Chicago: University of Chicago Press, 2008.

13 Q. Ismail, *Abiding by Sri Lanka: On Peace, Place, and Postcoloniality*, Minneapolis: University of Minnesota Press, 2005, p. 112.

14 Q. Ismail, *op. cit.*, p. 118.

15 I. B. Wells-Barnett, "Mob Rule in New Orleans," in *On Lynching*, Rpt. 1900, Amherst: Humanity Books, 2002. See also W. Ivy Hair, *Carnival Of Fury: Robert Charles and the New Orleans Race Riot of 1900*, Baton Rouge: Louisiana State University Press, 1976.

16 F. Turner, *Redemption, op. cit.*, p. 66.

17 *Ibid.*

18 B. Lopez, *Resistance*, New York: Knopf, 2005, p. 154.

11

Wordmaps

Howard Horowitz

In the middle of the night during the biggest snowstorm in 50 years, I had a vivid dream. Manhattan Island was a coherent image of words, with landmark places located where they belong on a map. This "visit from the muse" was so astonishing that I jumped out of bed, found the dusty pile of "Manhattan" poem fragments that had sporadically accumulated over several decades, and began to organize them geographically. After many inspirations and transformations, "Manhattan" was published on the Op-Ed page of *The New York Times*.[1] It was a cartographic poem in the form of the island that could be recited aloud or viewed as a map. It was a map made of words: a "wordmap."

That vivid dream of Manhattan Island as a cartographic text sparked a leap from writing shaped geographical poems to writing wordmaps. Early drafts, produced on my own computer, were rough approximations of the island's shape. Getting the island's shape precisely correct required professional assistance, which was generously provided by the superb cartographers Lawrence Andreas and Stuart Allan of Raven Maps (Medford, Oregon). As a writer, making the text fit into the map often seemed as slow as chipping away at a block of granite to create a sculpture. Whenever another landmark or historical event was added into the poem, adjustments were needed to make sure that other places were not knocked into wrong locations. In general, it was easier to establish correct north – south text locations than to maintain good east–west locations. Because Manhattan is such a long narrow island, adding places usually required adding exactly one full line of text, to avoid dislocating nearby places. As the poem grew in size and complexity, it began to resemble the densely packed congestion of the city itself.

The content of "Manhattan" includes geologic and human history, landmark skyscrapers and museums, popular culture, and critical infrastructure. Like a map, the poem's narrative moves freely through the geologic and historical past, but is limited by its date of production. In the case of "Manhattan," that date is 1997, so landmarks such as the World Trade Center and the Fulton Fish Market remain in the text, though they no longer exist on the island today.

In her essay *Poems Shaped Like Maps*, Adele Haft identifies visual poems that were written to be map-like, dating back to the "pattern poetry" of medieval and Renaissance times.[2] Various forms have been created over the centuries, some playful shapes and some concrete word patterns. Since the 1960s, a burst of map-like poems have been written, and Haft's article illustrates and discusses several, including "Manhattan."

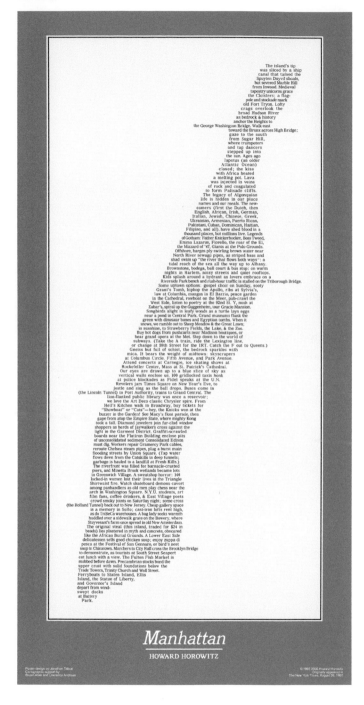

Figure 11.1 Manhattan poster.

Source: Copyright Howard Horowitz.

The riverfront was filled for barnacle-crusted
piers, and Minetta Brook wetlands became lots
in Greenwich Village. A sweatshop horror: 146
locked-in women lost their lives in the Triangle
Shirtwaist fire. Watch skateboard demons cavort
among panhandlers as old men play chess near the
arch in Washington Square. N.Y.U. students, art
film fans, coffee drinkers, & East Village poets
crowd smoky joints on Saturday night; some cross
(the Holland Tunnel) back out to New Jersey. Cheap gallery space
is a memory in SoHo; cast-iron lofts rent high,
as do TriBeCa warehouses. A bag lady seeks warmth
huddled over a sidewalk grate on the Bowery, where
Stuyvesant's farm once spread in old New Amsterdam.
The original steal (this island, traded for $24 in
beads) lies plastered in myth and concrete, obscured
like the African Burial Grounds. A Lower East Side
delicatessen sells good chicken soup; enjoy zuppa di
pesca at the Festival of San Gennaro, or bird's nest
soup in Chinatown. Marchers to City Hall cross the Brooklyn Bridge
to demonstrate, as tourists at South Street Seaport
eat lunch with a view. The Fulton Fish Market is
mobbed before dawn. Precambrian stocks bond the
upper crust with solid foundations below the
Trade Towers, Trinity Church and Wall Street.
Ferryboats to Staten Island, Ellis
Island, the Statue of Liberty,
and Governor's Island
depart from wind-
swept docks
at Battery
 Park.

Figure 11.2 Manhattan poster, detail.

Source: Copyright Howard Horowitz.

For me, years of writing geographically oriented poems involved a slow but steady progression towards the creation of wordmaps. The precursors were symbolically shaped to represent abstract geographical features; they were nature-oriented variants of the tradition known as "concrete poetry." The shaped poems became more geographically and historically specific. Two examples are "Idaho" and "Redwood Creek." "Idaho" is a map-shaped poem about planting trees in Idaho, but the text is not cartographic.[3]

"Redwood Creek" is geographically specific to Humboldt County with regard to narrative content. The voice of the story is that of the stream. Redwood Creek alludes to the area's history, from the Chilula Indians to the logging of the great trees with the consequent erosion, siltation, and decline of salmon runs. However, the shape of "Redwood Creek" is symbolic and abstract, rather than explicitly cartographic.[4]

"The Oregon Coast" maps a continent's edge: a place unstable and prone to extremes. The bedrock of words – erupted from within, or deposited by floods of events – is deformed and uplifted by impacts of geology and history, by the rigors of cartography, and by the personal touches of memory and love. The focus extends from the beaches and coastal dunes inland about 30 miles to the crest of the Coast Range.

Figure 11.3 "Oregon Coast" by Howard Horowitz, layout by Raven Press.
Source: Copyright Howard Horowitz.

The narrative includes rugged landforms, fog-shrouded old growth trees and clearcuts, human settlements and daily life in fishing towns and working forests. The most important waterways are delineated in blue, with critical cartographic assistance once again provided by Lawrence Andreas at Raven Maps. The major capes and coastal landmarks are named at their correct locations, and the north-to-south flow of the narrative is cartographically correct, from the mouth of the Columbia south to the California border. However, the need for clear vivid sentences limits the control of east–west word locations.

Wordmaps about rivers and watersheds pose special challenges. They are easier to read when the river flows from north to south, such as the Hudson River. It is harder when the river flows from south to north, as the Wallkill River does. Thus in my "Wallkill Watershed," the text follows the water's flow from headwaters to mouth, which means reading the lines from bottom-to-top. Reflecting the river's watershed, tributary streams are included as separate texts that flow into the main text.

Other wordmaps are focused on mountain ranges and physical ecosystems. The "Reading Prong" tells the story of the Appalachian Highlands region described by that name, which encompasses a swath of eastern Pennsylvania, northwest New Jersey, and the Hudson Highlands of New York. The forested ridges and agricultural valleys have an old geological and cultural history, which is the text of the poem. The words are stretched out in lines that correspond to the Appalachian ridges, and the text is folded in ways that approximate the folds and curvature of those ridges.

Wordmaps occupy a peculiar intersection of geography, poetry, and visual art. "Cooking them up" involves practice and the required ingredients: knowledge of the place being portrayed, laborious focus on the text, and a sweet spoonful of inspiration.[5]

Notes

1   H. Horowitz, "Manhattan," in *The New York Times*, August 30, 1997, p. 23.
2   A. J. Haft, "Poems shaped like maps: (Di)versifying the teaching of geography, II," *Cartographic Perspectives* 36, Spring 2000, 66–91.
3   H. Horowitz, "Idaho," in *Close to the Ground*, A Yew Book: Hulogos'i Press, 1986.
4   H. Horowitz, "Redwood Creek," in *Calapooya Collage*, Adrienne Lee Press, 1990.
5   A sample of wordmaps can be viewed at the website www.wordmaps.net.

12

Using early modern maps in literary studies
Views and caveats from London

Janelle Jenstad

In William Haughton's 1598 play, *Englishmen for My Money*, the moneylender's three daughters are courted by three unsuitable foreigners. The suitors are diverted from the moneylender's house in the fog by Frisco, a servant who gives them a spurious tour that supposedly takes them past London Stone in the heart of the city, Ivy Bridge Lane far west of the city, and Shoreditch, a neighborhood north of the city.[1] On the bare, unlocalized amphitheater stages of early modern London, dialogic mapping was necessary to establish place. Riffing on this dramatic habit, Frisco verbally maps out an absurd landscape. To the playgoer in 1598, the foreigners' urban ineptitude would have been a hilarious marker of their unsuitability as husbands for London maids. Readers now, however, like the three foreign suitors, are unlikely to know that London Stone, Ivy Bridge Lane, and Shoreditch are not proximate. The joke is lost on most of us unless we consult a map.

Foreigners to early modern London ourselves, contemporary readers may well miss many of the local references in texts deeply rooted in the urban environment that produced them. *Englishmen For My Money* is one of the first examples of a city comedy.[2] To understand even the basic moves of this dramatic sub-genre, which flourished from the 1590s to the 1630s, we need an intimate knowledge of the streets, buildings, and markets of early modern London. City comedies make a particular claim upon their viewers' cultural expertise, but other plays set in London have comparable expectations. For example, in Shakespeare's *1 Henry IV*, Prince Hal slums around Eastcheap with Falstaff and the tavern crowd, robs travelers on nearby Gads Hill, and is summoned to Westminster by his disappointed father. Likewise, the many poems, chronicles, pamphlets, and works of prose fiction that take up the matter of life of London assume a readership familiar with the space and denizens of London.

While most academics in the field eventually make a pilgrimage to London, as teachers we need to find a way of providing a comparable experience to students who may never walk its streets. Teaching editions often include rudimentary "wayfinding" maps.[3] They outline the city wall, major streets, and theater locations in enough detail that one can see, for example, the rationale behind the titles of *Eastward Ho* and *Westward Ho*, which invoke the cries of boatmen heading downriver or upriver respectively,[4] but they provide little sense of the space in which Londoners lived, worked, and played. My primary pedagogical tool has been the Agas map – a 6'2" long-view of London in the 1560s that gives a wonderful sense of a medieval city spilling over its ancient wall into the surrounding farmland.[5] The map depicts streets, waterways, houses, churches, watercraft, cranes on the wharves, gardens laid out in green spaces, and the crosses and conduits that dot the streets. Although the Agas map represents buildings as three-dimensional objects instead of flattening them into a ground plan, the perspective's nearly perpendicular angle makes most of the streets and buildings visible and identifiable, allowing it to function as a street plan as well as an image of London as it might have looked. Only such a large-scale map establishes a "local habitation"[6] – a phrase often quoted in my discipline to signal the material embodiment of an idea – for the action of plays like Thomas Middleton's *A Chaste Maid in Cheapside*.[7] Only a map with this much detail shows the irony of a title positing a virginal maid in the main market street where all things were for sale.

When I returned from my own pilgrimage to London in 1997 with photocopies of the Guildhall Library copy of the map,[8] I taped the eight individual sheets together and covered the resulting scroll with transparent adhesive shelf-liner. In my Shakespeare and English Renaissance drama courses, we use erasable markers to highlight salient features of the literary landscape: the places mentioned in the texts we are reading, the neighborhoods where theatergoers lived, and the Thames that shaped the city's growth and determined local travel. We trace the infamous transportation of the timbers of Shakespeare's Theater from Smithfield to its new site on Bankside, where it was reborn as the Globe. Plotting characters' movements in each play often reveals a geographic component to social transgressions, marriage options, and financial choices or limitations. In *The Shoemaker's Holiday*, Simon Eyre's initially low status is reflected in his wife's origins as a tripe seller in Eastcheap. He walks two blocks from his Tower Street workshop to conduct the shady deal at Billingsgate (a fish market and industrial dockyard) that makes him fabulously rich. His upward mobility is mirrored in his urban movements out of his immediate neighborhood. He is "sent for ... to the Guild Hal,"[9] made a sheriff and shortly thereafter Lord Mayor. His mayoral legacy is the building of Leadenhall Market on Gracechurch (a street that was hallowed by being part of the royal processional route) in the neighborhood of Cornhill, where his mayoral predecessor and rival Oatley lives. The map confirms students' sense that Eyre's rise happens in a social space where he is initially an interloper and eventually a founder.

Because these exercises have been so pedagogically useful, the Agas map has since become the platform for *The Map of Early Modern London* (*MoEML*), a digital humanities project that aims to provide a sense of the space of London circa 1550–1650.[10] The site offers descriptions, transcriptions of primary texts, and other resources, all linked to high-resolution scans of the map. The *MoEML* is open-source, which means that students can access it from home, the library, or a classroom, and teachers can devise assignments based on the map. One of my colleagues has students find the aforementioned settings of *1 Henry IV*, an activity that diagrams the extent of Hal's departure – both physical and moral – from his father's desires. The digital map brings playful possibilities to the map as well as serious scholarship. Visitors to the

Figure 12.1 The Map of Early Modern London. Used by kind permission of the Guildhall Library, London.

Source: available online at http://mapoflondon.uvic.ca.

site can zoom in to view details, at which level they often "become lost" in the city, electronically wandering the streets and marvelling at the architecture, according to informal feedback and user studies. They can zoom out to see the shape of the city, or locate themselves on the key map. They can pan across the map, turn layers of information on and off, and "query" the map to find out what information the website can provide about a neighborhood or building. In this way, the *MoEML* works a bit like a guidebook. One is meant to wander a bit, consult the book, turn down another street, and consult the book again. The digital map permits users to travel purposefully or spontaneously. The accompanying text and links make it possible to jump around in a non-linear fashion, exactly as Frisco does on his fanciful tour of the city in *Englishmen For My Money*, thus liberating users from the constraints of proximity if they wish, but allowing them to pinpoint their coordinates at any time.

One aspect of my research uses the map to recover the lost spatial component of mayoral pageants and royal entries. These are peripatetic forms of occasional theater, entailing static or mobile pageants (collectively known as an entertainment) performed in the streets on celebratory occasions such as the annual inaugurations of the new Lord Mayor on 29 October or the passage of a monarch through the city. Although the texts of the entertainments survive, it is hard to capture in print either the disjointed nature of the entertainment as a whole or the importance of specific locations to a pageant's meaning. No one spectator would have seen the entire entertainment. The crowd gathered at Gracechurch Street would see only the pageant performed there, for example, while the procession moved on to a new pageant and a new set of observers. The *MoEML* aims to capture something of this partial experience by attaching components of the printed books to the relevant streets. Visitors to the *MoEML* can read the entire text in a linear fashion if they wish, but they can also experience a single pageant through the "Pageant" link from the street, as if they themselves were lining the street at that point and having the procession arrive at their vantage point. As the number of texts in the library grows, so will the number of links to streets where pageants were annually performed. The section of Cheapside between St. Peter's and St. Paul's was so frequently traced by processions, parades, and these occasional entertainments that it came to have a sacred quality, as the density of hyperlinks to this street attests.[11]

The map has served well as the platform for a library and digital encyclopaedia of London, but it is also a text in its own right. To a scholar trained to read for fissures, inconsistencies, representational strategies, agendas, and imbrication of early modern literature in culture, the Agas map demands analysis and interpretation. Early maps are what Jess Edwards has called "a noisily rhetorical art," with multiple signifying strategies that serve various ideological functions.[12] Maps of London work to express a stable image of the city despite a 400 percent population increase and immigration both from abroad and from within England that challenged the ability of late medieval civic institutions to exercise control over the populace.[13] Edwards notes that "controversies around the borders of literary study have inevitably centered on the textuality of cartography: the extent to which maps themselves can and should be read like literary texts."[14] I spend the first meeting of my graduate course on Representations of London talking about the rhetoric of early maps, modifying literary critical questions to suit the "text" under scrutiny. Readers of maps and readers of literature alike must consider point of view (how "partial" is the view?) and fundamental questions of what is or is not represented.

With several early maps of London spread before us, we begin with point of view. Is the map's perspective stereographic, bird's-eye, or ground plan? The first (a panorama "depicting the subject as it presented itself to the eye of an observer at a point on the ground or not far above it"[15]) offers a view of the city from a point of approach. Such maps are aimed at the armchair traveler, the primary consumer of sixteenth-century atlases. Claes Jansz Visscher's 1616 long "View of London" is a close panorama that makes the city occupy most of the foreground. The perspective invites a sense of wonder. The bird's-eye view depicts the city "as seen obliquely from a more elevated point of vision," while the ground plan or "plat" is "drawn from a theoretically vertical viewpoint."[16] These last two viewpoints allow the map-reader to see the city in its totality. Such maps offer an alternative to viewing the city from one of the few platforms afforded by the architecture of the time (St. Paul's Cathedral tower and the upper floor of the Royal Exchange).[17]

Inclusions and omissions work in tandem with perspective to define the city. The Braun and Hogenberg ground plan map of London labels only the Tower and Westminster, thus containing the city – whose relationship to the Crown was always complex – within two signifiers of royal authority. The Agas map takes a far-seeing bird's-eye view that makes London an anomaly rising out of an agricultural space, unlike Visscher's city that dominates the landscape. Work and play happen outside the walls of the Agas London, while the city is oddly devoid of life, its empty streets cleansed of dirt and traffic. To the left of the Great Conduit in Cheapside are three large cans, for example. The apprentices whose job it was to fill the cans with water each day are nowhere to be seen. Given both the importance of the conduits as a gathering place and the persistent contemporary complaints about overcrowding, the absence of human figures provokes immediate question about the work of the map.

Historian Richard L. Kagan's analysis of urban maps in early modern Spain offers a paradigm for reading the Agas map. Kagan classifies maps as "communocentric"– realizing "the city as a human community or well-governed republic endowed with a character, history, customs, and tradition uniquely its own" – or "chorographic" – offering a "complete and comprehensible visual record of a particular place."[18] The communocentric map depicts the *civitas* (the people) and/or the *res publica* (the government), while the chorographic map depicts the *urbs* (the architecture) and sometimes the surrounding landscape. The maps in Braun and Hogenberg's *Civitates Orbis Terrarum* are often communocentric. The *civitas* is always rendered via large foreground figures in their native dress. The map of Antwerp in this collection celebrates the *res publica*, with soldiers in the streets, fortifications and cannons, and bars or gates limiting entry to the city; trees are planted in orderly lines, and the churches, exaggerated in scale, loom over the other buildings. The Agas map, on the other hand, offers a chorographic view of the *urbs*. It conceptualizes the city as an aggregate of buildings, not a community of people.

Kagan's paradigm, drawn from early modern thinkers, offers a powerful tool for reading London literature, which anxiously limits and defines the city in various ways.

In Shakespeare's *Coriolanus*, for example, a Tribune asks "What is the city but the people?" (an argument giving priority to the *civitas*).[19] Pageants and proclamations envision the city as its governing institutions (the *res publica*). John Taylor anatomizes the *urbs* in verse catalogs that list jails, inns, and other places.[20] Other texts give voice to the *urbs* by anthropomorphizing neighborhoods and structures. Thomas Dekker's *The Dead Term* is a dialogue between Westminster and London, each wondering if she is still a city when she is deserted by her inhabitants during the vacations.[21] Ben Jonson's mock-epic poem, *The Famous Voyage*, takes two hero-adventurers up the Fleet Ditch on a sewage barge, showing the disorderly underbelly of the *res publica* (prisoners), the worst of the *civitas* (prostitutes), and the backside of the *urbs* (human and industrial waste disposal systems).[22] In this way, terms from the history of cartography help us make sense of a body of literature.

True interdisciplinarity happens when two disciplines become mutually informing. The *MoEML*'s literary analyses augment and qualify the geography derived from the Agas map. As work progressed through many stages of design, encoding, redesign, and re-encoding, it became clear that the map alone does not give a complete sense of all the places that make up the "space" of London. There are practical problems of identification, perspective, scope, and time. The key to the numbered structures is now lost. Some places of cultural import, such as Tyburn, where executions took place, are beyond the map's borders. The theaters were all built after the map was drawn, while many buildings on the map were demolished as the city grew. Some of the narrower streets running east–west, such as Finimore or "fiue foote lane,"[23] cannot be seen behind the houses and other structures, all of which are drawn from the south. Every place label on the *MoEML* is therefore the result of extensive reading. Every street or site description draws on literary references to establish the cultural importance of the place, and most streets link to a dynamically generated archive of literary allusions.[24]

More of a challenge are the spaces – Foucault's heterotopias[25]– that cannot be mapped, such as the marketplace, the theater, the fairground, and the ship.[26] Jean Howard's recent analysis of city comedy is organized into "four city places that the drama turns into significant social spaces by its narration of them" – the Royal Exchange, debtors' prisons, bawdy houses, and ballrooms and academies[27] – three of which resist pinpointing but are nonetheless key to the construction of space in the early modern cultural imagination. It is mainly through the Topics pages and a network of links to ubiquitous concepts like the market that the *MoEML* will convey the importance of these spaces to the early modern cultural imagination. In a sense, then, the data associated with the map is as much an interpretation of the map as it is of early modern literature.

The literature of London is replete with references to streets, buildings, neighborhoods, wards, parishes, landmarks, and the natural landscape. The gap between the early modern Londoner's local knowledge and our own ignorance as metaphoric "foreigners" means that we not only miss the joke but miss the pleasure of recognition unless we become naturalized. Extensive reading helps us to understand the significance of place, but the Agas map and its digital avatar in the *MoEML* can take us into the environment in a way that mere words cannot. The map allows someone not already familiar with the intricacies of parishes, neighborhoods, and landmarks to see at a glance how London was laid out and how one place was connected to – or disconnected from – another. In gaining a sense of the relative location of streets, sites, neighborhoods, and watercourses, we acquire the local knowledge to navigate the range of London's literature and culture.

Notes

1   William Haughton, *Englishmen for My Money,* London, 1616, sig. G1v.
2   City comedies are characterized by a London setting, carefully delineated class boundaries, and an obsession with money and marriage; A. Leggatt, *Citizen Comedy in the Age of Shakespeare,* Toronto: University of Toronto Press, 1973, p. 4.
3   See, for example, the map of places mentioned in *Englishmen for My Money* in L. E. Kermode (ed.), *Three Renaissance Usury Plays,* Manchester: Manchester University Press, 2009, p. 350.
4   G. Chapman, J. Marston, and B. Jonson, *Eastvvard hoe,* London, 1605; T. Dekker and J. Webster, *Vvest-vvard hoe,* London, 1607.
5   The map, based on an earlier copperplate map, was attributed to the surveyor Ralph Agas, whose name remains inextricably linked to the map even though he cannot have been the cartographer; S. P. Marks, *The Map of Mid Sixteenth Century London: An Investigation into the Relationship Between a Copper-Engraved Map and its Derivatives,* London: London Topographical Society, 1964, pp. 21–22.
6   W. Shakespeare, *A Midsummer Night's Dream,* 5.1.17; quoted from D. Bevington (ed.), *The Complete Works of Shakespeare,* 5th edn, London: Longman, 2004.
7   *A Chast Mayd in Cheape-Side,* London, 1630. The play was performed in 1613.
8   The other two known copies are held by the Public Record Office and the Pepysian Library. All three are post-1603 prints of the 1560s map. The map is the basis of A. Prockter and R. Taylor (comps.), *The A to Z of Elizabethan London,* Kent: Harry Margary, 1979.
9   T. Dekker, *The Shomakers Holiday,* London, 1600, sig. E2v.
10   J. Jenstad (ed.), *The Map of Early Modern London,* online, available http://mapof-london.uvic.ca.
11   See also the map tracing all the processional routes in L. Manley, *Literature and Culture in Early Modern London,* Cambridge: Cambridge University Press, 1995, pp. 226–27.
12   J. Edwards, "How to Read an Early Modern Map: Between the Particular and the General, the Material and the Abstract, Words and Mathematics," *Early Modern Literary Studies* 9.1, 2003, 6.4, online, available http://purl.oclc.org/emls/09-1/edwamaps.html.
13   A. L. Beier and R. Finlay, "The Significance of the Metropolis," in Beier and Finlay (eds), *London 1500–1700: The Making of the Metropolis,* London: Longman, 1986, pp. 1–33.
14   Edwards, *op. cit.,* 6.3.

15 R. A. Skelton, Introduction, *Civitates Orbis Terrarum*, G. Braun and F. Hogenberg, 1572–1618, reprinted Cleveland: World Publishing, 1966, p. xi.

16 *Ibid.*

17 C. Stevenson, "Vantage Points in the Seventeenth-Century City," *The London Journal* 33, 2008, 218.

18 R. L. Kagan, "*Urbs* and *Civitas*," in D. Buisseret (ed.), *Envisioning the City: Six Studies in Urban Cartography*, Chicago: University of Chicago Press, 1998, pp. 76–77.

19 Shakespeare, *Coriolanus*, 3.1.202, in Bevington (ed.), *op cit.*

20 For example, *The Praise and Vertue of a Iayle, and Iaylers*, London, 1623; *The Carriers Cosmographie,* London, 1637.

21 *The Dead Tearme*, London, 1608.

22 See A. McRae, "On the Famous Voyage: Ben Jonson and Civic Space," in A. Gordon and B. Klein (eds), *Literature, Mapping, and the Politics of Space in Early Modern Britain*, Cambridge: Cambridge University Press, 2001, pp. 181–203.

23 J. Stow, *A Survey of London. Reprinted from the Text of 1603*, ed. Charles Lethbridge Kingsford, Oxford: Clarendon, 1908, vol. 2, p. 1.

24 See "Contributor Guidelines: Street or Site Short Essays," online, available http://mapoflondon.uvic.ca/guidelines_contributors.php and "Literary References" online, available http://mapoflondon.uvic.ca/index_litref.php.

25 M. Foucault, "Of Other Spaces," J. Miskowiec (trans.), *Diacritics* 16.1, 1986, 22–27.

26 See J.-C. Agnew, *Worlds Apart: The Market and the Theater in Anglo-American Thought, 1550–1750*, Cambridge: Cambridge University Press, 1986.

27 J. Howard, *Theater of a City: The Places of London Comedy, 1598–1642*, Philadelphia: University of Pennsylvania Press, 2007, p. 23; J. Dillon, in *Theatre, Court and City 1595–1610*, Cambridge: Cambridge University Press, 2000, discusses "the place of theatre," "the place of exchange," "the place of dirt," and "the place of accommodation."

13

"along Broadway 2009"

Robbert Flick

Broadway is a major street in downtown Los Angeles. It appeared on most early maps of the city, and was originally called Calle de la Eternidad (Eternity Street) since the original dirt road from the pueblo connected to a cemetery. In one of the first plans of Los Angeles after the Mexican–American War (1846–48), by Lieutenant Edward Ord in 1849, Broadway was identified as Fort Street. Soon after the street was renamed Broadway in 1890, the first of many large department stores opened as the city began to grow rapidly. Between 1903 and 1930, 19 major theaters and movie palaces joined the retailing businesses as well as several buildings that are now historic landmarks, including the Bradbury Building (designed by George Wyman in 1893), which became a favorite location site for Hollywood movies. Buildings owned by two major newspapers, *The Los Angeles Times* and the Los Angeles *Examiner*, were also part of the Broadway extravaganza.

In recent decades, Broadway has been transformed into LA's busiest shopping street for the local Latino population. The Broadway Historic Theater District contains the largest collection of vintage cinema and vaudeville theaters in the country, some kept alive by Spanish-language movie programs.

Sources

L. Pitt and D. Pitt, *Los Angeles A to Z: an encyclopedia of the city and county*, Berkeley: University of California Press, 1997.

C. Roseman, R. Wallach, D. Taube, L. McCann, and G. DeVerteuil, *The Historic Core of Los Angeles*, Charleston, SC: Arcadia Publishing, 2004.

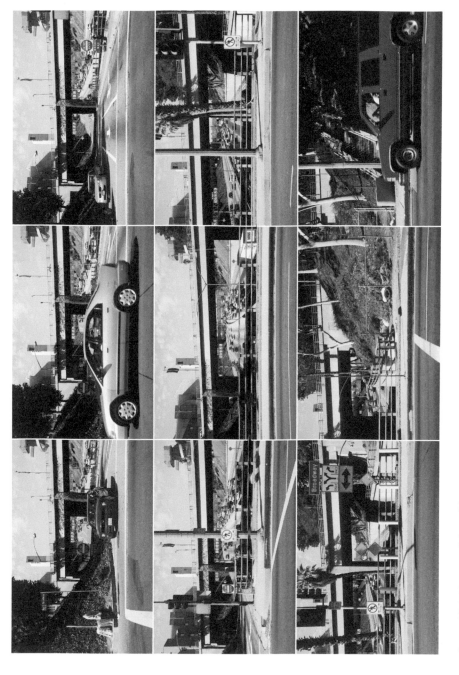

Figure 13.1 Photograph by Robbert Flick.

Figure 13.2 Photograph by Robbert Flick.

Figure 13.3 Photograph by Robbert Flick.

Figure 13.4 Photograph by Robbert Flick.

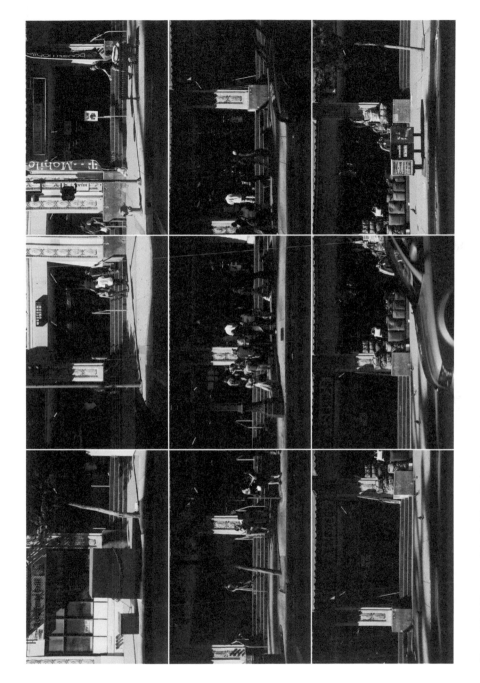

Figure 13.5 Photograph by Robbert Flick.

14

Thoreau's geopoetics

Sarah Luria

Henry David Thoreau, naturalist, essayist, and land surveyor – in short, nineteenth-century geohumanist – is an obvious subject for this collection. The common ground that this book seeks to describe Thoreau explored all his life. Thoreau earned most of his income from land surveying, which took advantage of his mathematical mind. But his heart lay with the creative writing he did in his journals and essays, which let his expansive imagination riff on the nature he observed during his surveys and his daily walks.

Thoreau is a particularly intriguing case to study as we seek to develop the field of geohumanities because he often complained that his two occupations – his surveying and his writing – were fundamentally at odds. In an entry to his journal on January 1, 1858, Thoreau writes:

> I have lately been surveying the Walden woods so extensively and minutely that I now see it mapped in my mind's eye – as, indeed, on paper – as so many men's wood-lots, and am aware than when I walk there I am at a given moment passing from such a one's wood lot to another's. I fear this particular dry knowledge may affect my imagination and fancy, that it will not be so easy to see so much wildness and native vigor there as formerly.[1]

Thoreau identifies several ways of seeing – as a property surveyor, a naturalist, and a poet – and contends that the surveyor's view threatens the others. In the summer of 1859, however, Thoreau did a lengthy survey of the Concord River that resolved this conflict and showed instead how compatible these views could be. His survey is full of "dry knowledge" that led his imagination to perceive in an intense way the "native vigor" of the riverscape. Indeed, the survey interweaves Thoreau's deep involvement in both the literary tradition and careful naturalist observation by creating a large visual and textual space where he can layer multiple points of view and, rather than have one occlude the other, see them all at once. Figure 14.1 shows a section of Thoreau's completed sketch of the river; the sketch in its entirety measures more than 7½ feet wide by 2¼ feet high.[2]

The Concord River Survey of 1859 is a culmination of Thoreau's life-long study of the Concord River. Through it he fulfills a project he outlined in his first book – *A Week on the Concord and Merrimack Rivers* (1849) – a wide-ranging account of a trip he and his brother took down the river in 1839. There he gave a tantalizing vision of

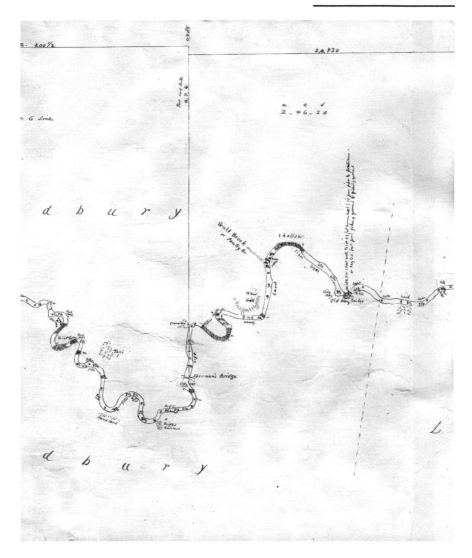

Figure 14.1 "Plan of Concord River from East Sudbury & Billerica Mills, 22.15 Miles, to be used on a trial in the S.J. Court, Sudbury & East Sudbury Meadow Corporation vs. Middlesex Canal, Taken by agreement of Parties, By L. Baldwin, Civil Engineer. Surveyed & Drawn by B.F. Perham. May 1834 [1859/1860] (rolled survey)" (http://www.concordlibrary.org/scollect/Thoreau_surveys/107a.htm).

Source: Courtesy Concord Free Public Library.

what naturalist description could be: as he and his companion "rested in the shade, or rowed leisurely along," Thoreau states, "we had recourse, from time to time, to the Gazetteer, which was our Navigator, and from its bald natural facts extracted the pleasure of poetry."[3] Thoreau's vision of a scientific poetry is inspiring indeed, but as Linck

Johnson has noted, many of Thoreau's descriptions of the river in *A Week* are drawn more from the pastoral poetry he had read in college than the actual facts of the river itself.[4] For example, Thoreau describes one segment of the journey: "Now we coasted along some shallow shore by the edge of a dense palisade of bulrushes, which straightly bounded the water as if clipt by art, reminding us of the reed forts of the East Indians, of which we had read."[5] In the Concord River survey of 1859, however, Thoreau succeeds at merging the elemental and synthetic points of view of the geographer-poet to at last realize "the pleasure of poetry" to be derived from "bald natural facts."

Today's interest in ecocriticism has led Thoreau scholars to note a gradual shift in his work, although never a complete one, from more "homocentric" to more "biocentric" writing.[6] H. Daniel Peck celebrates Thoreau's journal as a particularly fertile meeting ground between his scientific and literary self. The freedom of the journal, Peck argues, gave Thoreau the space to experiment and develop his style beyond the conventional language of romantic pastoralism into a more "elemental" style that could better "captur[e] … nature's processes." Since the journal had no limit to its pages (in print it fills 14 volumes), Thoreau could include drawings and minute descriptions of nature that could go on as long as he liked.[7] Lawrence Buell notes that the 1850s Journal "became less a repository for thoughts, quotations, anecdotal vignettes, and drafts, and more a record of regular, meticulous daily extrospection" and showed Thoreau's shift "from young transcendentalist literatus … to the middle-aged ruralist."[8] While Thoreau's journal has received much attention, critics are only beginning to turn to his land surveys.[9]

The Concord River Survey demonstrates the inseparability of Thoreau's homocentric and biocentric views. Rather than the survey getting in the way of his "fancy," one leads to the other and back again. The survey began when Thoreau was hired to measure the bridges on the river to help settle a lawsuit by farmers against the mill owners, whose dams the farmers claimed were flooding their fields.[10] Thoreau appears to have gone way beyond the task he was hired for and spent the whole summer making a full naturalist survey of the river and its banks. Ralph Waldo Emerson described Thoreau's efforts in a letter that summer: "Henry T. occupies himself with the history of the river, measures it, weighs it, and strains it through a colander to all eternity."[11] Thoreau measured the river's depths and breadth along its entire 22-mile length. He calculated its volume.[12] From his meticulous observations, Thoreau drew conclusions about how the riverbed and its banks were formed. He took notes in his journal of the changes in the river's flora and fauna he saw from day to day. In short he created just as Emerson claimed a "history of the river." In the end, Thoreau's view of the river surpassed the narrow objectives of the lawsuit. As Brian Donahue has suggested, Thoreau's findings were equivocal and were quite possibly never used in court.[13] One might conclude Thoreau's survey showed too much – through its multiplication of views (objective, subjective, bird's eye view, ground view, surface, depths), it revealed the ambiguity of the river's situation. Together, journal and sketch and Thoreau's pages of measurements produced a clear picture, not of property rights, but of the river's very being, and its uncontainable "native vigor."

Given the transdisciplinary nature of Thoreau's survey, how are we to read it? To simply consider the map in isolation is to divorce it from the story behind it, and to hence be less cognizant of the imaginative work it surely is. Most importantly, it is to miss a moment when Thoreau's geographical and literary imaginations productively converge. "There is a property in the horizon" Emerson states "which no man has but he whose eye can integrate all the parts, that is, the poet."[14] Thoreau's many ways of seeing demand a different sort of reading, one that reconstructs all the parts he integrated into it and considers them all at once. To that end, Figure 14.2 represents a start at connecting the river sketch to some of the texts and views – his calculations and tables of measurements, the taxonomic systems of the flora he identifies, the relevant journal entries, passages from *A Week,* and preliminary sketches – that went behind the notation of each specific site (Figure 14.2).

Much more data obviously could be attached. The point of Figure 14.2 is to give us a way of reading Thoreau's sketch of the Concord River, we might say, vertically, through the various texts and work that underlies it, rather than simply horizontally, taking it merely at face value, a "dry" picture of where the river is deep or shallow, its banks hard or soft. If we can see these connections between the river's "bald natural facts" we can begin to see the "poetry" Thoreau discovered in them. Figure 14.2 gives us a way to reread Thoreau's sketch, gloss its terms, explore its literary antecedents, much as we would a poem, in order to see how it works. Indeed, Thoreau's example suggests we might be advised to analyze any map through the various views, conditions (physical, political, historical), texts (notebooks, decrees, calculations), and life of the person who produced it. If we don't read a map vertically, then we miss the humanities story at the bottom of its geographical text.

Deep reading

Consider, for example, the layers beneath Thoreau's notations on the survey of the grassy sections of the river (Figure 14.3).

On the surface, these markings say the river is shallow and grassy, but in doing so they also link vertically to the opening of *A Week on the Concord and Merrimack Rivers*, which announces that the Indian name for the Concord River was "The Musketaquid, or the Grass-ground river." As recorded on the survey in 1859, the grassy sections confirm the history of the river Thoreau spelled out 10 years earlier in *A Week*, "To an extinct race it was grass-ground, where they hunted and fished, and it is still perennial grass-ground to Concord farmers, who own the Great Meadows, and get the hay from year to year."[15] It is the farmers' assertion that the mill owners' dams are flooding the river and ruining the hay on its banks that will cause Thoreau to be asked to survey the river's bridges. We do not read too much into the river sketch of 1859 by noting these connections. If they are in the map-makers' eye and his past writings about the same river, why not recognize them? By doing so we see how deep the survey's notations go – here for instance the grassy sections connect to the character of the river itself. The river is known to be so shallow and sluggish that grass can grow in it, hence its Indian name for its most distinguishing feature, one which Thoreau mentions frequently in his journal as when he records one farmer's remark that "he has seen a chip go faster up-stream there

Thoreau's Survey of the "Revolution" on the Concord River, Summer of 1859

Sarah Luria, College of the Holy Cross

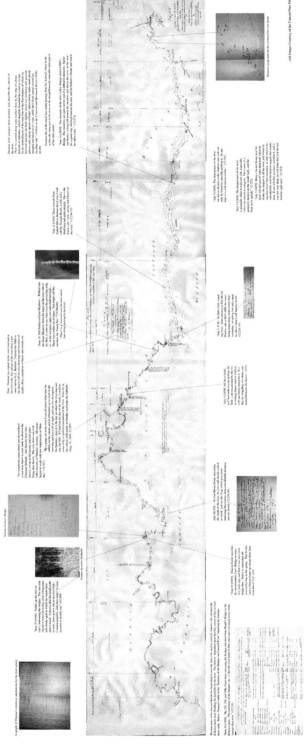

Figure 14.2 Poster of Thoreau's entire river survey with related documents, S. Luria.

Source: Survey image and documents Courtesy Concord Free Public Library.

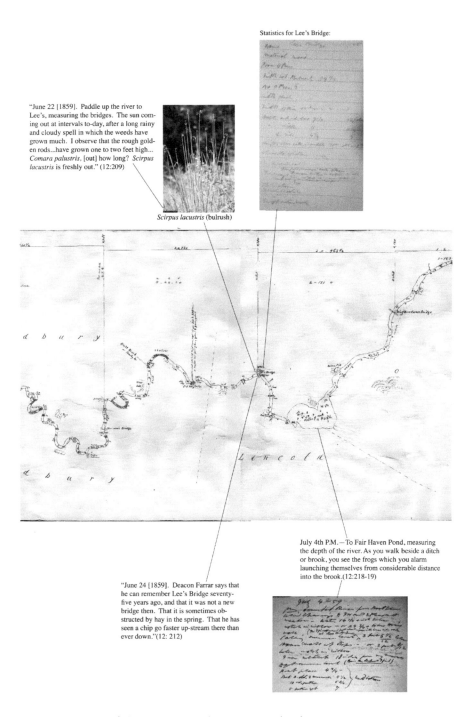

Statistics for Lee's Bridge:

"June 22 [1859]. Paddle up the river to Lee's, measuring the bridges. The sun coming out at intervals to-day, after a long rainy and cloudy spell in which the weeds have grown much. I observe that the rough golden rods...have grown one to two feet high... *Comara palustris*, [out] how long? *Scirpus lacustris* is freshly out.*" (12:209)

Scirpus lacustris (bulrush)

July 4th P.M.—To Fair Haven Pond, measuring the depth of the river. As you walk beside a ditch or brook, you see the frogs which you alarm launching themselves from considerable distance into the brook.(12:218-19)

"June 24 [1859]. Deacon Farrar says that he can remember Lee's Bridge seventy-five years ago, and that it was not a new bridge then. That it is sometimes obstructed by hay in the spring. That he has seen a chip go faster up-stream there than ever down."(12: 212)

Figure 14.3 Poster of Thoreau's Concord River Survey, detail, S. Luria.

Source: Survey image and documents Courtesy Concord Free Public Library.

[at Lee's Bridge] than ever down."[16] Through his notation that the riverbed is grassy, Thoreau touches upon the river's very being, and this reminds us of what the river used to be called and, more deeply still, the people, whom Thoreau imagines as "extinct," who once lived there.

Following these links shows the human connection to the river – the Native American name for it Thoreau prefers to use, the stories the current locals repeat about the river, the fact that the river was one of Thoreau's favorite subjects to write about – and how any survey of the Concord River is in part a survey of Thoreau's literary career. All these links erase the boundary between biocentric and homocentric writing because the way we see and know the Concord River is through a long history of human engagement with it and attempts to define its nature.

The more you know about Thoreau the more you can see in the survey. Thoreau's sketch includes numerous depth soundings, a standard feature of maps. But for an avid reader of Thoreau these soundings, however expected, recall his famous survey of Walden Pond. Printed in *Walden* (1854), the survey caps the story Thoreau tells of how he finally got to the bottom of the mystery surrounding the pond's true depths. Just as he debunked local legends about the pond's depth then (including its having "no bottom"), so too in his river survey Thoreau wields his scientific method to correct local superstitions, especially about the Concord River's "deep holes," which Thoreau claims "are so unfathomable and mysterious, not to say bottomless, to the swimmers and fishermen."[17] In his journal, Thoreau notes:

> July 5 [1859]. Many a farmer living near the river will tell you of some deep hole which he thinks the deepest in all the river, and which he says has never been sounded. ... It only need to be considerably over his head to acquire this reputation. If you tell him you have sounded it, and it was not very deep, he will think that you did not find the right spot.[18]

The making of the survey thus offers a familiar Thoreau story, his wry humor calling out the false wisdom that his stubborn neighbors cling to. The river's depths figure for Thoreau a challenge to, as Emerson might put it in transcendentalist terminology, go "beyond" its visible surface and discover what lies hidden beneath, and this in turn leads as we shall see to insight about the river's being. This trajectory resolves Thoreau's conflict between "dry knowledge" and his imagination and fancy. Sounding the river's depths doesn't distract Thoreau from the more poetical, imaginative ruminations he might have, they lead to it.

From surveyor to poet

At "Lee's Bridge" on the survey we see how Thoreau's work as a surveyor and writer elide. Thoreau's journal entry for the day he surveyed Lee's Bridge reads:

June 22 [1859]. Paddle up the river to Lee's, measuring the bridges.

The sun coming out at intervals to-day, after a long rainy and cloudy spell in which the weeds have grown much, I observe that the rough golden rods ... which have grown one to two feet high, have many of them ... immediately drooped their tops, hanging down five or six inches. ... *Comara palustris*, how long? *Scirpus lacustris* is freshly out.[19]

Thoreau surveys the bridge as hired to do, but he can't help but botanize while he is at it. Juxtaposed as he makes them here, we can see the similarities between these two kinds of looking. His bridge surveys involve exact measurements of the bridges' various parts, but even as he observes the golden rod, he focuses on its parts and gauges its height. Both activities point to the deeply mathematical and analytical nature of Thoreau's writings, whether he is recording the perimeters of a property lot in his surveyor's field book, or the species he encountered on his walks in his journal. In this instance, Thoreau makes a table for each bridge where he records 15 or so features including its height, number of piers, their widths, and the distance between them. Thoreau's identification of plants such as the bulrush by their Latin taxonomic name *Scirpus lacustris* invokes another rational abstract table of measurement that forms a verbal analogy to his measurements for the bridge. The parts of the plant are analyzed, categorized, and abstracted to arrive at an overall verbal taxonomic map of the plant "kingdom." As Thoreau narrates nature in his journal, he is always calculating and categorizing. Thoreau had a natural sense of measurement; he thought and saw mathematically, as Emerson noted "[Thoreau] could pace 16 rods more accurately than another man could measure them with rod and chain ... He could estimate the measure of a tree very well by his eye."[20] Thoreau is a natural surveyor; his feet serve just as well as a surveyor's rod and chain.

However artificial and rational they seem, Thoreau's measurements point to their own imaginary quality. The Concord River of course is not as contained as the privately owned lots Thoreau usually surveyed, and the more Thoreau measures the river, the more we appreciate the unnaturalness of the task. On the one hand Thoreau's tables of bridge statistics and depth soundings suggest that the dimensions of the river can be known and grasped, and its riverbed, often imperceptible to the human eye, can be visualized. So too, the river sketch is divided into sections, defined by the man-made bridges, with linear distances between them recorded across the top (Figure 14.3). The river, we learn from this, is just over 16 miles "direct" linear distance, but due to its windings is 22.15 miles in actual length, and is "contained within a breadth of two miles 26 rods."[21] In contrast to this imposition of rational measurement and taxonomies, the informal quality of the sketch and its doodle-like drawings show that the river is insistently irregular – its depths that vary inch to inch and its craggy shore remind us that there are no straight lines in nature and that there is nothing geometric and reducible about it. The river's infinite facts are to some degree beyond measurement. Emerson captures this tension in his claim that Thoreau measures the river "to all eternity," and the paired texts of survey and journal confirm the fact: the rather contained, limited sketch on which Thoreau can only enter so much information and still be clear, and the voluminous journal which over the course of the entire summer pursues the elusive task of getting more and more of the river down.

Part of what compels Thoreau so is that the river keeps changing and this makes him have to pay closer and closer attention to it. While observing the spread and drifting of bushes in the water just above Lee's Bridge, Thoreau notes that the bushes have grown, seemingly overnight:

> I notice a black willow top a foot above water, a dozen rods from shore ... where the water is ten feet deep by my measure, and it is alive and green. Yet one who was not almost daily on the river would not perceive this revolution constantly going on.

Even the person who might be thought to know the land best – the owner – Thoreau observes, may not notice how the bushes have advanced upon his land, the "transplanted plants look so at home there." In the daily observations and trips on the river, the journal and survey attempt to exceed human limits of perception and chart this "imperceptible ... revolution."[22] It seems almost a superhuman feat, one that recalls an earlier moment in Thoreau's first draft for *A Week on the Concord and Merrimack Rivers*, when Thoreau admired the bittern, the wading bird that he frequently saw on the river:

> By its [the bittern's] patient study by rocks and sandy capes had it wrested the whole of her secret from nature yet? What a rich experience must be its – standing on one leg and looking out from its dull eye for so long, on sunshine and rain, moon and stars! What could it tell of stagnant pools, and reeds and damp night fogs. It would be worthwhile to look into the eye which has been open and seeing at such hours and in such solitudes. I would fain lay my eye side by side with its.[23]

By imitating the bittern, Thoreau hopes to attain this bird's eye view at ground level, and by laying his "eye side by side with its" transcend the artificial gulf between human beings and "nature," and see nature as it sees itself. He sees nature not as an object but identifies with it as a subject, imitates it, enters into its very being. In the passage from *A Week*, this ideal radical project still exhibits the homocentric quality of Thoreau's writing with its hyper-literary style, its bookish "fain" – a word he frequently uses in his poetry, and his playing with the literary convention and pun of the poet, the bard, identifying himself with a bird, as Keats does in "Ode to a Nightingale." In the more biocentric river survey of 1859, Thoreau loses this literary self-consciousness and simply imitates the bittern with his attempt at a non-stop way of seeing through his daily (and nightly) efforts to track the river's conditions.

Indeed, we can say Thoreau exceeds the bittern's view. Through his extensive journal entries on the river in 1859, Thoreau's sights are so fixed on the daily changes he sees that the effect resembles a time-lapse film of the scene; plotted on the survey these journal entries accumulate to show much of the river over time at once. This multifocused attempt of journals, tables, charts, and sketch to "integrate all the parts" and show the river in both time and space, from both above and on and in the river, draws upon and in some sense improves upon the technologies of seeing of his day. A hot air balloon's "bird's eye view," the distancing view of the map, is modified by being brought down to the ground and seeing it more intensely through identification with the subject. The static image of the daguerreotype is made dynamic. Emerson complained that a daguerreotype's promise to capture one's appearance, and through one's expression perhaps even one's inner character, was so unnatural it must fail: in the strained effort to stay still for some minutes the "total expression ... escape[s] from the face" and you are left with "the very portrait of a mask and not of a man." The technology seeks to do the impossible: "Could you not by grasping it very tight hold the stream of a river ... and prevent it from flowing?"[24] Thoreau's sketch isn't a snapshot, or an extended exposure that kills the very life it hoped to render; instead it achieves a deep focus shot that keeps background and foreground in focus at the same time; it's a sedimentary, cumulative image, a palimpsest with the past of *A Week* and its theories showing through and fulfilled here.

In the past, the poet Thoreau would have ultimately been interested in what all these observations of nature tell us about human nature, but the survey and journal show his writerly imagination to be satisfied simply by the river itself. When Thoreau notes on July 5, 1859 that "The deep places in the river are not so obvious as the shallow ones and can only be found by carefully probing it. So perhaps it is with human nature," we get the move his earlier works including *Walden* often make.[25] Here, read in the context of the journal's devotion to recording the facts of the river, Thoreau's "so perhaps it is with human nature" seems perfunctory. Instead, the poetic pay-off of the journal's dry descriptions is not the story they build of the river itself and Thoreau's Herculean efforts to record it. As Buell notes, Thoreau "became increasingly interested in defining nature's structure, both spiritual and material, for its own sake, as against how nature might subserve humanity, which was Emerson's primary consideration."[26] In this grounded transcendence, Thoreau merges his daily life with that of the river and ends up not in the ethereal Emersonian realms of the Oversoul but contentedly and insightfully on the sandy bottom and hay-covered banks of the Concord River.

The river poet

Whatever the objective truth of Thoreau's scientific measurements and his notable accuracy, the river survey announces itself as deeply subjective; from Thoreau's perspective this makes it more valuable still. The sketch is personal, colored by local knowledge such as "swimming place," and "Old Hay Bridge hereabouts," where old timers recalled the spot. The river survey not only highlights Thoreau as the scientist but the anti-scientist: his philosophical conflict, as Nina Baym has pointed out, with the trend in the sciences of his day toward a belief in objective, de-personalized data.[27] His observations in the journal thus often begin: "*I* notice a black willow top a foot above water"– not simply "the black willows are a foot above water"– and "*I hear* now that snapping sound under the [lily] pads."[28] Thoreau makes sure to point out his own participation in the sighting. He was skeptical in fact of the scientific method, as he remarked in an earlier volume of his journal: "I think that the man of science makes this mistake, and the mass of mankind along with him: that you should coolly give your chief attention to the phenomenon which excites you as something independent of you, and not as it is related to you."[29] Thoreau's journal and sketch is anything but detached: he doesn't just appear to walk every inch of the river's banks, and float over every inch of its 22 miles of water, he immerses himself in it: "Bathing at Barrett's Bay," he writes "I find it to be composed in good part of sawdust, mixed with sand. There is a narrow channel on each side, deepest on the south."[30]

On the surface, Thoreau's sketch of the Concord River might seem to be unexceptional, but if we read it vertically, we can plumb its depths. The survey is as much a portrait of the river in 1859 as it is a portrait of Thoreau. The river survey serves as Thoreau's autobiography in that it shows us what Thoreau was up to in the summer of 1859. The notations of "hard" and "soft" on the map mark where Thoreau walked the river's banks and the depth soundings show where he stopped his boat. We have his life in river form. The stream of his days, recorded in the journal's loose form and organized only by time – "22 June," "5 July" – is plotted on the sketch, organized by space, upstream to downstream, to produce a clarifying, transcendent, bird's eye view. Thoreau would only live for three more years; he would die in 1862 from tuberculosis at the age of 44. It seems fortunate that he was able to produce this work, a tribute really, to the river that had figured so large in his short life.

Thoreau's geopoetic survey of the Concord River shows that in places he did resolve the tension between the two halves of his life. His surveyor's view, discernible in his gradually more "biocentric" style, led to what we value now as some of the best writing of his career, drawn from the plainest facts. If we consider his lifelong study of the Concord River, we might say his survey work fulfilled in a geographic way the romantic vision he sketched in *A Week* of a superior kind of poetry that emanated only from a deep involvement with the land. Thoreau liked to chide the narrow view of farmers, whose workaday greed prevented them from seeing the big picture – the poetry in their farms. But in *A Week* he draws upon romantic poetry's fascination with rustic life and decides it is the farmers who are the greatest poets of all. Such men were:

> greater men than Homer, or Chaucer, or Shakspeare [sic] only they never got time to say so; they never took to the way of writing. Look at their fields, and imagine what they might write, if ever they should put pen to paper. Or what have they not written on the face of the earth already, clearing, and burning, and scratching, and harrowing, and plowing, and subsoiling, in and in, and out and out, and over and over, again and again, erasing what they had already written for want of parchment.[31]

In the paid labor and labor of love of his meticulous survey of the Concord River of 1859, Thoreau had his ground and he had his parchment and he developed his way of seeing and writing in a uniquely geographical, transcendentalist style.

Ultimately, the Concord River Survey defines a poetics that challenges Emerson's acquisitive homocentric point of view. In contrast to Emerson's confident pronouncement that "There is a property in the horizon" that belongs to the poet "whose eye can integrate all the parts," Thoreau's survey discovers that there is *no* property in the landscape; the force of the river's "native vigor" overwhelms the attempts to produce a coherent view of it from which one can gain insights into human nature and achieve transcendence, or even establish one's proprietary rights to its resources. And that is the source of its poetry.

Acknowledgments

I am grateful to the Concord Public Library Special Collections for the use of the images of Thoreau's river survey. Leslie Wilson, Curator of Special Collections, provided generous assistance in my research of Thoreau's survey. Robin Bernstein, Renee Bergland, Michael Dear, Betsy Klimasmith, Jim Ketchum, and Laura Saltz provided helpful feedback on earlier drafts of this essay.

Notes

1 B. Torrey and F. H. Allen (eds), *The Journal of Henry D. Thoreau Vol. X*, Boston: Houghton Mifflin, 1949, p. 233.
2 Thoreau's sketch of the Concord River relies upon previous surveys of the river in 1811 and then again in 1834, as Thoreau's title cited here for the sketch notes. Indeed, Thoreau does not even put his name on this sketch. To the best of my knowledge so far, and in consultation with Leslie Wilson, Curator of Special Collections at Concord Public Library, what we see here is Thoreau copying the shape of the river from the previous surveys. The exact correspondence between Thoreau's journal and his notes show that much of the sketch is Thoreau's own work. For example, the depth soundings Thoreau plotted onto draft sketches of the river appear in the corresponding places on the larger sketch. The entire sketch appears to be drawn in Thoreau's hand. Thoreau seems to have filled in these older surveys with his own observations of the river in the summer of 1859.
3 H. D. Thoreau, *A Week on the Concord and Merrimack Rivers*, C. F. Hovde, W. L. Howarth, and E. H. Witherell (eds), Princeton, NJ: Princeton University Press, 1980, p. 90.
4 L. C. Johnson, *Thoreau's Complex Weave: The Writing of a Week on the Concord and Merrimack Rivers*, Charlottesville, VA: University of Virginia Press, 1986, pp. 7–8.
5 H. D. Thoreau, *A Week*, p. 44.
6 L. Buell, *The Environmental Imagination: Thoreau, Nature Writing, and the Formation of American Culture*, Cambridge, MA: Harvard University Press, 1995, p. 138.
7 H. Daniel Peck, "Unlikely Kindred Spirits: A New Vision of Landscape in the Works of Henry David Thoreau and Asher B. Durand," *American Literary History* 2005, 687–713, 690, 694.
8 L. Buell, *op.cit.*, p. 117.
9 R. Van Noy considers Thoreau's survey work in his *Surveying the Interior: Literary Cartographers and the Sense of Place*, Reno: University of Nevada Press, 2003, pp. 38–72.
10 For an excellent history of the controversy between farmers and mill owners over use of the Concord River, see B. Donahue, "Dammed at Both Ends and Cursed in

the Middle: 'The Flowage' of the Concord River Meadows, 1798–1862,"
Environmental Review Sumer/Fall, 1989, pp. 1–20.

11 Letter of Emerson to Elizabeth Hoar, August 3, 1859. Quoted in W. Harding,
The Days of Henry Thoreau, New York: Knopf, 1967, p. 411.

12 See *The Journal of Henry David Thoreau*, Vol. 12, p. 274. Further evidence that
Thoreau calculated the volume of the river is suggested by his pages of notes and
calculations for the survey. Many of these are owned by Concord Public Library
Special Collections. They contain pages of large multiplication problems drawn
from Thoreau's extensive tables of measurements, which suggest he was calculating the volume and not only the length and breadth of the river. I am grateful to
Avner Ash for helping me evaluate Thoreau's calculations. In his journal, Thoreau
notes "Rudely calculating the capacity of the river here and comparing it with my
boat's place, I find it about as two to one, and such is the slowness of the current,
viz. nine minutes to four and a half to a hundred feet" (*Journal*, Vol. 12, p. 274).

13 Both millowners and farmers changed environment in ways that defeat their
objectives. Farmers deforested river banks to take advantage of increased markets
produced by the railroad between Concord and Boston (see Donahue, pp. 17–18).
The court record of the case makes no mention of Thoreau or his survey. See the
*Report of the Joint Special Committee Upon the Subject of the Flowage of Meadows on
Concord and Sudbury Rivers, 28 January 1860* (Boston, William White, Printer to
the State, 1860), Special Collections, Concord Public Library.

14 R. W. Emerson, "Nature," in *The Norton Anthology of American Literature: Volume B*
7th edn, N. Baym (ed.), New York: Norton, 2007, p. 1112.

15 H. D. Thoreau, *A Week*, p. 5.

16 *Journal*, Vol. 12, p. 213.

17 H. D. Thoreau, *Walden*, J. S. Cramer (ed.), New Haven, CT: Yale University Press,
p. 276; *Journal* Vol. 12, p. 276.

18 *Journal*, Vol. 12, pp. 221–22.

19 *Journal*, Vol. 12, p. 209.

20 R. W. Emerson, "Thoreau," in *The Norton Anthology of American Literature: Volume
B*, 7th edn, N. Baym (ed.), New York: Norton, 2007, p. 1235.

21 *Journal* Vol. 12, p. 225.

22 *Journal* Vol. 12, pp. 209–10.

23 Johnson, *op. cit.*, p. 293.

24 Quoted in Sean Ross Meehan, *Mediating American Autobiography*, Columbia:
University of Missouri Press, 2008, pp. 65–66.

25 *Journal* Vol. 12, p. 222.

26 Buell, *op.cit.*, p. 117.

27 See N. Baym, "Thoreau's View of Science," *Journal of the History of Ideas* (1965),
211, 221–34.

28 *Journal* Vol. 12, pp. 209–10; emphasis mine.

29 *Journal* Vol. 10, pp. 164–65; quoted in N. Baym, p. 232.

30 *Journal* Vol. 12, p. 225.

31 H. D. Thoreau, *A Week*, p. 8.

3 VISUAL GEOGRAPHIES
Geoimagery

Jim Ketchum

Strange as it may seem, the critique of representation that so deeply affected a broad range of scholarly disciplines while spawning still others arrived late to academic geography. Strange, because geography has been practiced for so long and so widely, and because geography itself depends so heavily upon representation. Geo-graphy is, literally, *earth-writing*, and its time-honored practices of exploration and description have allowed it to carry forward, over the course of many centuries and from nation to nation, what is known about the Earth, it peoples, and its places, its many distant lands and far away events. As Denis Cosgrove so eloquently and simply expressed it, the endeavor of geography "lays particular claim to the globe. Its intellectual task is, by definition, to describe the globe's surface."[1]

Such descriptions depend however on the ability to fix what becomes known about the world through commonly accepted and comprehensible forms such as the written word, statistics, maps, charts, drawings, and photographs. By passing that knowledge down through time from generation to generation and also across space from place to place, readers and viewers who have not been to those far-flung places, met its distant peoples, observed its remote vistas, or witnessed its dramatic events can feel as though they have been there, or at least can feel justified in imagining how it would feel had they been there, in that place, at that time, able to see it through their own eyes.

As such, geography embodies a double-meaning. Commonly understood, it is either that which is there or those representations which bring what is there here before us, now, with clarity and accuracy. We frequently use the word both ways, often interchangeably, often without noticing the difference. Perhaps this is why the broad critique of representation, which shocked so many other disciplines into a deeply reflective posture, made slow inroads into geography: what geography was and meant was simply so self-evident that its descriptions hardly seemed like representations at all. Much as a window is merely a frame for seeing what is really out there in the world, a geographic representation seemed unlike, say, a painting, which is not a window through which we see worldly objects with our own eyes, but rather a creative reflection of that world produced, no matter how faithfully or accurately, through some other person's imaginative and interpretive act.

Thus art history and other visual disciplines such as cinema studies or architecture, or disciplines that clearly studied systems of representation, such as literature or anthropology, could be deeply affected by growing doubts about representation, which had previously been understood from a scientific viewpoint as value-free. Geography's own self-searching came a little late, but it did come. J. B. Harley warned about the potential maps held to act as tools of the power elite, and Mark Monmonier showed us that maps, if not always tools of the powerful, could still routinely mislead us.[2] J. M. Blaut explored the imaginative "colonizer's model of the world."[3] In that same vein, Lewis and Wigen published *The Myth of Continents*, deconstructing overarching conceptual premises for ordering our knowledge of peoples and events.[4] Likewise, landscape came under suspicion as an artificial construct that could be employed to serve the dark purposes of any number of hidden ambitions. New subfields were born such as critical geopolitics, which studies the taken-for-granted ways in which we represent nations, peoples, and other spatially conceived entities, and how those conceptions influence our public debates and function in the exercise of power.

The ways that geography represented knowledge of other peoples, places, and events to ourselves and to others through cartography, words, statistics, and images, of necessity all came under review. Lewis and Wigen argued that the point of all this was not academic navel-gazing, but something much more important: to reach a broader audience – since geographic concepts are so important to the public sphere, old geographic concepts had to be rethought, revolutionized, and in some cases simply exploded to make way for new thinking. A decade later, coupled with an emerging revolution in apprehending the world at distance through new, improved, or simply more ubiquitous technologies, we find that the revolution in geographic imagining for which they had hoped is in fact taking place on a broad scale. Spreading through academia but also beyond its boundaries, the revolution in geographic thought has entered the public consciousness. Geographic perspectives and its key metaphors and concepts, such as mapping and scale, have been rapidly adopted across a broad spectrum of society in efforts to explain and describe new perceived realities and experiences in the twenty-first century.

This has become particularly true in the realm of artistic production. Artists, it may be said, although not completely unfettered by convention and institutional constraints, do in many cases have more freedom to explore new territories and ideas, to create in new directions, and to blaze new trails than do the rest of us, and often admit to even training their sensibilities in the direction of the undiscovered countries of their imaginations. Laura Kurgan, the artist whom I interviewed and whose work I explore in one of the chapters that follows, revealed to me as much: that she is driven more by her own intellectual and artistic interests than she is by the need to sell artwork to collectors. In addition, artistic practice holds a special potential for exploring geographic concepts and perspectives because it still unfolds, in the tradition of the avant-garde, as a conscious and active resistance to received knowledge and conventions of representation, as if the juries selecting and approving new works of art to be allowed into the Paris Salon still existed and had to be actively resisted and exposed. The impulse to question authoritative content and its means of expression is alive and well in the world of art.

But artistic production offers fertile ground for the exploration of geographic perspectives and concepts for another reason: while academic investigation is still modeled on a dualism that separates the subject of study from the perceiving, objective thinker, artistic production allows and even celebrates the artist as the investigator whose experiences embody the act of the investigation and whose evidence or truth of that encounter is produced in and through his or her body. Think of Jackson Pollock, whose art resided as much in his confrontation with the canvas and his movements through space as it did in the individual paintings that resulted. Art is therefore ideally situated to explore questions now vexing much of the academic world – the exploration of the role the body plays in the production and reproduction of social life.

We can note that a previous cycle of academic critique studied representation for the ways it subtly served structures of authority and power, but we can also note that critique's subtle suggestion that once that veil were lifted and the processes beneath were exposed for what they were, that we would be able to exist in a kind of freedom, at liberty to remake the world in a more truthful, open, and transparent way. We should also take notice of that critique's distance from its subject, as if the perceiver could intellectually strip away the operations of representation that obfuscated truth and that created its own network of truths in its place. Perhaps it is the proliferation of images throughout society and the ubiquitous technologies that produce and circulate them, as well as the continuing human horrors they portray, that today has led us to a somewhat less hopeful stance with regard to the future, but also to new methods of exploration and resistance: we may not be able to escape the grid and all it implies, and we may not be able to free ourselves from the discourses that continue to produce war, environmental degradation, and social oppression, but we may be able to effect a deeper rupture in those discourses by recognizing that we are not above and beyond the structures of power but within them, that they operate through us enabled by all the social spaces, places, landscapes, nations, and scales we unceasingly recreate from moment to moment in the everyday world.

That we can, in fact must, use our bodies as the locations for a continuing and perhaps more complex critique of representation and its networks is the fundamental realization that makes the works in this section new and different from anything that could have come before. The eight chapters grouped in this section are all either written by artists or by critics discussing an artistic work. Aitken and Dixon discuss the director Paul Anderson's film *There Will Be Blood* and the ways it simultaneously inhabits and deconstructs landscape conventions. Caren Kaplan explores a work of photographic art by the French artist Sophie Ristelhueber, who carefully deconstructs notions of scale and the ways they work through our bodies in our habits of viewing. Phil Govedare is a painter concerned with environmental degradation and the ways we experience and accept it. Stephen Young is a geographer whose artistic work in creating gallery exhibitions of the Earth as remotely sensed from the distances of space explores the potential that abstraction holds for confronting viewers and intervening in their received ideas about the Earth, nature, and environmental change. Lize Mogel is an artist who carefully manipulates cartographic conventions to awaken us to the ways that maps accomplish the objectives of political agendas. In my own chapter, I explore Laura Kurgan's artwork responding to September 11 as a skillfully crafted experience for the art viewer meant to intervene in the discourses and images that help manufacture a public consent for war. Ursula Biemann's discusses her own artworks, which look at the role of surveillance in the spatial politics of migrations in sub-Saharan Africa. Finally, Norma Iglesias-Prieto describes a broad project involving the creation of an animated film, a documentary, and a survey of children's attitudes in both Tijuana and San Diego about the border between those two cities and its many meanings, and the hopes those children have for the future.

Taken together, these eight chapters are not meant to represent a balanced survey of new directions in artistic practices taking place at the intersection of geography and the humanities. This zone of production is in fact so diverse, so dynamic, and so fertile that such a survey would not really be possible, and certainly not by exploring just eight creative works. Rather, what these chapters do express is the broader attempt to explore the human individual as the location for the production of meaning in an increasingly networked society, and consequently as the potential location for interventions in and resistance to the production of those meanings. What these pieces all share is a common recognition that the work of culture in producing meaning through us never stops, that there is no outside to culture. At the same time however, there is a recognition that if the political truly is personal, then we always have at our disposal the opportunity to act and react in a positive, if oppositional way, to those forces always subtly at work in causing those moral outrages which we perceive as happening at a distance through the technologies we hold so close at hand.

Notes

1 D. Cosgrove, *Apollo's Eye: a Cartographic Genealogy of the Earth in the Western Imagination,* Baltimore: John Hopkins University Press, 2001.
2 See the chapter "Maps, Knowledge, and Power" in J.B. Harley, *The New Nature of Maps: Essays in the History of Cartography* Baltimore: The Johns Hopkins University Press, 2001; and M. Monmonier, *How to Lie with Maps,* Chicago: University of Chicago Press, 1996.
3 J. M. Blaut, *The Colonizer's Model of the World: Geographical Diffusionism and Eurocentric History,* New York: The Guilford Press, 1993.
4 M. Lewis and K. Wigen, *The Myth of Continents: a Critique of Metageography,* Berkeley: University of California Press, 1997.

15

*El otro lado de la línea/*The other side of the line

Norma Iglesias-Prieto

Transborderism

Among the world's frontiers, the boundary line between the United States and Mexico, particularly the Tijuana–San Diego section, is the one with the highest levels of inequality, but also of interaction, integration, and economic interdependence. This interaction is expressed in the multiple exchanges and flows between the two cities. Unfortunately, the high level of exchange and economic interdependence does not necessarily translate into equivalent levels of awareness, understanding, acceptance, or commitment towards the neighboring city and its inhabitants. In this complex border zone, the level of interaction is determined by what I define as the degree of "transborderism." This term refers to the frequency, intensity, directionality, and scale of crossing activities; the type of material and symbolic exchanges; and the social and cultural meanings attached to the interactions. A higher level of transborderism is associated with greater cultural capacity and richness, increased complexity in the ways people perceive the border, as well as richer concepts of self-identity. It is often said that "everything depends on the mirror in which one looks," but it also depends on the characteristics of *who* is doing the looking and the *place* from which one observes. For this reason, it is essential to analyze social concepts of the border from the viewpoints of the people and places in the border zone's different social and cultural milieux. The main distinctions are among those who cross the border and those who do not, the reasons for crossing or not crossing, and whether crossing is from north to south or south to north.

For those who cross the border, there is an enormous diversity of interactions. In the border zone between San Diego and Tijuana these may be grouped into four major types, which in turn generate four levels of complexity in social imagery. Relationships range from the most basic, impersonal, and superficial to the most intense and complex. First, there are relationships of a cold, temporary, impersonal character, arising from sporadic crossings and interactions of a *commercial nature* between people who do not know one another, who do not necessarily behave according to local cultural norms, and are not trying to go beyond a client relationship. Here, we are speaking primarily of relationships formed in micro-spaces of the city, generally designed for consumerism and tourism, and concluded in relatively little time and in confined areas in the city. In Tijuana, such zones are principally in the vicinity of Avenida

Revolución and three or four adjacent streets where one finds bars, restaurants, arts and crafts businesses, and, more recently, pharmacies. Such places are visited mostly by tourists from the United States but not by Tijuana residents. Visitors are not acquainted with the urban, social, and cultural dynamics of the city and its inhabitants. Their Tijuana landscape is built around exotic mental images the tourists have acquired prior to their arrival (including donkeys painted like zebras, piñatas, brightly colored walls, pictures of bullfighting, margaritas in big cups, charro hats, etc.). In contrast, when people from Tijuana cross over to San Diego, their commercial interactions generally occur in large shopping centers and recreational areas. These areas are not exclusively for foreign tourists (unlike Avenida Revolución in Tijuana), but for a much broader clientele that includes local residents as well as tourists. The images of the border deriving from these client relationships tend to remain in the realm of stereotype, appearance, and the superficial; that is to say, focused on the most visible characteristics of the border zone. Such first impressions may surprise and startle visitors, pleasantly or unpleasantly, simply because they are different.

Second, there are interactions that require *periodic encounters and personalized relationships* but not emotional connection. These include crossings for regular purchases of medicine, food, and other products, as well as visits to doctors and dentists. This type of interaction presumes a greater familiarity with the urban area, since the stores, medical offices, and hospitals are located in various parts of the city. It also suggests a level of cultural understanding that comes from confidence needed in the context of medical and other services. For people traveling from San Diego to Tijuana, knowledge of the Spanish language is important but not essential, since all of these services are bilingual. While the purchase of medications does not necessarily follow an ethnic-cultural pattern, the use of medical services does. This personalized type of interaction implies a more sophisticated and abstract perception of the border cities. For people coming to San Diego from Tijuana, encounters typically relate to places in different parts of the city that offer more specialized products and services beyond the shopping malls.

Third, some transborder *relationships characterized by warmth and emotion* are associated with crossings to visit family, friends, or significant others. They may be infrequent and limited to special events such as weddings and funerals, but they can also involve frequent crossings at regular intervals undertaken for other purposes such as recreation, cultural events and dining, medical care, and other products and services. People in this category have a broad knowledge of the urban spaces on both sides of the border. The interactions from north to south are dominated by visits by Mexican and Mexican-American residents of California.

Finally, we have the emotional, intense, and deeply engaged relationships of the *"transborder" citizens* who personify the border zone's diversity and dynamism. They include people with dual citizenship; with experience of having lived, studied, and worked on both sides of the border; who are bilingual and bicultural; and who have profoundly personal ties on both sides. The transborder citizens typically cross frequently, even on a daily basis to work or take children to school. They possess the cultural ability and awareness that allow them to move independently on both sides. Most interestingly, they tend to have the most complex understanding of the border: they recognize it as a kind of fracture; a boundary where one feels the exercise of power, abuse, and suffering; but also as a space of multiple opportunities and great cultural riches. Typically, they are conscious of the variety of problems and challenges associated with the border, but they also have a great commitment to their place in the transborder metropolis.

In summary, transborder people develop a variety of abilities and more complex behavioral patterns that enable and promote a binational existence. Transborder lifestyles necessarily involve expanded capacities and activity patterns. Transborder citizens are often more critically self-aware of the everyday realities of border living. The future of the borderlands lies precisely in the involvement and potential of such transborderized citizens.

Schoolchildren's views of the border

In the summer of 2008 I had the opportunity to coordinate – together with Yvon Guillon of the University of Rennes (France) – a project entitled "The Other Side of the Line: Views from Tijuana and San Diego," which involved 48 people from Mexico, the United States, and France. The French presence in this project was the result of participants' technical and pedagogical expertise on children's animation workshops, as well as their cultural interest in border issues developed in previous projects such as the 28th Douarnenez Film Festival (in 2005), which focused on Mexicans in the United States and the US–Mexico Border. The project consisted of four phases:

1   The creation of two short animated films by children from Tijuana and San Diego based on the theme "The other side of the line." The cartoons that resulted were the work of 22 children between the ages of 11 and 13, in two workshops offered by the French animated-film experts.
2   A training workshop on model animation for the young visual artists together with graduate and undergraduate students from San Diego and Tijuana.
3   The production of a documentary film about the project and border life.
4   A research project addressing notions of "otherness" through children's imagery of the two sides of the Tijuana–San Diego border.

Through the workshops, children from each city produced a short animated film about the children on the other side of the border. The two groups of children discussed the way they think about and portray the other side. Subsequently, the children came to consensus regarding the story, characters, situation, context, sets, and so on. They also wrote the script, built the set and characters out of paper and fabric, shot the entire sequence of scenes, recorded the soundtrack, and edited their respective film animations.

The 22 children who participated in the project were of mixed gender and possessed a variety of experiences of the "other side." This experience ranged from those who had never crossed the border (usually because their parents had never wanted to, in the case of San Diegans, or because they lacked a US visa, in the case of Tijuanenses);

those who crossed regularly for reasons of family or school; and those who at some time had lived on the other side/*el otro lado*. The two short animated films (5 minutes each) produced by the children may be found at http://delotroladodelalinea.word-press.com/.

The children of Tijuana

One of our first questions was: "What is the border like?" All the children in Tijuana identified and associated the border with the metal fence that divides the two countries. They recognized its function was to halt the flow of people from south to north, "so the Mexicans can't go to work in the fields over there." All the children knew about the fence and the great majority had seen it. For them, this was without doubt the most graphic and obvious perception of the border. They were also aware (including those who had never crossed) of the difficulty of crossing the border and of obtaining a visa. Some children defined the border in more abstract, even sophisticated terms, saying for example that "the border is a place that separates two places"; or that it is a "boundary of place and time."

This high degree of awareness about the separation of the boundary line led to a more specific question: "What is the difference between Tijuana and San Diego?" One of the most common answers was the higher level of civic-mindedness attributed to San Diego: "The laws are respected there," or "People are more educated." The answers also revealed a preoccupation with the lack of safety in Tijuana relative to San Diego: "The police there do help people"; "In the United States, they have safety and alarms"; and "Nobody steals, even if they wanted to."

The difference in socioeconomic levels of prosperity between the two sides of the border was very clear to the children of Tijuana. For example, they noted that: "Over there, there aren't any homeless people"; "The houses are bigger and more modern"; "The toys are way cooler"; "They have a lot of parks." In contrast, about their home town Tijuana, the children observed: "There are houses without any roof."

In addition to their verbal comments, every child drew a house in San Diego and one in Tijuana. Their sketches revealed the huge contrast in the amenities and size of the dwellings in both countries. The San Diego houses appear with swimming pools, big gardens, many bedrooms, spacious kitchens, garages, and several automobiles. Home security systems made a large impression on the children given the levels of insecurity in their own city. The drawings of Tijuana houses, in contrast, tend to be much smaller and more modest, although there were many variations that revealed awareness of the diversity of styles and socioeconomic levels in that city.

The Tijuana children were asked to collaborate in a drawing of San Diego. They showed the city as a huge shopping center and amusement park with almost no houses, and where institutions such as schools, hospitals, government offices, churches, and so on were absent. The United States was always associated with big businesses and retail brands such as McDonald's, Costco, Wal-Mart, Target, and so on. In addition, the drawing powerfully drew attention to a dollar tree, few people in the streets (a handful are shown sunbathing on the beach), and large public restrooms. The girl who drew these last items commented: "Over there, everywhere there are bathrooms, not like here where you can't find one anywhere".

The Tijuana children, as much in their pictures as in their conversations, also noted other differences aside from the material, as for example family and community dynamics. They recognized that in Tijuana – in spite of the difficult economic conditions and lack of security – there is a lot of support through family members, friends, and neighbors. In general, Tijuana was represented as a space full of human feeling and social relationships, something that was absent in their imagery of San Diego.

Finally, the children were asked: If the United States were an animal, which animal would it be? One of the girls said: "A bald eagle"; another said: "A lion, because it wants to be the king of the jungle"; and a third suggested: "A donkey, because they are stubborn and foolish." Such characterizations relate to common Mexican perceptions of the asymmetry of power relations between Mexico and the United States.

After many debates and discussions, the characters and story for the animated film were at last defined. The film was entitled *Wacha el Border*, alluding as much to the geopolitical frontier as to the border's most colloquial form of cultural expression: Spanglish (which is very familiar to transborderized people). There are four protagonists: two humans (Cynthia, from Tijuana; and Max, from San Diego) and two animals (Chester, the cat from Tijuana; and Paulina, the panther from San Diego). The story refers to ordinary daily activities such as preparing to cross the border, looking

Figures 15.1 Artwork from *The Other Side of the Line.*
Source: Photography from "The Other Side of the Line" Project Archive.

for passports, leaving on time, waiting in long lines for the chaotic crossing, interacting meanwhile with street vendors, experiencing the coldness of the immigration officer, driving around on the busy freeways, and so on. It also incorporates unusual events such as the rescue of a panther at the San Diego Zoo; or the destruction of the wall dividing the two countries in the year 2019 (commemorated in the film by a new monument at the site of the former international boundary).

The panther turns out to be the film's most enigmatic character, since in spite of living in a beautiful, clean, well-maintained place, she is still a captive longing for freedom and food other than healthy vegetables. In San Diego the protagonists frequently experience harassment: first, the panther is mistreated by some children at the zoo; later all are pursued by law enforcement officers, a link with the children's perception of social oppression in the United States. Thanks to Cynthia and Max, who represent the children of both sides of the border, the panther is liberated and gets to enjoy meat: "This is the most delicious I've tasted in a long time!" Then they all flee in a giant bubble to the Mexican side of the border. Their deed is reported in the newspaper: "Children cross the border in a bubble," along with another note alluding to violence: "Violence deforms human beings." The story ends with a twist, very much in the style of Mexican soap operas, when Cynthia and Max turn out to be siblings. The film's message is one of hope, presenting "an ideal world, without borders, where there are neither winners nor losers," but instead everyone belongs to one large family.

Figures 15.2 Artwork from *The Other Side of the Line*.
Source: Photography from "The Other Side of the Line" Project Archive.

Figures 15.3 Artwork from *The Other Side of the Line.*
Source: Photography from "The Other Side of the Line" Project Archive.

The San Diego children

In the work with the San Diego children, little or no reference was made to the fence dividing the two countries. The great majority of the group had not seen it, and some even did not know of its existence. These perceptions reflect the asymmetry of the cross-border experience that makes it impossible for one side (Tijuana) to forget the existence, power, and control of the north, while the other side (San Diego) essentially ignores its identity as a border city and the nearby international boundary with its metal fence. The San Diego children generally represented Tijuana as an "arid", "dusty" place, of "orange" and "brown" colors – referring to the "clay" and "soil"; with the taste of "tacos" and spices; with the sound of "trumpets" and "mariachis"; surrounded by "cactus" as well as "donkeys," "pack mules," and "dogs"; people with "sombreros," "ponchos," and "maracas" in a landscape of "churches" and "missions." Such stereotypical images are common in movies, television, and advertising. They fail to capture the more complicated realities of Tijuana. In a rare instance of more nuanced understanding, one transborder girl colored Tijuana green because it reminds her of "cilantro" and "nopalitos," images drawn from her personal experiences.

In the children's drawings of Tijuana, there was enormous diversity and contrast. The images tended to be more heterogeneous than the images of San Diego drawn by

Figures 15.4 Artwork from *The Other Side of the Line*.
Source: Photography from "The Other Side of the Line" Project Archive.

the Tijuana children. First, there were drawings by children who have never been to Mexico and have no cultural connection with Mexico: they depicted Tijuana as a desolate place, dehumanized and dangerous; or associated it with elements or products they have seen in Mexican culture by way of the American marketplace, such as Taco Bell. A second group of images were by children with no experience of crossing the border, but who have learned and retained positive notions of some elements of Mexican culture; they tend to represent Tijuana in romanticized form as a small, traditional town, with adobe houses, churches, and markets. The last group of drawings was by children experienced in border crossing and with strong family ties in Tijuana. They represented the city as a modest place in material terms, but full of color, fun, and emotion associated with their strong relationships with family and friends.

The San Diego group included two very influential students: one was a girl with transborder experience, who knew about "Mexican things." She embodied the frontier: "I am the border, for me living here or there, switching from English to Spanish is as normal as breathing." She always questioned the more stereotypical and negative images of Tijuana. The second was a boy who advanced a very negative vision of the other side, characterizing Mexico in general as "a problem." These two opinion leaders not only differed on the positive/complex versus negative/stereotypical versions they espoused, but also on the kind of story they proposed for the film. The girl, who

Figures 15.5 Artwork from *The Other Side of the Line*.
Source: Photography from "The Other Side of the Line" Project Archive.

was supported by half the group (mostly girls), wanted to make a film about ecological problems along the border, a topic that emphasized the border as a human construction and a natural ecosystem that recognizes no political boundaries. The boy (supported mostly by the boys) wanted to make a police action film about "gang members" and "the mafiosos" of organized crime, emphasizing the border's function to protect San Diego from dangers originating on the other side.

As in the Tijuana workshop, after many debates the San Diego children arrived at consensus on a film entitled "Beyond the Border." It is the story of a young North American woman reporter who – assisted by a Mexican mosquito – discovers the illegal activities of an American businessman who is making a lot of money disposing of industrial and urban waste on the Mexican side of the border. It's a film of ecological drama mixed with police action footage that incorporates a series of Hollywood movie conventions, including an American hero who saves everyone else. In the animation, the themes of garbage and violence (including crimes, kidnaps, robberies) are discussed. Allusion is made to corruption, to complicity along the border, and to the underworld of organized crime which is associated with cholo graffiti and slang. The film also touches on the difficulties of finding work and on powerful figures such as the boss who is always threatening to fire the protagonist. The film makes a concession by identifying the "kidnappers" as Anglo-Saxon Americans, thus conceding that

Figures 15.6 Artwork from *The Other Side of the Line.*
Source: Photography from "The Other Side of the Line" Project Archive.

"we are part of the problem" and avoiding the perception that borderland difficulties are attributable solely to Mexico. The San Diego children make no reference to every-day life or to real relationships between equal human beings on the other side, although the film ends by suggesting that the two cities should cooperate. The crime story appears on TV (the great generator of reality for the children) but only on the American channel, not the Mexican one. The San Diego children's film emphasizes cooperation among unequal neighbors (even though differences are accentuated more than similarities), but it lacks the vision of an emerging integrated transborder metropolis.

Representing cultural differences

Understandably, the children of Tijuana and San Diego told stories that reflect points of view that are intimately tied to their degrees of transborderism. In comparison with the San Diegans, the Tijuana children made greater efforts to incorporate "the other" as part of their reality by involving main characters from each country, using Spanish and English languages, and incorporating perspectives from both sides. The Tijuana children – throughout the project – always took the position that the border was a human construction that impedes the natural flow of people, animals, and natural resources. For this reason, in creating their film they constantly made comments questioning and resisting the very existence of the boundary fence and of borders in general. "Animals don't know that there are borders"; or "Cats don't use passports," they observed. Significantly, the Tijuana film ends by suggesting that the wall will be removed and transborder communities, represented by Cynthia and Max's family, will be reunited. The film's social diagnostics derive strongly from a populist, everyday viewpoint.

For the San Diego children, the border was something natural and unquestioned because it was "necessary." In general, the children seemed convinced that the border "should exist" because "it protects us." Although the transborder girl suggested that the border be demolished, the rest of the children would not agree because: "It isn't realistic." The girl responded: "But, is a talking mosquito realistic?" Ultimately, however, her argument did not prevail, and the San Diego film remained resolutely separated from Mexico, even though in reality the borderline is only minutes away.

More generally, the dynamic of the workshops, the exercise of expressing and debating different perspectives and life experiences along the border, and the process of negotiation to produce films about "the other" generated positive attitudes about the border. Furthermore, the French presence in the workshop was useful; by bringing a third culture and language to the work, French participants functioned as neutral elements that facilitated in-depth discussions of difference and similarity between Mexico and the United States. The children were better able to explain to "outside" collaborators not only how they see and represent others, but also how they see and represent themselves. The debates acted to humanize those on *el otro lado* (the other side) because participants realized that beyond the material differences they also shared many things, such as aspirations and dreams.

Much remains to be done in terms of the awareness and commitments among neighbors along the US–Mexico border. Through their film-making, the children of Tijuana and San Diego revealed strikingly different attitudes and conceptions of each others' cities. But their collaborations also showed that taking an early interest in and working through children's imagery of the borderlands, it is possible to generate more positive commitments to a collective future in an increasingly integrated and diverse world.

Acknowledgment

Translated from the Spanish by Myra Bailes and Michael Dear.

16

The space of ambiguity
Sophie Ristelhueber's aerial perspective

Caren Kaplan

> Governments at war use the truth whenever it suits their purposes; it is ambiguity they find intolerable.
>
> George H. Roeder, Jr.[1]

Since the advent of World War I, views from the air have become part of the iconic imagery of war in Western modernity. As aviation advanced throughout the twentieth century, often via the initiatives required by the waging of war from the air, so, too, did the technologies of sight and imaging. If the "bird's eye view" has become part of the way terrain and landscape become recognizable in contemporary Western culture, it has also come to constitute the ground truth of cartography and to serve as a vital tool in the modern art of warfare.[2] Throughout the twentieth and into the twenty-first centuries, aerial perspective, the vertical view from above, mechanically reproduced through both analog photography and digital imaging, has produced realist aesthetics as well as militarized geospatial practices.

Realism's ground truth – What happened? Who saw it? When? Where? – demands witnessing in the place of denial, facts instead of sanitized views. This passion for certainty is, as John Taylor has argued, "at the heart of reportage on war."[3] Such realism meets the unceasing need of governments and militaries for controlled imagery of war and helps to justify the underlying national interest in the ways in which war and its effects are portrayed. Thus, one of the pressing questions in a time of war is how to address imagery that is both iconic for war photography – grief, violence, destruction – and too graphic or "realistic" for mainstream media or inconvenient for the government. Although "realistic," aerial photography can appear to be too remote or too objective for the intimate or sentimental point of view associated with some iconic images of war. But if we want to think through the ways that war photography works to serve the interests of governments we might choose to avoid the recuperation of nationalism inherent to the heroic model of war-time photojournalism in favor of another visual strategy: abstraction and ambiguity.[4]

In this nexus of representational critical engagement, the 1990–91 First Persian Gulf War[5] serves as a meaningful example. Celebrated by Western militaries and governments as the perfect apotheosis of airpower and contemporary information technologies, it was the first US war to incorporate fully satellite navigation and guidance systems. The conflict also remains notable for the imagery of "precision" bombing that many of the armaments themselves – the so-called "smart bombs" – produced as they were deployed to targets.[6] This new generation "TV war" brought a seeming realism to those watching at home: simulated real-time action with spectacular effects. But there was very little "on the ground" reporting due to both US and Iraqi policies that strictly controlled Western journalists' access to the conflict. Estimates of Iraqi war dead number in the hundreds of thousands.[7] In the West, at least, there were few "close-ups" of this carnage.

This war, unlike the war with Vietnam or World War II, presented Westerners with a different mediascape: one of absence and distance rather than proximity and sensation.[8] The quandary for representational critique in this instance, therefore, could be located on the line between realism and abstraction, between reportage and art. A war was fought, the US-led coalition forces attacked Iraq by air and by ground: What is the relation between seeing these actions and knowing anything about their effects? What use can be made of the knowledge of war that we receive through visual media?

The imagery produced by Sophie Ristelhueber, a French artist, troubles the quest for realism, certainty, and documentary sentimentality that so often characterizes the representational practices of modern warfare. At the close of the First Persian Gulf War, Ristelhueber traveled to Kuwait's border with Iraq and photographed areas that she was able to reach by helicopter and on foot. These photographs, exhibited as *Fait* in Europe (as *Aftermath* in the USA), cite with great subtlety and complexity the conventions of landscape and aerial imagery: ruins and reconnaissance, architecture and target, vistas and maps. The distancing effect of this kind of vertical and oblique aerial photography places viewers in an aesthetic zone that takes some time to engage. Ristelhueber's photography emphatically avoids the representation of embodied carnage and the usual realist conventions of war photography by drawing on a different set of conventions to produce an aerial perspective; one that critically engages both vertical and oblique or horizontal angles. Situating Ristelhueber's photographs in relation to scale can alert the viewer to the interpretive exigencies and possibilities of aerial perspective.[9]

Aerial perspective and the production of scale

The relationship between vertical and horizontal views in the cultural history of Western modernity has been structured as always already oppositional. Therefore, like all binary categories, the distinction between the vertical and horizontal has been naturalized as foundational and mutually exclusive. Each axis is produced by invested disciplines and fields; for example, Eyal Weizmann argues that modern geopolitics produces a "flat" discourse that "looks across" landscape, mostly ignoring the vertical dimension.[10] However, since vertical views help to establish the geometrics of terrain that make possible a horizontalizing discourse, this "cartographic" flattening cannot be "seen" without the vertical view from above. Although the oppositional aspect of

Figure 16.1 Installation view of New Room of Contemporary Art exhibition, *Sophie Ristelhueber: Fait*, April 1998.

Source: Photo courtesy of the Albright-Knox Art Gallery, Buffalo, NY.

vertical and horizontal perspectives is vigorously asserted in various disciplines and fields of study, in some visual practices the opposition is overstated and even untenable.

In this regard, aerial perspective may be understood to deconstruct the geographical concept of scale. I am using the term "scale" here in its more expanded sense, displaced from its origin as a mathematical relationship foundational to cartography, and more closely linked to inquiries into the ways in which knowledge of space is produced.[11] The omniscient or universalizing gaze from above articulates neatly with geometric scale if it can be conceptualized as hierarchical or "nested," moving vertically in a graduated progression from smaller to larger or from specific to general. Visually, this kind of scale can be imagined as a cone: from a point on the ground expanding out-ward and upward or from the "God's eye" on high radiating out and downward.

Vertical scale is complemented by horizontal scale – a perspective that is presumed to provide a more "grounded" or "human" viewpoint. Linked to the Renaissance "discovery" of linear perspective and the establishment of the vanishing point, the "realistic" claims of linear perspective are foundational to landscape painting as well as to cartography and navigation. From the European Enlightenment on, then, the proportional relationships that structure Western visual culture constitute scalar qualities that are quantifiable, rational, and yet also naturalized – one can hardly imagine "seeing" otherwise.

The primacy of aerial perspective as a way of seeing in war can leave the impression that the ground is only something to be viewed at a distance, as space to be measured or analyzed en route to a target. But the immediacy and proximity of the ground is always a part of vertical knowledge – the air and the ground, like all binaries, make each other possible. The view from the air, vertical or oblique, as the target is approached, has become one of the primary ways in which war can be "seen" by those at a distance.[12] As we engage in this operation, the precision, speed, and power of the weaponry is mythologized as the promise of objective truth offered by the vertical scale of war waged from the air.

A more dialectical approach to the binary of vertical and horizontal, already care-fully elaborated in cultural geography in relation to movements of capital and global-ization[13] can be applied to a theorization of aerial perspective to avoid the rigid separation of categories into mutually exclusive entities. But the argument made by geographers Sallie Marston, John Paul Jones III, and Keith Woodward against the concept of scale in favor of a "flat ontology" pursues a more deconstructive approach.[14] Marston, Jones, and Woodward's objections to scale are multiple, complex, and linked to disciplinarily specific debates.[15] Deconstructing the opposition of vertical and hor-izontal, their argument for a flat ontology helps us to engage the aerial perspective at work in *Fait*. When neither the ground nor the air holds primacy, it becomes possible to view differently the visual conventions of war and to make a space for the more complicated evidence of ambiguity.

The work of flat facts: *Fait's* aerial perspective

While much has been written about the convergence of technology and warfare that produced certain kinds of spectatorial subjects during the 1990–91 air war, less attention has been paid by cultural and visual studies scholars to the short but intense ground war that began in January 1991.[16] At the start of the war, Iraq was perceived to have the fourth largest army in the world.[17] With recent experience in an eight-year war with neighboring Iran, the Iraqis had an extensive trench system and large numbers of tanks and fighting vehicles.[18] Although both sides prepared for a ground war, the anticipated "classic tank battles" did not occur.[19] Largely defeated by the air war, many Iraqi soldiers perished in trenches and ground vehicles. A war that was continually characterized as a spectacularly successful demonstration of high-tech air-power had little space for disturbing images of dead bodies on the ground.[20] Rather, the military offered news outlets plenty of views from guided weaponry, planes, and satellites – images that cited the iconic history of aerial reconnaissance photography.

Aerial reconnaissance photography has been part of the conduct of war since the inception of the technology in the nineteenth century – first, via balloons and then, in the twentieth century, as part of the apparatus of the airplane. Reconnaissance images are examples of what Allan Sekula has called "applied photography"; apparently "free from 'higher' meaning in their common usage."[21] Yet, these utilitarian images are hardly neutral. Although aerial reconnaissance images are believed to be marginal to the interpretive and evaluative practices of both high art and commercial photography, they can be "read" in many ways. Certainly, they can be understood to convey the point of view of air power – distanced and abstract, on the one hand, and time, as well as site, specific, on the other. The interpretation of such images leans one way or the other depending on context of use and, therefore, the meaning can be de-linked from any intention of the image-maker.

Ristelhueber's deliberate incorporation of aerial photographic techniques and references to views from the air might be understood to recuperate the dominant imagery of a war that most of us only saw from a distance via digital and analog media. But in *Fait*'s rigorous engagement with the distancing aesthetics of aerial reconnaissance photography, as both Jim Ketchum and Marc Mayer have argued, no discernible narrative is established and it is even difficult to tell what it is that we are meant to see.[22] The indeterminacy of the imagery in *Fait* is not simply an experiment in form (although it can be appreciated as such). Cheryl Brutvan and Charles Merewether point to the "shifting planes," "absence of recognizable scale," and "loss of perspective" in the photographs that constitute *Fait*.[23] The evacuation of photojournalist conventions – no blasted bodies, no keening figures – forces the viewer to try to make sense out of abstractions and ambiguities, to figure out not only what is there to be seen but scale as well, and thus, to reflect on the disappearance of "news," "information," and "evidence," all of which are believed to be foundational to the public sphere in a democratic society.[24]

The ground truth of aerial perspective in the era of long wars and securitization references a grim history of spatialized violence. Counter to Le Corbusier's dictum that "when the eye sees clearly, the mind makes a clear decision"[25] – a rationale for precision bombing as well as modernism – *Fait* offers ambiguity and loss of scale often to the point of "flattening" the objects that appear in the image. If *Fait*'s aerial perspective evades the fascist sensibility that accompanies the sublime spectacularization of war as well as the sentimental nationalisms that are masked by the realist conventions of war reportage, it remains committed to negotiating distance and destabilizing relations of scale. As Ian Walker argues, *Fait*'s "post-reportage" strategy "suggests not what photography cannot do, but what it can."[26]

In French, *fait* means "fact" and it also connotes the completion of an action or "something that is done" ... *c'est fait*. The aerial perspective at work in *Fait* is not a dialectical alternative to militarized ways of seeing. Rather, in engaging the scales of vertical and horizontal, it reveals their historical construction, their ongoing invested exclusions, and their effects. Views from the air can be seen to be highly mediated, even ideological, rather than innocently "realistic" or objectively factual. Indeed, *Fait* shows us the ground as a fact, but it reveals a cartographic nightmare of evacuated meaning. In these photographs, flattened landscapes of violent acquisition resonate with a long history of colonial discourse in general, and decades of air war over Iraq in particular, as authoritative maps figured empty deserts or barren land at stake in contests for territorial control of resources like oil.[27] *Fait* causes us to lean in closer, to try to discern what has happened, to see what the representation of certainty has wrought in modernity's endless wars.

In *Fait*, what do we see? A moonscape? A line of fortifications? Grids of landscape? Toxic pollution? Tracks in sand? This photography flattens its views and disturbs the relations of scale to the point of loss. Not so much empty (because the photographs are chock full of interdisciplinary, historical, and cultural references), *Fait* nevertheless refuses to offer a specific, bounded national, or even regional viewpoint. But these images are not just floating in time and space, offering a nomadic alterity. Rather, Ristelhueber's photographs are so full of possible meaning that the loss of scale and flattened ontology makes possible any number of new connections. Through aerial perspective, *Fait* offers the flattened logic of the fact – a way to begin to understand what we have done in the space and time of Iraq.

Notes

1 G. H. Roeder, Jr., *The Censored War: The American Visual Experience During World War Two*, New Haven, CT: Yale University Press, 1993.

2 For a comprehensive account of this world view see D. Cosgrove, *Apollo's Eye: A Cartographic Genealogy of the Earth in the Western Imagination*, Baltimore, MD: Johns Hopkins University Press, 2001; and B. Newhall, *Airborne Camera: The World from the Air and Outer Space*, New York, NY: Hastings House, 1969.

3 J. Taylor, *War Photography: Realism in the British Press*, London: Routledge, 1991, p. 1.

4 See the ongoing work of filmmaker/artist/public intellectual L. Sachs: *I Am Not a War Photographer*, www.lynnesachs.com (circa 2007), as well as an extended engagement with the notion of "geoaesthetic trace" in J. A. Ketchum, *Journey to the Surface of the Earth: The Geoaesthetic Trace and the Production of Alternative Geographical Knowledge (Sophie Ristelhueber, Laura Kurgan)*, Ann Arbor, MI: University of Michigan Press, 2005.

5 In the Arab world and elsewhere, the Iran–Iraq war of 1980–88 is referred to as the "Gulf War" but, in the United States, the war between the United States and Iraq

in 1990–91 is either referred to as the "Gulf War" or the "first Persian Gulf War" and the occupation that began in 2003 as the "second Gulf War." The confusion over what to call these conflicts is a form of cognitive dissonance that masks invested points of view.

6 The "smart-bomb" footage has been discussed widely; for a comprehensive discussion see D. Kellner, *The Persian Gulf TV War*, Boulder, CO: Westview Press, 1992. I have written on this topic in C. Kaplan, "Precision Targets: GPS and the Militarization of U.S. Consumer Identity," *American Quarterly* 58, 2006, 693–714.

7 Middle East Watch, *Needless Deaths in the Gulf War: Civilian Casualties During the Air Campaign and Violations of the Laws of War*, New York, NY: Human Rights Watch, 1991.

8 See J. Baudrillard, *The Gulf War Did Not Take Place*, Bloomington, IN: Indiana University Press, 1995.

9 In this very brief version of a longer work-in-progress, I am not intending to recuperate aesthetic modernism as a strategy of resistance to realist politics. Rather, I am arguing that in specific contexts a critical engagement of the visual tools of war may undermine the government's ideological practices by revealing complex ambiguities rather than simplistic certainties. I do not intend to celebrate or monumentalize ahistorical or purely aesthetic abstraction.

10 E. Weizman, "The Politics of Verticality," *Open Democracy*, 2002, www.open democracy.net/conflict-politicsverticality/article_801.jsp

11 See, for example, D. Harvey, *Justice, Nature, and the Geography of Difference*, Oxford: Blackwell, 1996; H. Lefebvre, *The Production of Space*, Oxford: Blackwell, 1974; R. B. McMaster and E. Sheppard, "Introduction: Scale and Geographic Inquiry," in *Scale and Geographic Inquiry: Nature, Society, and Method*, E. Sheppard and R. B. McMaster (eds), Malden: Blackwell, 2004, pp. 1–22; and N. Smith, *Uneven Development: Nature, Capital, and the Production of Space*, Oxford: Blackwell, 1984.

12 C. Kaplan and R. Kelly, "Dead Reckoning: Aerial Perception and the Social Construction of Targets," *Vectors: Journal of Culture and Technology in a Dynamic Vernacular*, 2:2, 2007, www.vectorsjournal.org/index.php?page=7&projectId=11

13 See for example, A. Amin, "Spatialities of Globalization," *Environment and Planning A* 34, 2002, 385–99; N. Brenner, "Between Fixity and Motion: Accumulation, Territorial Organization and the Historical Geography of Spatial Scales," *Environment and Planning D: Society and Space* 16, 1998, 459–81; and S. Graham, "Vertical Geopolitics: Baghdad and After," *Antipode* 36:1, 2004, 12–23.

14 S. A. Marston, J. P. Jones III, and K. Woodward, "Human Geography Without Scale," *Transactions of the Institute of British Geographers* 30, 2005, 416–32, and J. P. Jones, III, K. Woodward, and S. A. Marston, "Situating Flatness," *Transactions of the Institute of British Geographers* 32, 2007, 264–76.

15 Their engagement of "flat ontologies" engages explicitly or implicitly the work of G. Deleuze and M. Delanda (among others): see, for example, G. Deleuze, *Difference and Repetition*, London: Continuum, 2004; and M. Delanda, *Intensive Science and Virtual Philosophy*, London: Continuum Books, 2002. See also the

responses to Marston *et al. op. cit.* that were published in *Transactions of the Institute of British Geographers* circa 2006–7.

16 See, for example, K. Robins and L. Levidow, "The Eye of the Storm," *Screen* 32:3, 1991, 324–28; P. Virilio, *L'Ecran du desert*, Paris: Editions Galilee, 1991; R. Stam, "Mobilizing Fictions: The Gulf War, the Media, and the Recruitment of the Spectator," *Public Culture* 4:2, 1992, 101–23; D. Gregory, *The Colonial Present*, Malden: Blackwell, 2004; and N. Mirzoeff, *Watching Babylon: The War in Iraq and Global Visual Culture*, New York, NY: Routledge, 2005.

17 A. Finlan, *The Gulf War 1991*, Oxford: Osprey, 2003, p. 17.

18 R. A. Schwartz, *Encyclopedia of the Persian Gulf War*, Jefferson: McFarland, 1998, p. 170.

19 M. R. Gordon and General B. E. Trainor, *The Generals' War: The Inside Story of the Conflict in the Gulf*, Boston: Little, Brown, 1995, p. 417.

20 Photojournalist Peter Turnley was one of the first reporters at the scene at the infamous "Highway of Death" – an air attack on retreating Iraqi ground forces on the last evening of the war. US media did not choose to circulate images such as those Turnley posted at "The Unseen Gulf War" (circa 2002), www.digitaljournalist.org/issue0212/pt_intro.html. For a discussion of the absence of violent war images during this time period see W. J. Mitchell, *The Reconfigured Eye: Visual Truth in the Photographic Era*, Cambridge, MA: MIT Press, 1994, p. 12.

21 A. Sekula, "The Instrumental Image: Steichen at War," in *Photography Against the Grain: Essays and Photo Works, 1973–1983*, Halifax, NS: The Press of the Nova Scotia College of Art and Design, 1984, p. 28.

22 Ketchum, *op. cit,* p. 200; Marc Mayer, *Sophie Ristelhueber: Fait*, Buffalo, NY: Albright-Knox Art Gallery, 1998, p. 1.

23 C. Brutvan, *Details of the World*, Boston, MA: Museum of Fine Arts, 2001, p. 151; C. Merewether, "After the Fact," *Grand Street* 16:4, 1998, 1.

24 Ristelhueber has described how she was drawn to an aerial photograph in *Time* magazine of an attack by French fighter jets on Iraqi Republican Guard ground troops due to its abstract composition and lack of scale (Ketchum 2005, 56).

25 Charles-Édouard Jeanneret-Gris ("Le Corbusier"), *Aircraft*, New York: St. Martin's Press, 1935, p. 13.

26 Ian Walker, "Desert Stories or Faith in Facts?" in *The Photographic Image in Digital Culture*, ed. Martin Lister, London: Routledge, 1995, p. 240.

27 Ella Shohat, *Taboo Memories, Diasporic Voices*, Durham: Duke University Press, 2006; Priya Satia, "The Defense of Inhumanity: Air Control and the British Idea of Arabia," *The American Historical Review* 3:1 (2006): 16–51.

17

Counter-geographies in the Sahara

Ursula Biemann

To engage in a discussion of why geography has been so important for the develop-ment of my artistic practice, I need to think back to the moment when I entered the professional world, for it encapsulated the early conjunction between geography and visual culture. By the late 1980s, when I graduated from art school, discourse on art was already considerably "contaminated" by other theoretical currents – such as eth-nography, urbanism, cultural and media studies, post-colonial criticism, and feminist theories – which not only represented new content but also provided instruments for reformulating the domain of symbolic production. It had become evident that art theory would no longer be the sole frame of reference for an aesthetic practice which would now have to position itself in relation to other terrains of knowledge produc-tion. This important discursive expansion coincided with the vigorous onset of glo-balization processes and a turn, in the arts, towards content-oriented work, enabling precisely this connection of diverse strands of critical interpretation. I sensed the necessity of developing an aesthetic practice that could respond to this complex and rather unique condition.

My videos and writings of the last 10 years chronicle two parallel processes: first, the process of discerning a political and geographical area of interest for my art prac-tice – for example, the gendered, international labor division along the Mexican border – and, second, the process of tracing out a research field at the juncture of dif-ferent forms of knowledge production where this practice could be situated. My simultaneous engagement with the geopolitical and social transformations that have been induced by globalization, and with the form in which these could be addressed in the expanded aesthetic field, are conceptually related. It doesn't surprise that the deregulation of an entrenched world order profoundly troubles the categories and methodologies through which this order has been established and maintained. The emerging protocol for this new world disorder is the unfettered transgression of bor-ders between genres and beyond the boundaries of conventional disciplines. These two ongoing processes – with parallel effects on geography and knowledge regimes – are connected and hinge, in my work, on the concept of the border.[1]

The proliferation of academic and visual work on border issues witnessed the begin-ning of geography turning into a major reference for organizing these new and com-plex contents. Many artists expressed a certain fascination with reinforced border regimes that emerged first along the US–Mexico border, later along the eastern and southern borders of the EU. From the beginning, I have been somewhat suspicious about the utility of documenting border fences and impressive surveillance technologies. It seems to me that even from a critical perspective, the focus on the line and its militarization cannot help but reproduce and reinforce the divisive force of the border as a concept. My various border videos are geographic projects in the sense that they engage in a process of visualizing spatial relations. When geography is understood as a spatialization of the dynamic social and economic relationships connecting local systems to the transnational, it becomes clear why border geographies are the site of extreme compression at all levels. Border areas, like the Spanish–Moroccan borderlands I document in the video *Europlex* (2003), are given their cultural meaning predominantly by being traversed: by container ships en route from West Africa to the Mediterranean, by boats transporting migrants on their perilous nocturnal journeys, by helicopter patrols keeping watch, by radio waves and radar lines, by itinerant plantation workers who pick vegetables for the EU market, by commuting housemaids going to work for the señoras in Andalusia, by border-guard patrols along the mountain paths, by buses transporting Moroccan women to Tangier where they peel Dutch shrimps to be shipped back to Holland, by pirates who procure goods from China,

Figure 17.1 Smuggling activities around the border of the Spanish enclaves in Morocco, *Europlex*, 2003.

and by women smugglers who tie these goods up under their skirts and carry them into the medina. This is the mobility I am concerned with in this video — the everyday mobility lived out on a local level, to produce micro-geographies that are deeply intermeshed with one another while reflecting a global schema.

Territories of transit

In later projects I have attempted to develop this notion of geography both as social practice and organizing system. *Sahara Chronicle* (2006–7), for instance, is a collection of videos on the modalities of migration across the Sahara; it chronicles the sub-Saharan exodus towards Europe as a social practice embedded in local and historical conditions. The project introduces the migration system as an arrangement of pivotal sites, each of which have a particular function in the striving for migratory autonomy, as well as in the attempts made by diverse authorities to contain and manage these movements. Video documents include the transit migration hub of Agadez and Arlit in Niger; Tuareg border guides in the Libyan desert; military patrols along the Algero-Moroccan frontier in Oujda; the Mauritanian port of

Figure 17.2 Migrants using the iron ore train along the border between Mauritania and the Polisario Front in Western Sahara, *Sahara Chronicle*, 2006–2007.

Nouadhibou on the border to the Polisario Front; the deportation prison in Laayoune, Western Sahara.

With its loose interconnectedness and its widespread geography, *Sahara Chronicle* mirrors the migration network itself. It does not intend to construct a homogeneous, overarching, contemporary narrative of a phenomenon that has long roots in colonial Africa and is extremely diverse and fragile in its present social organization and human experience. No authorial voice, or any other narrative device, is used to tie the carefully chosen scenes together; the full structure of the network comes together solely in the mind of the viewer who mentally draws connecting lines between the nodes at which migratory intensity is bundled.

This text is not primarily intended to interpret these videos; rather, it is a place for making some further reflections about the politics of visual practice with regard to migration, with a particular emphasis on illegal migration. It is also a place for offering some of the connections and insights acquired in the field about the nature of this sophisticated migration network, the intersection of resources and migration routes and the entanglement of migration and sustainability in the Sahara.

As part of the massive economic and political diaspora of our world of transnational capitalism, migrant workers uniquely embody the condition of cultural displacement and social discrimination. But, the task of a political aesthetics today is not to capture an image that best symbolizes our times; rather than positing the ultimate image, the task is to intervene effectively in current flows of representation, their narratives and framing devices. In some instances, the accepted story needs to be undone and we

Figure 17.3 Sahara Chronicle installation at Arnolfini, Bristol UK, 2007.

should not get anxious about reassembling it into another story too soon. The preferred mode of signification in *Sahara Chronicle*, therefore, is fragmentation and disassembly.

The project contains an undefined number of videos, which are never shown all at once, since there is always something unknown, hidden, and incomplete about clandestine migration. My preferred way of showing them is in the form of an installation, whereby some videos are projected and others can be viewed on monitors, creating a multiperspective audiovisual environment that can be inhabited by viewers, in much the same way that migration space is inhabited by the actors depicted.

Imaging clandestinity

The Western media have a very peculiar way of representing clandestine migration to Europe. They direct its spotlight on the failure of the stranded migrants (the "Naufragés") and celebrate police efforts that successfully apprehend transgressors; victorious passages go undocumented. The media seem to succumb to every temptation of condensing reality into a symbol, thrusting the whole issue into discursive

Figure 17.4 Main figures in the clandestine migration system channeling West Africans through the desert. Site: desert truck terminal in Agadez, Niger. *Sahara Chronicle*, 2006–2007.

disrepair. In a perpetual loop, television clips capture the state of being intercepted, caught in a process of never reaching the destination, a freeze-frame of the *Raft of the Medusa* drifting off the shores of Senegal. In cinematographic language, this fixed spatial determination is simply called "a shot," suggesting that the real is no longer represented but targeted. In the staccato of television news, this particular shot becomes the symbol that encapsulates the meaning of the entire drama. It is evident that complex social relations are not negotiated in this frantic manner. Apart from the time compression, which creates an immense discrepancy between representation and social reality, there is something seriously inadequate about this robotic viewpoint when it is directed at the shifting and precarious movements of life.

But the mundane truth behind the trauma-like recurrence might be that these images are not the outcome of intense aesthetic reflection but the convenient product of current media politics under the strain of growing competition. Since their mission is to cover events rather than explain conditions, news channels do not see why they should send out expensive camera teams to remote desert towns in the Western Sahara or Niger, unless some drastic event makes these places internationally newsworthy. So we are likely to be presented with the lazy and less costly version of the story that only covers the most visible end points of a long journey.

Evidence and artifice

Sahara Chronicle includes a number of records of the more or less successful efforts at keeping the fluctuating migration currents through Morocco, Mauritania, and Libya in check, by means ranging from off-road patrols in border terrain to aerial surveys by propeller planes and high-tech surveillance drones. Engaging with this politics of containment sucked me right into the gigantic visualizing apparatus and made me a part of it.

One of the records follows the border brigades in the Algero-Moroccan frontierland, where they half-heartedly poke around popular hiding places for clandestinos near the train tracks. Nobody was found that day, but the colonel in charge of the area was pleased to demonstrate the efforts made by the royal brigades in impeding migration flows to Europe. As their budget is barely enough to cover one surveillance flight per week in the vast desert areas around border cities like Oujda or Laayoune, I didn't want to initiate an extra flight for aerial filming that would risk the detection of a group of clandestine migrants hiding in the dunes.

The police were willing to give me the photographs they had taken on previous tours; these pictures have a different status from the frames I would have shot from the same plane, functioning as evidence for use within the confidential circuits of police investigation. They capture the moment between recognition and possible disciplinary action. A simultaneous role as witness and record endows these images with a juridical effect, providing evidence of infringement and the occasion for moral judgment and deportation. However, the scrolling text introduces a thriving solidarity between the transiting migrants and the local populations. Moroccan carpenters have started to prefabricate boat kits, which can be quickly assembled by migrants in their desert hideouts. Distanced judgment is baffled here by a sense of local complicity.

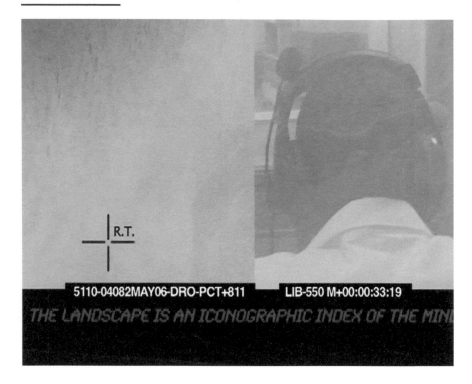

Figure 17.5 Digital montage of a surveillance desert drone image, *Sahara Chronicle*, 2006–2007.

Another video is dedicated to some of the most high-tech surveillance technologies currently being deployed on military missions, from the war in Iraq to the Saharan desert front. Libya has received the newest models of unmanned airplanes from Germany, in return for their active demonstration of hindering migration flux to Europe. These drones glide over the desert borders, transmitting televisual data back to a remote receiver in real time. Other observation machines are equipped with night vision and thermal cameras, extending surveillance into realms invisible to the human eye.

Colonel Muammar Kadhafi's military department was not as cooperative as the Moroccan brigades in handing its visual intelligence to me, but we can safely assume that the images produced by these drones are no longer film-based photography, like the ones used in Moroccan aerial reconnaissance, but computer generated. These digital technologies have created new ways in which an image can be linked to an actual object; the indexical linkage required in previous concepts of documentary realism has been traded for new methods of attaining and validating empirical knowledge.[2] The drone images are simulacra used as representation.

Lack of source material meant that I had to artificially construct it from high-resolution satellite images of the Libyan desert. The soundtrack is composed of many layers of recordings from Saharan and Middle Eastern radio and TV stations, mixed with electronic sounds, music fragments, and winds. This artificial videography addresses the important fact that migratory space cannot simply be documented by conventional video-making on the ground. We need to enter the more ethereal strata of signal territories created by the streaming of images and the diffusion of sounds and information – territories with a relentless and excessive meaning production.

The abstraction of these images is offset in yet another video, with sequences of the hard reality experienced by those who have no visa to the borderless world of signs. The overcrowded deportation center in a former colonial prison in Laayoune, Western Sahara, offers a sight that propels you back 200 years into a somber past. Close your eyes and you can hear the chains jangle. The main light source is a barred skylight, a hole in the roof through which a harsh stream of sunlight pierces the sweaty gloom, making every mosquito and every grain of dust dance in front of your eyes. Slowly getting used to the scene, you see starvation, weakness, disease, and sun-scorched eyes; none of this matters when the goal is in sight, but it is excruciating to bear when hope has slid away. The only traces of the migrants' trajectories are the fragile architectures they had built in the remote desert dunes during the days and weeks of holding out while water stocks were running low. The aerial photographs show that, around some of these shelters, an area is marked by stones like the outline of a garden or a place for prayer, as if the deadly expanse was a place too vast to comprehend.

A network run by a transnational tribe

The core of *Sahara Chronicle*, however, is set in one of the truck terminals for desert crossing in Agadez.[3] The town, at the heart of Niger, is the southern gate to the Saharan basin for the main routes coming from west Africa; it is a major trans-Saharan trading center, and capital of the Tuareg. Saharan people live in open space, mobility is everything in this geography. They have developed different methods of mastering the terrain. Of necessity, life is lean. And portable. Tuareg culture has worked out a system of information, a specific topographic literacy, with itineraries and means of communication. They are GPS embodied. In this environment, orientation makes all the difference between drifting and traveling, between fate and destination. In their minds, prosperity and power are located in movement rather than bounded territory.

The video documents the great departure of the "Exodés," those many young men and few young women from West Africa on a quest for a better life in the Maghreb, or, in a more distant, blurry vision, in Europe. In contrast to the images of failed arrival, these scenes show the moment of potentiality at which anything seems possible. The excitement about the risky outcome of their adventure is very tangible among the passengers. What unites them is the common goal of accessing the labor markets in the north.

In joining this greater venture, they contribute to an elaborate system of information exchange, routing, and social organization that spans the immense Saharan region and, in doing so, creates a translocal space that will exist for as long as these social practices last. What we witness is a large-scale geographic reconfiguration, activated by growing practices of migration which are highly flexible – proficient at rerouting, reorganizing, and going covert in record time. It is in this guerilla fashion that the geography is made productive, by those players defined by global capitalist logic as immobilized: the poor and the deprived. The focus is on the unrepresented, rebellious, and obstinate local practices of space, which resist and circumvent any attempts to discipline them.

If we want to understand what makes this emerging migration system work, one of the things we need to look at is the historic condition of the region. For it is the conceptual difference between nomadic and colonial politics of space that lies at the heart of the Sahara being turned once again into a contested zone mobility. The immense Saharan territory of the Tuareg tribes was split in five by the Empires at the Berlin Conference in 1884. Since then, their space of mobility and livelihood has made up substantial areas of Algeria, Libya, Mali, Niger, and Chad. Denied a proper state, the Tuareg constitute a minority within these national cultures and are granted fewer civil rights than native citizens. Nonetheless, as a distinct linguistic and cultural entity, they maintain their identification as a people across the boundaries. Tuareg territorial structure is, by definition, transnational; it provides the framework for social and economic, if not political, organization. The role of these nomads is central to the transnational process of repurposing their old caravan routes as highways for illicit migration. Their unique topographical expertise and tribal ties are in high demand as a steady flow of sub-Saharan migrants passes through Agadez and Arlit.

The Tuareg rebellion in Niger in the mid-1990s, which made another attempt at consolidating their tribes into a nation-state, was directly linked to uranium mining in Arlit and the exclusion of the Tuareg from the wealth found on their territory. In the remote capital the revenue from uranium extraction was shared among the French owners and the Nigerian elite who recruited miners from other ethnic groups from the south. The rebellion ended with a peace treaty that promised better social integration.

I interviewed former Tuareg rebel leader, Adawa, who is current head of the clandestine transportation operations in Arlit on the Algerian migration route. For lack of better opportunities, the returning former rebels saw a possibility of making business with the transit migrants. Transportation services were needed; besides, Arlit, like Agadez, is a desert gate that can be controlled and taxed, but the desert border is a vast terrain and roving border patrols are few and far between. Some passengers were documented, many not. Deploying rebel tactics, they swarm out in jeeps at night and bypass the border checkpoints with their full charge of migrants before melting into the dark dunes.

The regional authority of Agadez saw the need to intervene in these opportunistic developments and formally mandated Adawa to manage the semi-legal transport of migrants in an organized fashion. The local authorities may have welcomed the fact that this locks him into a criminalized position which compromises any further rebellious plans. Semi-legal, yet authorized, the business keeps the rebel pacified while generating extra income and power for the officials: a well-planned, if precarious, balance. This solves two problems at once: putting an experienced man in charge of logistics and keeping him occupied and accountable. Should Adawa ever prove to be uncooperative, the authorities can put him away without much ado. He understands that he has been taken hostage and that his status as a semi-citizen of Niger is directly linked to his guidance of more and more people into a terrain of bare survival in which citizenship is suspended.

What these transit and border recordings aim at is not the consolidation of a national unity, as media reports on border defense inevitably attempt, but its opposite: the permeability and constant subversion of national space. To some extent, television reports on clandestine boat passengers do this too; yet, importantly, the shadowy and potentially subversive circumstances of such border passages are assimilated all too quickly into a disciplined national order in which the interventions of state officials play a leading part. Yet documenting reality today also means to recognize that the massive departures mark the beginning of a migratory existence for a great number of people whose lives will not integrate in a single space and whose interests will no longer be served by one nation-state. Images of border passages allow us to cultivate an alternative imaginary to national culture, one that is based on cultural practices that harness and play with national boundaries.

Concluding, I wish to return to some central concerns regarding my artistic practice in the context of this book. First, I want to emphasize that my videos *are* geographies, if by geography we mean a visual form of spatializing territorial and human relations. What makes these videos a distinct geographic practice, rather than say a documentary tool for an earth scientist, is their essayist form. The video essay typically has a non-linear narrative structure and follows a subjective logic that doesn't shy away from loops and discontinuities. It could end at any point or continue beyond its end and it certainly doesn't follow a particular line of argument that would assume a proposition, conclusion, or deduction. It is not conceived as a sequence in time but as constructed coexistence in space. Not unlike the transnational subjects of my research, the essay practices dislocation, it sets across national boundaries and continents and ties together disparate places through a particular logic, arranging the material into a particular field of connections. In other words, the essayist approach is not about documenting realities but about organizing complexities. This very quality makes the audiovisual essay a suitable genre for my investigation of a subject matter like globalization. In this field of knowledge, many issues around economy, identity, spatiality, technology, and mobility converge and are placed in a complicated relationship to one another. The attempt to draw these layers together leads inevitably to the creation of an imaginary space, a sort of theoretical platform on which these reflections can take place and be in dialog with each other. In every work, essayists install this kind of space. We can think of it as an imaginary topography, on which all kinds of thoughts and events taking place in various sites and non-sites experience a spatial order.[4] Beyond being extremely suitable for my concerns, the essayist genre has directed my work continuously towards working in the human geographies of migration and global labor as visual-spatial configurations. The fact that this method does not confine my results to an objective logic has allowed me to integrate connections that seem unrelated, such as the encounter with prostitutes in the Turkish border town Trabzon in the context of a research on the Caspian oil geography entitled *Black Sea Files* (2005). At first sight, there is no immediate causal relation between the massive capital flows generated by the west-bound oil and the trafficking of women in the Black Sea basin, but their economies and trajectories are intricately connected through international visa agreements, reviving cultural affiliations and regional post-socialist histories. The technological adventure of pipeline building is juxtaposed here with the intimate experience of forced prostitution. These links are fragile but nevertheless very important if we want to convey the complexity and precarity of contemporary human geographies. The inclusion of the prostitutes reflects my ongoing concern of introducing a gendered perspective in the visual investigation of globalization processes and migration systems. Although we see the existence of a significant body of academic literature on women and space, geography, migration, and transnationalism, this gendered sense of geography has not been counterbalanced by a significant boost in the domain of aesthetic production. A spatial perspective on migration harbors the great potential of making a move from the individual and experiential zoom on subjectivity formation to a wider structural and systemic understanding of migration. We are in fact in the process of opening up a field of investigation in which a great deal of visual experimentation has yet to occur.

Notes

1 My first video essay, *Performing the Border*, 1999, is an analysis of the US–Mexico border as a place performed by gender, mobility, and labor.
2 M. J. P. Wolf discusses this shift of the indexical linkage in his essay "Subjunctive Documentary: Computer Imaging and Simulation," in *Collecting Visible Evidence*, Jane M. Gaines and M. Renov (eds), Minneapolis/London: University of Minnesota Press, 1999, pp. 274–91.
3 For a detailed discussion of this, see U. Biemann, "Agadez Chronicle – Post-colonial politics of space and mobility in the Sahara," *The Maghreb Connection*, Barcelona: Actar, 2006, pp. 43–67.
4 For a more extensive discussion on the relation between transnationalism and the video essay see "Performing Borders: The Transnational Video, Stuffit – The Video Essay in the Digital Age," Dordrecht: Voldemeer/Springer, 2003.

18

Laura Kurgan, September 11, and the art of critical geography

Jim Ketchum

Acts of violence at the scale of warfare have recently increased in clashes between "the West" and "the Middle East," a phenomenon infamously predicted by Samuel P. Huntington in his internationally excoriated thesis.[1] This violence continues to produce new landscapes from Afghanistan to Manhattan. Of course, these are not the carefully planned landscapes normally evaluated in scholarly journals or lauded for their design features; neither are these the haphazard accretions of the everyday landscape typically investigated by cultural geographers. Nevertheless, new landscapes of devastation continue to emerge in the contemporary world. We should not ignore them simply due to their disturbing nature. Indeed, "the landscape of devastation is still a landscape" writes Susan Sontag, and as she convincingly argues, their investigation remains a moral imperative for politically engaged individuals.[2]

As images through which we may apprehend the world from a distance at the touch of a button appear, multiply, and circulate with increasing ease, violently *unbuilt* landscapes become ever more important as cultural signs, serving as arbiters of social messages affecting our thought, judgment, and ultimately our behavior. As Denis Cosgrove argues, the social importance of *geographic* representations (those that produce public knowledge of other peoples, places, and events from a distance through fixed forms such as sketches, photographs, maps, satellite images, and written texts) is that they possess significant "agency in shaping understanding and further action in the world itself."[3] If the world we know did indeed change with September 11, as has so often been stated, it did so in part because of the way we were capable of coming to know such events, their devastated landscapes, and the people and forces that cause them through the wide availability of new kinds of visual technologies and the easy circulatory speed through which they multiply and travel, but also because of the way that geographic and historical interpretations and understandings are embedded in those images and technologies.

This essay discusses these understandings by interpreting one artwork by the artist Laura Kurgan as a particularly skillful artistic engagement with the complex effects of visual images and their technologies, especially as they mediate the ways we comprehend large-scale acts of violence such as the attack on the World Trade Center. "I see my work as very experimental with spatial technology," says Kurgan. Although some of her work is "quite poetic, it's poetic with these strange kinds of new tools."[4] "My work becomes more a combination of conceptual art and geography and mapmaking" she says,[5] and becomes an intervention in the way geographic knowledge is produced and reproduced to help us make sense of horrific events such as September 11.

New York, September 11, 2001, four days later ...

On the morning of September 11, 2001, Laura Kurgan was at work in her New York studio on a project for a major international art exhibition, *Control_Space: Rhetorics of Surveillance from Bentham to Big Brother*, which would bring together the work of 54 artists at the ZKM Center for Media and Technology in Karlsrühe, Germany. Invited to participate in *Control_Space* and given a budget to produce an artwork based on her interest in satellite imagery, Kurgan thought the best way to spend that money would be to acquire a satellite image that would also be of use to a non-profit organization. She found officials at Global Forest Watch receptive to the idea. "I worked with them to look for an illegal road in a rainforest in Cameroon," says Kurgan. Proving its existence would be the first step in shutting down the illegal logging and rainforest destruction it facilitated.[6]

Kurgan originally processed the order for the rainforest image with Space Imaging Inc. in March of 2001, but clouds had prevented the satellite from recording a useful image in the intervening months. The continued delays caused growing concern for Kurgan, especially as the October 2001 unveiling of *Control_Space* drew closer. "Three weeks before the show my image still hadn't arrived," says Kurgan. While still waiting, the World Trade Center came under attack. Kurgan quickly contacted Space Imaging Inc. to alter her order. "The same satellite company is also in the habit of when a major event happens you know that they're going to try and take that picture because they're also really interested in publicizing their technology." Space Imaging, Inc. provided Kurgan with an image of Manhattan produced four days after the attack, captured by the IKONOS satellite. From this satellite image, Kurgan created a work of art (Figure 18.1), *New York, September 11th, 2001... four days later. ...*[7]

Kurgan enlarged the satellite image to produce a work roughly 15 feet wide by 48 feet long. Printed in sections on pre-laminated paper, the pieces were then assembled and mounted on a plywood platform installed on the gallery floor, a process similar to the production of billboards. Visitors to the *Control_Space* exhibition were encouraged to walk on the image in order to gain a closer look. This of course involved viewers touching the work of art – normally a taboo in museums – but the transgression of this taboo was one key element of Kurgan's artistic strategy for the piece. The artwork also includes a strip along the right edge that repeats the left hand edge of the image and slightly enlarges it, but in a different color palette and with a dramatic increase in contrast. To produce this effect, Kurgan manipulated the satellite data by using the image-processing software to choose a pixel in the middle of the ground zero site and asking the software to isolate similar pixels. The result was that the software "tended to equate rubble with buildings," thus doing nothing to bring clarity to the scene. The work appeared in one of many extremely large gallery spaces at the ZKM Center

Figure 18.1 Laura Kurgan, *New York, September 11, 2001, Four Days Later...* (2001). Digital print on pre-laminated paper from Ikonos satellite data of September 15, 2001 by Space Imaging, Inc. 1 pixel = 1 meter, 669" x 236", installation view ZKM/Center for Art and Media, Karlsrühe, Germany. Used by permission.

Source: Photo © Jens Preusse, Vienna.

in Karlsrühe, which included two overhanging balconies placed at one and two stories above the gallery floor. From these balconies, gallery visitors could view the work from a distant perspective in a way not typical in standard museum spaces. Needless to say perhaps, Kurgan designed *New York, September 11th, 2001... four days later...* specifically for this space, utilizing its peculiarities to produce an effective – and unusual – work of art.[8]

Transgressing the world picture

Kurgan's *September 11th* piece depends on an art historical legacy that begins with Dada and can be seen as a part of a larger movement in art practice that critiques the social manufacture of a public consent for war through mass media. Some aspects of this critique can be seen as having key geographical aspects.[9] Elsewhere in this book, Lize Mogel describes the recent explosive growth in critical cartography by artists using maps, a phenomenon that also depends on that same art historical legacy from Dada on. But Kurgan's work is also an extension and revision of those practices, as she goes beyond engaging with systems of signs displayed across the flat picture plane to investigate the social space around it. For her, satellite images of the aftermaths of acts of war provide something radically new in the field of visual culture. Not yet fitting comfortably into established patterns of comprehension, in their strangeness lay their potential.

"Satellite technology is another way of defining the globe and picturing the globe" she says, but her real interest in these images involves the current "revolution in our experience of space and time brought on by new technologies."[10] In her September 11 piece, Kurgan uses a single image to produce a complex experience that might provide an opening into what Martin Heidegger would call, in one of his most famous and often-quoted essays, a moment in which the possibility of "genuine reflection" into the age of the "world picture" – an era in which knowledge is limited in scope through the application of advancing technology and all that it implies – might begin.[11]

Krugan's threefold strategy

Many have argued that attitudes of critical reflection and creative questioning are often initiated when one is placed in a position of inbetweenness – when one becomes aware of his or her presence in a place, yet simultaneously feels like a stranger in that place.[12] In constructing *New York, September 11th, 2001 … four days later …*, Kurgan created a site-specific work of art capable of opening up this experience of inbetweenness for the art viewer by foregrounding the active cultural production of geographic knowledge taking place through viewing images of devastated landscapes.

How might people begin to understand the complex relationships between their own lives and distant events? Kurgan's strategy for initiating the possibly critical geographic reflection through a specific kind of "inbetweenness" is threefold. By working against the art viewer's expectations, *September 11th, 2001* opens the possibility that the viewer may transgress her own art-viewing expectations in three simultaneous ways, each of which reinforces and encourages the other. In fact, Kurgan's *September 11th* is highly attuned to the ongoing debate in geography over conceptualizations of scale, particularly to Marston, Jones, and Woodward's concern that giving agency to larger forces at the nested scale of the globe disempowers individuals in ways that makes both globalization and lived, everyday experience puzzlingly abstract.[13]

First transgression: embodying the eye

The ZKM Center for Media and Technology features a number of extraordinarily large gallery spaces, each of which includes two progressively higher balconies directed inward toward the gallery. In this case, the balconies provided the viewer with the chance to take in the artwork all in one view, a near impossibility from the gallery floor. "It was a large museum and certainly there aren't many venues where I could have done such a large piece, but part of the idea was that when you were looking at the piece you couldn't get the overview, so it was a satellite picture but you couldn't get the satellite view."[14] Satellite views such as this typically appear on a computer screen where they remain at the mercy of the viewer who may manipulate them in multiple ways or simply click ahead to the next image. In this case, however, Kurgan reversed the power relationship between the viewer and the viewed. The reversal produces some anxiety in the viewer, who no longer controls it. This reversal points out the power these images have over us, when normal use in everyday life makes it seem as if it is we who control the image.

Kurgan's *September 11th* also drew the museum-goer to it as the only object present within a single gallery space of enormous size – a space which at first glance perhaps better resembles an airplane hanger than an art gallery. In addition to the size of the gallery, the effect of being drawn to the work is increased by its being mounted on the floor instead of the wall, a situation that prevents the visitor seeing the work's formal qualities, at least at first. Instead, the visitor's viewing of the work is initially blocked due to the oblique angle of the picture plane as it lies on the floor at some distance from the viewer. The combination of the large though obscured object, the presence of the gallery guards surrounding it, and the other visitors gathered around the work – all set at a considerable distance from the point at which the visitor enters the gallery – work to produce a desire, a drawing toward the object, an interest in knowing what the object contains.

"The ideal gallery subtracts from the artwork all cues that interfere with the fact that it is art," argues Brian O'Doherty in his classic interrogation of the modern gallery space.[15] The modern, white, sterile art gallery is a "fictional space" in the sense that it cues social expectations. It is a technology for viewing, but just as with a scientific instrument such as a microscope or telescope, that technology requires a particular subject to look through it and a certain type of object at the other end of the lens. The result is a supposed intensity of viewing, not a practice altogether different from everyday life but rather a practice altogether more concentrated. The result is that the gallery space functions to evoke a set of concepts which seem natural: that art is properly looked at by a viewer engaged in a particular kind of disengaged, contemplative perspective; that art appropriately contains only certain types of meaning; and an implied message that the best kind of knowledge is intellectual, thoughtful, and not of the body, but despite it. The comfort of being immersed in the act of viewing is contrasted with the discomfort of being in an "empty" space with nothing to see. Kurgan uses the unaesthetic qualities of the space – the hard concrete surfaces and their unrelenting grayness, the sharp echoes caused by the hard surfaces and the distant walls and ceiling, and the contrast in scale between that of the human being and that of the enormous room – to enhance this effect of discomfort versus luxury, of void versus presence, of stark empty space versus the *fully narrated space* provided by O'Doherty's "white cube."

Kurgan also chooses to place the object on the floor when, in fact, it could have been hung upon the wall in a style more typical for the viewing of images. This metaphoric maneuver removes the "elevated" art object from its pedestal and places it instead within the three-dimensional realm normally inhabited by bodies (as opposed to viewers) in everyday life. If the museum encourages visitors to appreciate the preciousness of the art object as distant and elevated, out of the mainstream of everyday life and its concerns, Kurgan's *New York, September 11th, 2001 … four days later …* brings us back to the subject of contemplation and the network of complications that produce it. Rather than create a single, irreplaceable precious object out of materials typically associated with high art (such as oil paint or bronze), Kurgan uses more prosaic materials (copy paper, plastic laminate, plywood). She increased this effect of deflating the object's preciousness by combining the formal content of the image with the object's materiality, using current events as the subject of her work rather than a more timeless high art subject such as the beauty of the human figure or the peace of a bucolic landscape.

In addition, Kurgan removes the prohibition against touching the art object—normally enforced in museum galleries all around the world by uniformed guards, impact-resistant glass enclosures, and the presence of surveillance cameras and alarm systems. In fact, Kurgan not only removed the prohibition but encouraged visitors to walk upon the work of art, something most of them felt uncomfortable doing even with the encouragement of the museum guards.[16] This practice emphasized the prosaic materials out of which the artwork was made: printer's ink and photographic paper, plywood and plastic laminate. Kurgan's use of everyday materials assists in intensifying the contrast between her work and those serving more traditional art discourses. Taken together, Kurgan's numerous tactics emphasize the material presence of the artwork as a transgression of one's usual art *viewing* expectations.

Second transgression: limiting vision

Kurgan's *September 11th* arrests the way users normally experience satellite images by taking it out of the constant flow and circulation of images common in everyday life. In contrast to the constant flow of data and information in contemporary society, by freezing one image and enlarging it to the immovable scale of a monument *New York, September 11th, 2001…* creates a sense of stillness and silence compared to the way daily life is often experienced: as a constant overlapping flow of noise and fluctuation. Instead, the artwork brings the stimulation normally produced by our image culture to a halt, producing the inbetweenness many have argued is necessary for critical reflection.

This effect of stillness is magnified by the subject matter under view. Certainly, the September 11 attack on the World Trade Center was experienced by a great many people around the world *only* as a series of images. In fact, a sudden profusion of images of the attack and subsequent destruction was the event's main product. However, Kurgan's *September 11th* contains that sense of flow and turns it into an abrupt stoppage through its monumental size, but also by turning an image that normally flickers past our eyes quickly in the conduct of everyday life into a still, silent abstraction. While engaged in their normal venues, images such as these provide us with the assurance that they tell us something meaningful as views from prosthetic eyes set in space. When removed from those technologies and frozen as still images, the pictures that normally flow through our vision become strange, abstract, and uncomfortably untethered to the computer or television screen. At the same time, the work's enormous size also encourages that sense of reverence and awe normally reserved for monuments, and which seems so inappropriate to the images that pass by us so fleetingly and insubstantially as digital images.

As a flat picture, Kurgan's *New York, September 11th, 2001…* is also typified by the formal qualities of abstraction. Certainly, it is not easy for a viewer to grasp the image all at once, as she would be able to do if looking at a computer or television screen. Of course, the gallery space at the ZKM Center in Karlsrühe where the work was exhibited had two balconies, allowing the viewer to draw back and gain the advantage of distance. Even so, it remained difficult to grasp the image "normally"– in a single all-encompassing view. As Kurgan says of the piece, "part of the idea was that when you were looking at the piece, you couldn't get the overview, so it was a satellite picture but you couldn't get the satellite view."[17] The difficulty of grasping the overall view was contrasted with the difficulty of penetrating the mysteries of the image's details. If contemplation of the image from one of the balconies left the viewer dissatisfied, the close-up view encouraged by Kurgan left the viewer equally flummoxed.

The expectation at work is that by stabilizing the eye upon a single expanse of the earth's surface, an intensity of seeing is thus achieved which enables the viewer to see more and, of course, know more. As such, each technological image participates in a highly structured act of looking in which we imagine that the empirical world is fully present and transparent to our gaze, and that visual technology is only an extension of the natural act of visual perception and consequent knowledge. Kurgan's *New York, September 11th, 2001 … four days later* participates in that narrative but also subverts it. By making the image abstract and alien, Kurgan at once produces the expectation that we should be able to know more, while pointing out to us that the image actually limits out vision. As Kurgan explains, "It wants to represent the comprehensible magnitude of the event," yet the image "offers only a certain kind of evidence."[18]

Third transgression: the corpse vanishes

As Allan Sekula notes, a photograph merely represents the *possibility* of meaning. Only as a result of its "embeddedness in a concrete discourse situation, can the photograph yield a clear semantic outcome."[19] Fixing the meaning or range of meanings of any photograph depends upon resorting to a limited range of available cultural scripts. This meaning depends upon the "reality effect." That is, the belief that a photograph is a window onto a self-evident, empirical world.[20] However, the "importance of war photographs lies less in whether or not they are 'real' and more in the reality they construct."[21]

Kurgan is in fact well aware that her work of art pictures a place where thousands of corpses exist at the moment the image is recorded. Hers is a landscape in which thousands of dead lie buried where they fell, a fact that resonates with the typical appeal of war photography. Yet, unlike Alexander Gardner's photographs of Antietam or, for that matter, any work in the long history of war photography, Kurgan's artwork curiously avoids picturing even a single corpse. As Kurgan says, "The image makes us witnesses: it is imperative that we look at it. The satellite's sensors capture a mass grave, what remains of a crime or an act of war. Nothing can justify or rationalize what happened here."[22] She continues that the image itself offers no instructions regarding how we should understand the event.

Of course, it is not the image itself but the *genre* to which her work refers that offers such instructions. War photography typically evokes humanist narratives that reduce any particular war to the universal of all wars, and reduce the individual corpse to "the pre-eminent sign of war" as the cost of historical progress. That sign helps make War's meaning universal; at the same time, the sign of the corpse within those photographs makes the act of war tragic but inevitable, for we have learned that the forces of historical progress are larger than any individual.[23] Rather than depict the dead in ways that would elicit emotions commonly associated with war photographs of the tragic dead, Kurgan produces an emotional vacuum by playing upon audience expectations queued by awareness of the genre of war photography. Kurgan makes war's pre-eminent sign disappear in the very place we expect it to appear. For Kurgan, picturing the bodies would only draw on a timeworn cultural script for understanding what the event means. By failing to depict the bodies in the space of a picture where the viewer knows – through everything that was happening in the mass media at that time – that more than 2,000 persons lie dead, victims of an act of violence at the scale of warfare, Kurgan calls into question the authority of both cultural narratives and the latest advanced image technologies that continually circulate them.

Conclusion

While technologies for viewing the world from great distances at ever-increasing resolutions give the impression that the act of seeing is becoming limitless, Kurgan's piece emphasizes the limits to images produced through new technologies. Her work points out that neither the encompassing satellite view nor examination of its details really tells us any more about the subject. In fact, the real subject – the circumstances surrounding the *event* of violence – becomes perhaps even more obscure due to the proliferation of satellite images. As the question "Why?" repeated itself endlessly in the American media in the days and weeks following the attack, Kurgan's work of art presented the technological image as not only failing to respond to such an important question, but as complicit in the production of the very event it was unable to explain. Complicit because, as the geographer J. M. Blaut convincingly argued, the myth of the West's superiority over all the other cultures and nations continues to validate the West's political and economic ambitions. For him, this myth was deeply interwoven with beliefs regarding technological superiority. He argued that technologies contain a hidden message: that those possessing advanced technologies demonstrate their superiority as individuals and collectively as cultures by using those technologies to exercise power over others. Thus, the technologies themselves become a sign that those who possess them must be superior, and those who do not must be inferior. For Blaut, "In the end, technological arguments end up being arguments about the inventors, not the inventions," and the arguments they make are always geographical as well as historical.[24] It is this myth, present in the narrative structure enacted in viewing satellite images of the attack on the World Trade Center, that Kurgan so skillfully deconstructs in *New York, September 11th, 2001*.[25]

Notes

1 See S. Huntington, "The Clash of Civilizations?," *Foreign Affairs* 72 (3), 1993, 22–49, and *The Clash of Civilizations and the Remaking of World Order*, New York: Simon & Schuster, 1997.

2 S. Sontag, "Looking at War: Photography's View of Destruction and Death," *The New Yorker*, December 9, 2002, 82–98.

3 D. Cosgrove, *Apollo's Eye: a Cartographic Genealogy of the Earth in the Western Imagination*, Baltimore, MD: John Hopkins University Press, 2001.

4 Both quotations are from a telephone interview I conducted with Laura Kurgan on August 30, 2003.

5 This quotation is from a telephone interview I conducted with Kurgan on June 4, 2003.

6 Information here is from the two interviews referred to previously; quotation is from the August 30 interview.

7 *Ibid.*

8 This description is drawn from the above mentioned interviews with Kurgan and her former website at www.princeton.edu/~kurgan/sep15/text.html accessed August 19, 2003.

9 J. Ketchum, *Journey to the Surface of the Earth: Contemporary Art, the Geoaesthetic Trace, and the Production of Geographic Knowledge*, dissertation completed at Syracuse University, 2005.

10 The above quotations are from my August 30, 2003 interview with Kurgan and her website biography, respectively, website accessed August 19, 2003 at www.princeton.edu/~kurgan/about/bio.htm.

11 M. Heidegger, "The Age of the World Picture," in W. Lovitt (trans. & ed.), *The Question Concerning Technology and Other Essays*, New York: Harper Torchbooks, 1977, pp. 115–54.

12 In fact, this has been the underlying assumption for much of the socially concerned artwork of the past century, from Dada to Guy Debord and the Situationists to the "Theater of the Absurd" to Gerard Genette's *Narrative Discourse*, Cornell, 1980, although this is perhaps better illustrated in literature by Alain Robbe-Grillet and the *nouveau roman*. In academic geography, Edward Sonja's "third space" is one example of a similar claim in the book, *Thirdspace: Journeys to Los Angeles and Other Real-and-Imagined Places*, Oxford: Basil Blackwell, 1996.

13 See S. Marston, "The Social Construction of Scale," *Progress in Human Geography* (2000) v.24, pp. 219–42, and S. Marston, J. P. Jones III, and K. Woodward, "Human Geography without Scale" (2005) v.30, pp. 416–32.

14 Quotation from my interview with Kurgan on June 4, 2003.

15 B. O'Doherty, *Inside the White Cube: the Ideology of the Gallery Space*, Berkeley: University of California Press, 1999, p. 14.

16 Author's interview with Kurgan, June 4, 2003.

17 *Ibid.*

18 Quotation taken from Kurgan's online description of the artwork, accessed August 19, 2003 at www.princeton.edu/~kurgan.

19 A. Sekula, "On the Invention of Photographic Meaning," in V. Burgin (ed.), *Thinking Photography*, London: Macmillan, 1982, pp. 84–109.

20 R. Barthes, "The Reality Effect," in T. Todorov (ed.), *French Literary Theory Today*, Cambridge: Cambridge University Press, 1982, pp. 11–17.

21 G. Klingsporn, "Icon of Real War: A Harvest of Death and American War Photography," *The Velvet Light Trap*, n. 45, 2000, 4–19.

22 Quotation taken from Kurgan's online description of the artwork, accessed August 19, 2003 at www.princeton.edu/~kurgan.

23 G. Klingsporn, "Icon of Real War: A Harvest of Death and American War Photography," *The Velvet Light Trap*, n. 45, 2000, 4–19.

24 J. M. Blaut, *The Colonizer's Model of the World: Geographical Diffusionism and Eurocentric History*, New York/London: the Guilford Press, 1993.

25 Felix Driver (see for example "Imaginative Geographies" in P. Cloke, P. Crang, and M. Goodwin (eds), *Introducing Human Geography*, London: Arnold, 1999, pp. 209–16) has also illustrated that, from a Western point of view, the West has typically represented itself as a progressive force at the forefront of history, while non-Western nations and peoples have typically been represented as behind the West in terms of modern progress, thereby producing an imaginative picture of the world as a mosaic of differing moments of cultural progress – of different moments in time – all manifested simultaneously. These differences often appear as the subtext of landscape depictions, and it is just such a geographical subtext that Kurgan works so carefully to deconstruct in *New York, September 11th, 2001…*

19

The Earth exposed

How geographers use art and science in their exploration of the Earth from space

Stephen S. Young

Art and science are often viewed as diametrically opposed. Science seeks the true meaning of nature through a specific methodology while art recreates nature through myriad ways of seeing. Science is done in a fashion that allows it to be replicated by others, whereas art is often unique and many times cannot be replicated even by the same artist. Despite this seemingly wide gulf between the arts and the sciences, there are areas where the two interact productively. There are scientists who utilize art to assist them in their work and artists who use science in their creative endeavors. In their study of the Earth, geographers employ both art and science in exploring and explaining the world. One way is through the use of satellite imaging, or remote sensing.

Seeing the Earth from space provides a new viewpoint for humans to understand and appreciate the world. From space we can see the results of powerful geologic forces that lifted the Andes high above the Pacific Ocean. We can also view the immense dry territories of the Tibetan Plateau and Sahara desert, as well as the lushness of the Amazon Basin or the Indonesian archipelago. These images can transform education into an exciting discovery of our planet as a complex connected whole; they can also be of breathtaking beauty, reminding us of our incredible planet.

In an exhibit *The Earth Exposed*, imagery was selected from a variety of airborne and satellite-based sensors, some a few kilometers above the Earth, others from over 35,000 kilometers aloft. These stunning visuals demonstrate how science and art are involved in the creation of imagery and the extraction of information. A variety of artistic perspectives and techniques can be incorporated. First and foremost is the visual power of the imagery itself. The technique of collage is also used to show interconnections in the world, and to display extra meaning through the juxtaposition of images. Some art shocks the viewer, especially when used to visualize potential futures. For instance, global climate change and potential sea-level rise are serious environmental issues, and through art we can visualize future sea-level rise and confront viewers with the possible consequences of global warming. Finally the artistic perspective of abstraction alters imagery to open the mind to new ways of seeing.

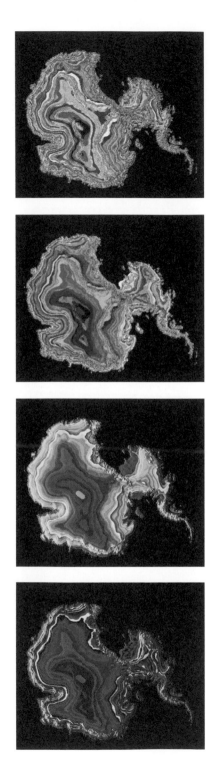

Figure 19.1 *Antarctica* by Stephen S. Young.

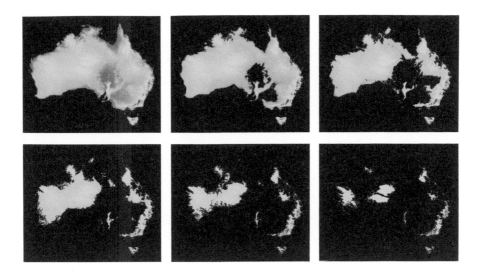

Figure 19.2 The Making of the Australian Archipelago by Stephen S. Young.

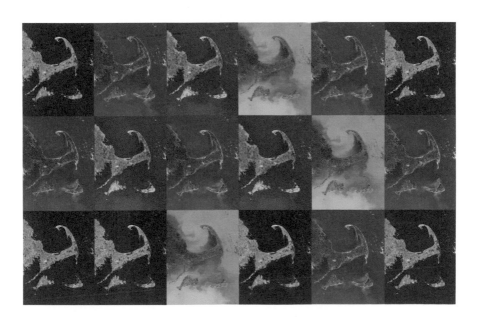

Figure 19.3 Capes as captured by the Warhol satellite by Stephen S. Young.

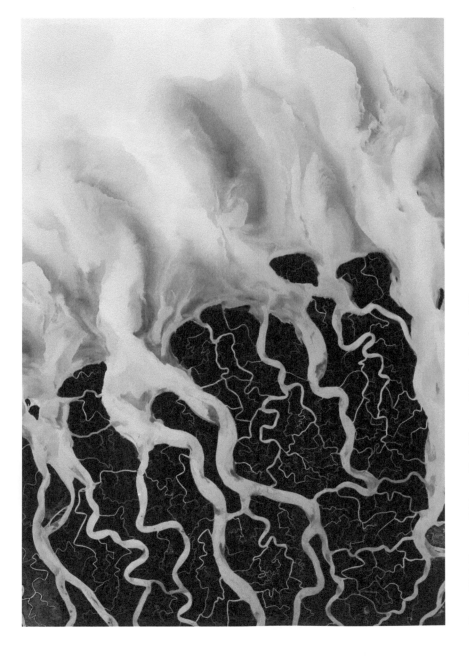

Figure 19.4 Ganges : White-on-Black/Black-on-White by Stephen S. Young.

20

Disorientation guides
Cartography as artistic medium

Lize Mogel

Maps have become part of a pop culture kit-of-parts within the cultural sphere, used as a form, an aesthetic, or a methodology. The rising number of college-level art, architecture, and design courses that teach "mapping" is a testament to this, as is the number of art exhibitions about maps concentrated in the last few years.

Why maps, and why now? This mapping impulse results from a convergence of a number of shifts in the way we think about representation and space. This trajectory has art historical roots in movements ranging from Surrealism to land art to institutional critique to interventionist practices. The way we visualize spatial and personal relationships has also radically changed – the absolute centrality of the internet to metropolitan citizens, saturation of electronic communication, and increased mobility have taught us to understand information as embodied in map/networked form.

For artists and designers, maps are a highly aesthetic form, able to articulate and spatialize complexity. However, much of contemporary art and design using maps only pays lip service to a key aspect of maps – their inherent politics. This work merely *represents* the political, but does not *produce* it.

Several recent cultural projects are able to successfully unleash the political aspects of the map. *Million Dollar Blocks* maps criminal justice data onto urban centers; the *disOrientation Guide* to the University of North Carolina-Chapel Hill provides a new representation of the "global" university; and *An Atlas of Radical Cartography* presents different artistic and cartographic strategies for political education. These three case studies share common elements. All are interdisciplinary projects, merging art and design, geography and cartography, and an activist intent. All set out to engage multiple publics. All are examples of critical or counter-cartography, making and using maps as a form of resistance.

Million Dollar Blocks

Project Team: Eric Cadora, Justice Mapping Center; Laura Kurgan, Columbia University; David Reinfurt, Graphic Designer; Sarah Williams, Columbia University.

THE NORTHWEST PASSAGE

or Can a   dia n   In   terrial
Waters. Be   c au   se of global i wa
rning, the p   as   sage is cle
at of ice for a p   sig   io d
of ti   me, mak   ing it possible for the e
ventu   al u se of the route for global shipping and
military   maneuvers. The sovereignty of the passag
e is in qu   e stion— Canada claims that it is part of its
conti   nent   al shelf and has taken steps to mark it
s   United States, claim that it traverse
g the   ational waters and should not
s   intern   ed by Canadian interests. If th
be controll   passage was opened, it would be a
ovide a faster ro   ute from west to east and would be a
ble to accommodate   larger vessels than the Panama Ca
nal can currently manage. It would also create potent
ial environmental problems— what would happe
n if an oil tanker run aground in these pristine
water s?   THE 1915 PANAMA PA S
an Fr CIFIC INTERNATION anci
isco hosted t AL EXPOSITION his W
orld's Fair, which celebrated the age
ring of the Panama Canal, a strategic, e
conomic, and engineering triumph for the United
States. The World's Fair displaced global geography
onto a local map- offering pavilions from twenty-tw
o nations as well as five ethnographic side-show dis
plays of Asians and Africans in "native" habitat. Th
e World's fair put San Francisco "on the m   n e wi?
as a viable port city. A 1916 map titled t (the Unite
the "Exposition City" focuses not on the Fai 09. This ar
hey rgrounds, but on the busy ship pi n helped engine
that secured   g piers along the eas p term e anama's ind
la. The Canal Zone s age of the city. ov ou n ded the Pan
ted in 1914 by the U.S. aft   er an initial unsuccessful a
ests. Fredrick Wiseman's 1977 documentary, "Canal Zone," pictu
community —there is little that reveals the location as Central
ns were excluded from most jobs in the Canal Zone, and Pana
e not permitted to fly (although several were planted in act
60's), American military bases were co n   s
rs. The Canal and the Can
to Panamanian control in 1
wements must be m ade to ke
he increasing size of
d the enormous amo
f traffic throu
the Canal.

THE MOTHBALL FLEET   Suisun
The   etimes kno
Bay Reserve Fleet is som
wn as the "ghost fleet" o r the "mothball fleet." Ap
proximately 96 American military ships— tankers, ba
ttleships, and aircraft carriers dating from World War II to
the Vietnam War are stored in Suisun Bay, near San Franci
sco. These are part of the National Reserve Defense Fleet, o
verseen by the U.S. Navy and the Maritime Administra ti
on. Other NRDF sites are located in James River,
Virginia, and Beaumont, Texas. Some ships are m
aintained in readiness, to be deployed in civil or
military emergencies. Others await the scrapyar
d in developing countries such as Bangladesh— wh
ere contractors take advantage of low-wage labor co
sts, and less enforcement of environmental, lab
or, health, and safety regulations. Ships a
re also donated to individu al st   at
es to be sunk and turned   The Canal Zo
into artificia   as a sovereign territory o
l fish re d States from 1903 until 19
eefs.   ea was given to the U.S. after t
er the "bloodless revolution"
ependence from Columb
ama Canal itself, comple
tempt by French inter
res a typical American.
American flags wer
manian flags wer
s of protest in the 19
trusted, and the Scho
there for almost forty yea
al Zone were restored
999. However, impro
ep p ace with t
uni o.
gh

Million Dollar Blocks is a mapping of data sets onto American cities that reframe questions of criminal justice and place. It is a collaboration between Eric Cadora and the Justice Mapping Center (JMC) and Laura Kurgan and the Spatial Information Design Lab (SIDL) at Columbia University's Graduate School of Architecture, Preservation, and Planning, and others.

The concept of "million dollar blocks" looks at the cumulative amount of government spending to incarcerate and reincarcerate individual residents at the scale of a single city block to an entire neighborhood. Criminal justice advocates suggest that there should be more accountability for this spending, often in millions of dollars for a single block. They propose that these millions should be invested directly into improving the social conditions in that neighborhood in order to decrease the amount of incarceration.

Urban crime maps most often show that crime is spread out throughout the city. Cadora asked "What would it look like if we instead mapped where people lived – how would that change the conversation [around criminal justice]?" The million dollar blocks maps first created by the JMC in 1998 and subsequently with the SIDL showed that incarceration is concentrated in very few areas, usually the lowest income areas with the highest percentage of people of color.

The JMC used million dollar blocks maps to help persuade state and local legislators and government agencies to initiate new practices in prisoner re-entry, sometimes with marked success.[1] The collaboration with Kurgan and the SIDL was meant to expand this work by creating a more finely honed set of tools for communication, and to reach another audience – the architecture and planning fields, and the general public. This happened in part through exhibitions of the maps in architecture and art venues, including in New York at the Architecture League (2007) and the Museum of Modern Art (2008), in art exhibitions in Syracuse (2007) and Los Angeles (2006), and at the Venice Architecture Biennale (2008).

Additionally, a day-long "Scenario Planning" workshop at the Architecture League activated the map even further. The workshop brought together professionals from architecture and planning, education, prisoner advocacy, housing advocacy, homeless and family services, and the Department of Corrections. They analyzed and compared geographic data for a single block in the New York City neighborhood of Brownsville, including location of social services, demographics, and prison and jail admissions, and created scenarios for the "reprogramming" of urban space to alleviate the concentration of incarceration and poverty.

By reframing the discussion in terms of the design of urban space, the *Million Dollar Blocks* project creates an awareness of the social and political conditions of that space. For example, "The Pattern," a SIDL publication on the project, zooms in on individual blocks in Phoenix, New York City, Witchita, and New Orleans. These share common characteristics including a large percentage of people of color and living in poverty, a large number of prison admissions, and certain physical characteristics that create social and physical isolation in a neighborhood – such as adjacency to a major highway, and visible disinvestment in the area in the amount of vacant land and deterioration of housing stock.

The challenge posed by the project to the people and agencies who design the physical components of the city, its infrastructure, and policies, is not only to understand how incarceration deeply affects the social fabric of the city, but more so to "initiate regenerative, even if incremental, forms of urban change."[2]

disOrientation guide

Project Team: Counter-Cartographies Collective: Maribel Casas-Cortes, Sebastian Cobarrubias, Craig Dalton, Liz Mason-Deese, Tim Stallman.[3]

The *disOrientation Guide* to UNC-Chapel Hill was produced by the Counter-Cartographies Collective (3Cs) in 2006. The collective, made up of politically active undergraduate and graduate students from geography, anthropology, and math, came together initially to discuss and organize around labor issues on campus. The *disOrientation Guide* came out of their interest in creating a different kind of political awareness – one that was not focused on a singular issue and using a particular activist language, but one that created more of a sense of the "big picture" and self-reflexiveness on the part of university students, and the university itself.

Disorientation guides have been popular on US university campuses since the 1960s, when they were used to protest university-wide investments and involvement in the Vietnam War. Today, many progressive student groups on campuses produce these guides, in zine, booklet, or newsletter form. Unlike orientation materials that help new students find basic services in the university community, the disorientation guides draw attention to institutional racism and sexism, conservative curricular agendas, and university ties to global finance, gentrification, and defense research.

The UNC *disOrientation Guide* diverges from the historical form and function of other guides. First, it takes the form of a map, spatializing information and the reader's relationship to it. Second, it eschews activist vernacular in favor of a more theoretically informed questioning of students' and the university's subjectivity. This guide is political, but in a way that is inclusive rather than exclusive in terms of language and perceived intent.

The front of the folded *disOrientation Guide* features an antipodal projection in which Chapel Hill surrounds or contains the rest of the world. Unfolded, this visual metaphor is furthered by the use of world maps which riff off of the idea of the "global campus" and its relationships to other places – not just in where students are from and where they visit during a study-abroad semester; but also in how the world is produced for students through course titles and areas of study.

The *disOrientation Guide* reads like a kind of Situationist *dérive*[4] through UNC-Chapel Hill and its global interests. It looks at the university from a number of scales and from a number of subject positions. The 3Cs describe the organizing principles of the front of the map as based on three theoretical poles – Deleuze and Guattari (the university as "a functioning body ..."), Marx ("a factory ..."), and Foucault ("... producing your world"). While conventional design employs an information hierarchy and is meant to communicate a singular, clear message, the design of this project is anti-hierarchical. It reproduces the methodology and multiple voices of the collective.

The map simultaneously considers aspects of the university's representation, infrastructure, and its environs from labor to curriculum to research dollars to fair

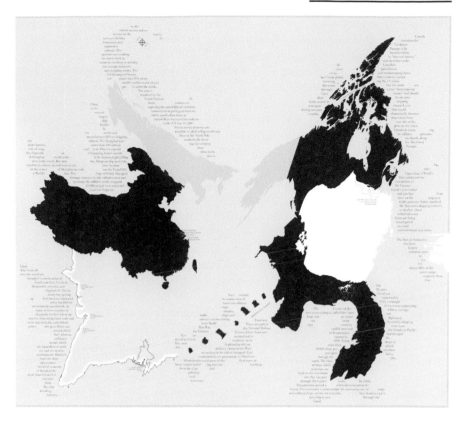

Figure 20.2 Mappa Mundi, 2008, Lize Mogel.

trade coffee shop locations. Counter to UNC-Chapel Hill's own image-making, the *disOrientation Guide* reveals the university as a complex and conflicted organism.

Over 1500 *disOrientation Guides* were distributed, many to students in geography and other departments at UNC-Chapel Hill and elsewhere. It is used as a teaching tool by graduate students and faculty throughout the USA. The most visible impact however is the interest in the guide from students and groups in other cities who want to create their own disorientation map. Even certain offices of the university have embraced the guide to an extent. For example, the study abroad office finds the specific maps about the international student body useful to visualize where the gaps in their study-abroad programs are. And while the Guide is critical of the university, the university also uses it to demonstrate its own openness.

An Atlas of Radical Cartography

An Atlas of Radical Cartography is a counter-cartographic project in three parts, organized by myself and artist/writer Alexis Bhagat.[5] The intent of the entire project is to discuss maps as an inherently political form, and to draw attention to the issues described – including migration, globalization, extraordinary rendition, surveillance, urban ecology, and waste management.

This grassroots project began with a book of 10 essays and 10 maps about social and political issues, published in December 2007. The contributors include artists, designers, geographers, architects, scholars, community organizers, and collectives who all have an activist practice. "Map" is defined broadly to include diagrams, plans, and drawings. The 10 maps are unbound, to be further distributed or displayed as posters. An internationally traveling exhibition, titled *An Atlas*, showcases these maps and others, and has allowed us to expand the book project in a number of directions. The maps are reproduced as large prints, giving the viewer a different physical relationship to the information. The exhibition has also included participatory mapping projects, audio tours, and documentary films. A third, discursive aspect of the project evolved as we presented the project in dozens of lectures and workshops to disciplinary specific audiences from art to geography to activism, and the general public. The book is also used in high school and college classrooms to teach socially engaged design, cartography, and geography, and the political ramifications of spatial practices.

The project as a whole presents 10 different conceptual and design strategies for visualizing social justice issues using maps. These 10 projects naturally form thematic linkages to each other, creating a broader conversation about the issues at hand and the political uses and limits of mapping. The underpinning "data" is sometimes very different from that of the other projects described in this essay. It is produced by local knowledge, personal experience, and collective action, as well as by more traditional research methods. The maps use recognizable geography, but also conceptual image-making, experiments with cartographic form, and thought diagrams. Several of the maps in the collection employ cartographic conventions as a strategy to critique existing power structures and the map itself. The three maps described below exemplify different strategies of representation and function.

The map "Selected CIA Aircraft Routes, 2001–2006" is by artist/geographer Trevor Paglen in collaboration with activist graphic designer John Emerson. The data was generated through Paglen's research into covert government projects. He identified dozens of planes leased or owned by the CIA through front companies that were used in extraordinary rendition flights. Many of these flights were further linked to the CIA program via news stories about citizens who were mistakenly rendited and eventually released. These "torture taxis," as Paglen calls them, are used to transport people to countries where they are interrogated using methods that are not legal in the USA.

The map makes visible a geographic aspect of the covert extraordinary rendition program, one that is not commonly seen in the mass media coverage which focuses more on human rights issues. It diagrams a specific set of relationships between places such as Washington DC, Guantánamo Bay, Cairo, and Kabul. By depicting these data as a kind of airline route map, Paglen and Emerson reveal just one small part of a sizable transportation infrastructure, but one that is entirely secret – a shadow aeronautics industry.

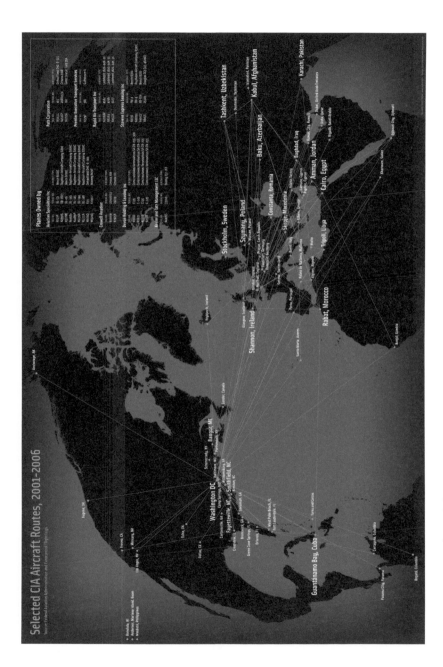

Figure 20.3 CIA Rendition Flights 2001–2006, 2006, Trevor Paglen and John Emerson.

This map was also produced as a public artwork, installed on a billboard on a major Los Angeles boulevard, part of a series of billboards presenting alternative and critical images of the effects of the war in Iraq, commissioned by the organization Clockshop.

Artist Pedro Lasch's "Guías de Rutas/Route Guides" is a map of North and South America, pictured as a singular landmass with no marked borders, imprinted with the words "Latino/a America." This map names, and thus symbolically claims, the entire American continent for Latinos and Latinas. This act of naming is aggressive, although in keeping with historic practices of cartography linked with colonization. It also recognizes a growing demographic shift as massive northern migration continues, and a shifting *Latinidad* identity.

Lasch created this project while working as a cultural educator with migrant populations in Queens, NY, and elsewhere. He gave two of these maps each to 20 individuals who he knew were crossing the US–Mexico border for work, commerce, immigration, leisure, and other purposes. The few that were returned to him bear traces of their journey from coffee rings to dirt stains. In *An Atlas of Radical Cartography*, Alejandro de Acosta describes Lasch's project as:

> a synthesis of the idea of an abstract, conceptual map and the concrete and ephemeral ones that are always being drawn in our lives, in our most minute moments. Lasch's maps are, as maps, quite useless. None of the wanderers that carried them could have used them to guide their movements. Instead, they suggest a spiritual aspect of the journey.[6]

The maps mark the physical and politically symbolic passage of a wide range of migrants. In exhibition, the maps are displayed with excerpts from interviews about the process of migration and how that affects one's own and cultural identity by the people who carried them.

My own contribution to the project, "From South to North," reorganizes the familiar territory of the world map so that it better describes the processes of globalization. This work layers Canada, San Francisco and its 1915 World's Fair site, Panama, and a "mothballed" navy fleet in Suisun Bay, east of San Francisco. The geographic forms are mirrored by text that relates shared narratives of sovereignty, shipping, and displacement.

"From South to North" relates how global geography was relocated onto a local map for the World's Fair in San Francisco. The Fair put its host city "on the map" as a key US port, and celebrated the opening of the Panama Canal a year earlier. The Panama Canal created a shipping route through the South American isthmus, changing the course of the global shipping industry. Panama was bisected by the Canal Zone, which was a US territory from 1903 until 1979. The Northwest Passage is a potential new shipping route that bisects the Canadian Arctic, now navigable in the summer because of climate change. A dispute over whether it lies in Canadian internal waters or international waters will determine its future use. The Mothball Fleet represents the end of the life cycle of a ship. The fleet consists of more than 70 decommissioned US Navy ships waiting to be scrapped or recycled, many in shipbreaking yards in South Asia where labor conditions are deplorable. This "geography lesson" describes a new set of geopolitical relationships that a standard world map cannot contain. However, as globalization brings part of the world closer together, the map becomes more difficult to read, and more disorienting.

The *disOrientation Guide*, *Million Dollar Blocks* maps, and *An Atlas of Radical Cartography* shift between art/design, geography, and activism. Their relationship to these fields can be simultaneously uneasy and expansive. These projects look back to a time when cartography was an art form, only now these maps serve to rebalance power structures instead of reinforce them. They also work against the value system of art and design in which the singular, collectible object is a prize above all else. Instead, they operate within a larger context of interpretation and action – which is what makes them politically relevant. These projects' hybrid nature helps them to move beyond mere representation of place or politics, and to actively seek out and produce social change.

Notes

1 Counter-Cartographies Collective, *disOrientation: your guide to UNC-Chapel Hill* (Counter-Cartographies Collective, 2006). http://www.counter-cartographies.org.
2 J. L. St. John, "A Road Map to Prevention," *Time*, March 26, 2007, 56.
3 Graduate School of Architecture, Planning and Preservation, *Spatial Information Design Lab: Scenario Planning Workshop,* New York, NY: Columbia University Press, 2008, p. 4.
4 The Situationist International was a European avant-garde cultural and political group (1957–72) whose activities and writings are highly influential in contemporary art and architecture fields. The *dérive* [French: *drift*] is a Situationist technique, where a group or individual wanders through the city without apparent aim, the course influenced by the city's "psychogeographic effects" for the purpose of "studying the terrain or to emotionally disorient oneself." G.-E. Debord, "Theory of the Dérive," *International Situationniste* 2, 1958, tr. by Ken Knabb, http://library.nothingness.org/articles/SI/en/display/314
5 A. Bhagat and L. Mogel, (eds) *An Atlas of Radical Cartography*. Los Angeles: Journal of Aesthetics and Protest Press, 2007.
6 A. de Acosta, "Latino/a America: A Geophilosophy for Wanderers," in A. Bhagat and L. Mogel (eds), *An Atlas of Radical Cartography*, Los Angeles, CA: Journal of Aesthetics and Protest Press, 2007, p. 73.
   Paglen, Trevor, and A.C. Thompson, *Torture Taxi: On the Trail of the CIA's Rendition Flights*. (Hoboken: Melville House Publishing, 2006).
   Graduate School of Architecture, Planning and Preservation, *Spatial Information Design Lab: Architecture and Justice*. (New York: Columbia University, 2008).
   Graduate School of Architecture, Planning and Preservation, *Spatial Information Design Lab: The Pattern*. (New York: Columbia University, 2008).
   Graduate School of Architecture, Planning and Preservation, *Spatial Information Design Lab: Scenario Planning Workshop*. (New York: Columbia University, 2008).
   Interviews with: Tim Stallman, Craig Dalton, Liz Mason-Deese, Laura Kurgan, Eric Cadora.

21

Avarice and tenderness in cinematic landscapes of the American West

Stuart C. Aitken and Deborah P. Dixon

Blood on the land

Director Paul Anderson's *There Will Be Blood* (2007)[1] is a brooding, critically acclaimed epic that harkens back to vast constructions of avarice and greed witnessed in classics such as *Citizen Kane* (1941)[2] and *Giant* (1956).[3] Like *Citizen Kane*, it charts the insidious manner in which the power to direct people and things corrupts the human soul. And, like *Giant*, it tracks the emergence of a Western landscape punctuated by empire builders who want to capitalize on the exploitation of human and physical resources. The complex acts of rendering, attending, viewing, and reading films of this kind require a nuanced appreciation of intertextuality that stipulates an understanding of the connections between geography and the humanities. Landscape features hugely in the ways that geography and art coexist, but often these connections are taken for granted. With this essay, we want to dig a little deeper into the politics of landscape representation by engaging its filmic contexts as emotionally charged movements.

The plains and peaks of the American West, when presented as grandiose cinematic spectacle, are things we are drawn into because they are, by and large, rendered without an accompanying science or geography that explains their engagement with a particular social or cultural order. This aesthetic is consumable and thereby open to manipulation. The modern consumption of images involves, for example, a mobilization of anaesthetizing strategies, what Giorgio Agamben describes as the destruction of experience. "Standing face to face with the great wonders of the world," he argues, "the overwhelming majority of people have no wish to experience it, preferring instead that the camera should."[4] To use Michael Rogin's provocative terminology, consumable images may be thought of as a "spectacle of amnesia in imperial politics."[5]

It is tempting to consider how Western landscapes are deterritorialized through the camera lens only to be reterritorialized with ideological intent through a variety of film-making techniques, from juxtaposition, "point-of-view" and "establishing" shots to lighting, sound, and narrative. Indeed, this has been a preoccupation of much of film geographic analysis, whether in the form of semiotics, psychoanalytic theory, or discourse analysis.[6] But therein lies a cinematic illusion, because Western movies are, in many ways, *also* and *at the same time* an ironic deconstruction of patriarchal America, carefully working to strip away the amnesiac brutality of just such an imperial mapping. It is the effective power of this process that we seek to uncover in this essay by following another route, one laid out by what Giuliana Bruno calls *carte du pays de tendre*,[7] or tender mappings. This particular mode of encounter has the potential to reveal the non-representational, highly contingent contexts of cinematic spectacle without reducing the amnesia in imperial politics.

Cinematic landscapes, motion, and emotions

Movies may look like reality and they may feel real when emotions are engaged and churned by what is on the screen, but they are something more. bell hooks argues that movies are like magic because they change things; they take the real and make it into something else. Audiences rarely get a dose of reality from movies, they get precisely the opposite; they get a recreated, reimagined reality. She writes: "the world of movies constitute a new frontier providing a sense of movement, of pulling away from the familiar and journeying into and beyond the world of the other."[8] To this we may add, with what effect?

The work of Agamben, Rogin, and other theorists suggests a reactionary politics to representational forms in cinema that often takes the form of subtle spatial framings that seek to deny movement, to fix the identity of people and things, and the meaning of their particular presence or absence. Film-makers often rely upon previous modes of spectatorship, such as paintings and music hall, to facilitate the "ease" with which audiences engage with cinematic viewing. They also reference their content, such that the meaning of people and place on film are interpreted in accord with established frames of reference.[9] In the process, these framings work to foreclose the possibility of alternate interpretations via an emphatic rendering of "naturalness" or "inevitability"; they may even, as Rogin[10] argues, attempt to forestall the cognitive act of interpretation itself. Indeed, the stable ordering of meaning through the manipulation of filmic landscapes and space relates to Rogin's[11] idea of cultural amnesia as it is fomented through spectacle, which is, "the cultural form for amnesiac representation, for specular displays are superficial and sensately intensified, short lived and repeatable ... Spectacles colonize everyday life ... and thereby turn domestic citizens into imperial subjects." This is not to suggest that amnesia is produced at the expense of affect and emotion but rather, as Thrift observes, a particular kind of emotion is produced through imperialism.[12] This is the heightening of a particular kind of emotion that sidesteps or forgets (through amnesia) the horror of, for example, war and slavery, and is pulled primarily towards the heroism of characters and contexts.[13]

Mike Crang notes that "[o]bservation is not just optical but haptical – a practice of grabbing hold of, reaching out, apprehending and touching."[14] It follows, then, that the production of film landscapes is intimately connected to movement and the production of other kinds of spaces, those associated with the practice of viewing. The attenuation of all of our senses within the darkened interiors of theaters is a special configuration and practice of a viewing: it sets up the possibility of the illusory eye/I following the camera. Giuliana Bruno theorizes our experience of cinema as traveling through space. From her classic *Streetwalkers on a Ruined Map* (1992)[15] to her epic *Atlas of Emotion* (2002),[16] she develops ideas of motion and emotion through a medium that is a montage of views co-constructed by the cinema viewer (whom she calls the voyager). With hooks, and 1930s Russian film-maker Sergei Eisenstein, she argues that "[t]here is a mobile dynamics involved in the act of viewing films, even as the spectator is seemingly static. The (im)mobile spectator moves across an imaginary path, traversing multiple sites and times. Her fictional navigation connects distant moments and far-apart places."[17] Bruno argues against the usual view of film space as a homogeneous field understandable from a unified perspective, favoring a viewing field that travels through and connects with ideologies, architectures, and cultures.

The imperial subject is by no means a passive, malleable creature. The production of identity is, in the case of Rogin, a matter of a co-construction of psychological and state apparatuses, while for Bruno it emerges through a process of intertextuality that is predicated on the possibility of slippage. Ultimately, an emphasis on intertextuality prompts certain skepticism in regard to surety, stability, rootedness, and order. Bruno engages the contexts of mapping in architecture, travel, design, housing, planning, and film.[18] She takes the history of mapping and contextualizes it in the arts, in desire, and in tenderness. Bruno calls her reworking of cartographic themes against prevailing imperial hegemony a "sentimental geography."[19] Her *Atlas of Emotion* moves in, between, and through seventeenth-century cartographies to twentieth-century films.

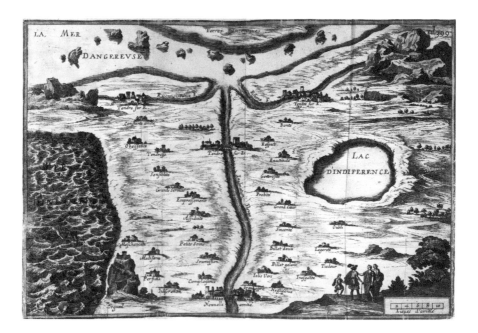

Figure 21.1 Carte du Pays de Tendre (Map of the Land of Tenderness) by Madelaine de Scudéry, which accompanied her novel *Clélie* (1654). Public domain.

At its heart is a map designed by Madelaine de Scudéry (Figure 21.1) to accompany her novel *Clélie* (1654).[20] Bruno argues that Scudéry's *Carte du Pays de Tendre* is a celebrated allegory for the female association of desire with space, and an exemplar of the ways in which cartography is inextricably linked with the shaping of female subjectivity.

Tom Conley suggests that the map in *Clélie* might have been drawn in opposition to contemporaneous military cartographies, inaugurated by neo-Cartesian engineers under kings Henry IV through Louis XIV, who redrew the defensive lines of France and designed fortified cities at a time when new siege technologies were changing the ways of war.[21] Scudéry's map quite literally highlights important passages and mobilities away from lakes of indifference, dangerous seas, and *terra incognitae* to favorable villages and towns of tenderness, large hearts, reflection, and sympathy. The map incarnates a narrative voyage that:

> visualizes, in the form of a landscape, an itinerary of emotions which is, in turn, the topos of the novel ... Scudéry's map effectively charts the motion of emotions – that particular landscape which the "motion" picture itself has turned into an art of mapping.[22]

Bruno's purview includes and goes well beyond maps, paintings, and films; it moves away from patriarchy and imperialism, away from surety, to a consideration of a tender geographical imagination and indeterminacy.

Western landscapes as imperial spectacles

Albert Bierstadt's paintings are often credited as a forerunner to the *mise-en-scène* of the classic Hollywood Western. In the late nineteenth century, Bierstadt moved to California and began a series of paintings that established his fame as the founder of the "California School of Landscape Painting." With *The Rocky Mountains, Lander's Peak* (1863)[23] – the painting that made him famous – and iconic works such as *The Yosemite Valley* (1868)[24] and *Mt. Whitney* (1875),[25] the populace of the Eastern Seaboard was treated to a particular rendering of the American West: a spectacle described by some as arousing Wagnerian aesthetics. Through his paintings, Bierstadt articulated some profound connections between landscape and the imperious manifest destiny of a young, mobile nation seeking a sense of self.

A "consumable empire" was created in the sense that Bierstadt's paintings invited settlement and conquest, calling to would-be pioneers with their idealized wide-open spaces and unoccupied, fertile lands.[25] For those less inclined to venture west, the landscapes could be bought and named. Bierstadt's proclivities towards high society propelled him to use large canvases, which excited well-heeled clients with walls to fill. At the height of his fame, Bierstadt went so far as to offer naming opportunities over western mountains for those willing to pay extravagant prices for his paintings.

Bierstadt's ability to conjure a manifestly destined nation is all the more surprising because his work is, for the most part, a fiction. The elements of his paintings (often repeated) may have existed in actuality but the totality of the scenes he painted were almost always fictive creations. It was a fiction that fooled no-one. Herman Melville (1857), for example, expressed displeasure at what he thought were propagandist leanings in the art of the Hudson Valley School (of which Bierstadt was a part), suggesting that, "[the paintings] look not only for more entertainment, but, at bottom, even for more reality than real life can itself show."[26] But then, these were very much "spectacles" and were received as such. The larger vision is a falsification, but the specific scenes draw us in as we willingly suspend our disbelief in the larger work. Bierstadt's paintings represent another world, and yet it is one to which we *feel* a tie.

Spectacle as illness and remedy

Lee Clark Mitchell, in his wonderful book *Westerns: Making the Man in Fiction and Film*, argues that, "[p]art of what critics have persistently thought of as a 'problem' of Bierstadt derives from his socially transgressive, metaphysically transcendental yearning – the unappeased craving."[27] Bierstadt's work is, to use a phrase from Brian Massumi,[28] both "illness and remedy" for the imperial subject. Bierstadt's positing of the American West as national icon, the creation of a hunger for manifest destiny through Western landscapes, and the conception of an imagined community out of false landscapes are more than mere frameworks or conventions. These connections draw us into an emotionally and politically rendered place. As with manifest destiny, it is a hunger to move beyond places that confine, and it is part of the imagined community that breaks territorial boundaries, but also reforms them.

Bierstadt achieves this sense of connection by creating a "lawless world that evokes too much emotional fervor, too much astonishment at the 'sensational', the 'lurid', and the otherwise extravagant."[29] This emotional fervor, then as today, is central to the mass appeal of spectacle. That is, spectacle must exceed science, history, and geography so that it can reterritorialize it. The success of US Western landscapes as spectacle lies in the way the constructors of those landscapes, from Albert Bierstadt to John Ford to Paul Anderson, enable viewers to forget the moment, to absorb the specificity of a scene, to embrace the falsity (or at least suspend disbelief), and to allow themselves to become an imperial subject. This is the illness, the part we connect to that implies complicity with imperialism through the consumption of spectacle.

But, there is another way to look at filmic landscapes, the part that is remedy to imperial imposition. This implies a re-territorialization that is not imperial in the sense that it lends itself to a politics of amnesia, but is rather connected to important ways in which particular emotions are mapped on a moving terrain. Through Bierstadt and Scudéry, we can work back to pre-filmic constructions of images and maps. Bruno offers a mapping which graphically embodies the idea that motion produces emotion and that emotion contains movement. Scudéry's *Carte du Pays de Tendre* reintroduces motion and emotions to the study of film and creates a place that is unreachable from semiotic theory and discourse analysis. Although she never defines "emotion," Bruno's tactile, tender mapping is nonetheless an allegory for female association of desire with space that is not referential to cognitive or psychoanalytic theory. It moves discussion away from the binary oppositions of home and travel, sedentarism and movement, the problematic connections between home and wellbeing, and the nurturing woman "in-her-place." Maria Walsh sees Bruno concentrating on a "mapping where exterior and interior collapse on the terrain of a singular topology."[30] Walsh goes on to point out that Bruno's reluctance to define emotion does not detract from an argument that presupposes an important intertextuality among the museum, the art gallery, the building, and the movie theater, which simultaneously recognizes its phantasmagoric precursors. These precursors relate to memory sites that affect and connect us. Herein lies a remedy to amnesia and the illness of the imperious subject.

Desire and landscapes of spectacle

A tender mapping of *There Will be Blood* begins with a palpable connection to landscape. If Scudéry's *Carte du Pays de Tendre* is a celebrated allegory for the female association of desire with space, then *There Will be Blood* suggests this connection from a male perspective, in a different but no less touching way. The first 20 minutes of the film has no dialogue as we follow, in a painful and fragmented way, Daniel Plainview's beginnings as a silver prospector and then oilman. The movie opens with a brief panorama of dry, desolate rolling hills and then takes us directly into those hills. The images are of men at toil, muscles straining in dark, humid, telluric spaces. Bones are broken in falls; skulls are crushed by falling machinery. Men emerge from narrow holes as if the earth were birthing them. The oil that covers them is dark, viscous, blood-like.

This visceral, embodied connection to landscape is transferred – with the first dialogue in the movie – to an embodied connection between father and son. Plainview gains his son as a baby from a miner who dies in one of his oil pits. The movie's dialogue begins as voiceover from a meeting of townspeople in an oil-rich region of Texas at which Plainview is presiding. The scene behind the voiceover is Plainview playing with the infant on a train trip. In taking H.W. as his son, he uses him as an innocent face beside his grizzled determination to add a semblance of family life while he pitches his oil project to the Texas town-folks.

The confining spaces of Plainview's beginnings are contrasted with the barren landscapes of early twentieth-century California that dominate the balance of the film. If the beginning of the film symbolizes Plainview's birth from the earth – an earth that breaks his body and molds it to "her" wishes – then the balance of the film is about control of the earth and the peoples who inhabit its surface. The face of Plainview is the face of an empire builder, a self-absorbed pioneer who traces a raging capitalist enterprise across the "undeveloped" landscapes of California. Of particular note is the facial intensity that actor Daniel Day-Lewis creates in his slightly forward leaning, almost surging figure of Plainview. His face brings alive the single-minded greed and avarice of Plainview, with a simultaneous hint – the remedy alongside illness – of human compassion and need. In *There Will Be Blood* that compassion and need are positioned most forcefully in Plainview's relationship with his son.

Plainview's face affectively parallels the façade of the wooden oil derricks, new town buildings, pipelines, and, finally, his echoing lonely mansion overlooking the

Pacific Ocean. The face and the façades are contextualized beside barren, forlorn land-scapes so untypical of popular representations of California and the American south-west. In *There Will Be Blood*, director Paul Anderson creates a landscape that forecloses upon any transformation other than into a corporate empire. This is by no means a landscape of manifest destiny, in the Bierstadtian sense, but it is nonetheless a land-scape that is consumable. The landscape mirrors, and is mirrored in, the face of Plainview. After contriving a suspect deal with an unwitting community, Plainview sets up his drilling operation. Rather than a spectacle of the American West, the viewer is treated to a dystopian and barren landscape upon which the new wood of the oil derricks provides welcome relief.

The structures of desire and repulsion in *There Will be Blood* are matched in struc-tures of fantasy and their relations to identity and landscape. Linda Williams suggests that fantasies "are not … wish-fulfilling linear narratives of mastery and control lead-ing to closure and the attainment of desire. They are marked, rather, by the prolonga-tion of desire, and by the lack of fixed position with respect to the objects and events fantasized."[31] Fantasy is a place where interior and exterior – conscious and uncon-scious, part and whole – collapse onto a singular plain. Bruno likens this to a "map-ping impulse … an architectural mobilization of affect and memory … an intimate geography, a body-city on a tender map."[32] Encountered in this way, Plainview's face and body are not only part of the consumable empire that Anderson creates in a dis-posable Californian landscape (illness), they also reflect his desire to connect with telluric spaces, landscapes, communities, and H.W. (remedy). *There Will Be Blood* is an outward journey of conquest, collusion, and single-minded empire building, but also simultaneously an inward journey to sentiment and the work of fathering. If the film's dialogue-free beginning feels like a metaphorical birthing that is a violent con-test between men's bodies and the earth, it is simultaneously a material birthing of H.W. Given that Plainview adopts H.W. as his son to further his ambitions, we none-theless glimpse intimacy between father and son. Daniel Day-Lewis holds Plainview's face in a visage that is simultaneously about desire for, and repulsion from, intimacy. The latter wins out when H.W. is struck deaf in a drilling accident and Plainview loses his lifeline with intimacy. It severs their relationship (H.W. is sent to a special school in San Francisco) and shakes Plainview's already tenuous hold on reality. The desire for intimacy becomes an unappeased craving that Plainview fills in other ways. His greed, avarice, and murderous ambitions mount over the years as he strives to build a pipeline from his oilfields to the coast.

Another narrative strand runs through *There Will be Blood*, represented by the angelic, boyish face of Eli Sunday (Paul Dano). Eli is pastor of the community against which Plainview struggles. Attempting to gain advantages for himself and his evan-gelical ministry, Eli sets himself up in opposition to Plainview. At one level, Eli's community represents the old world that Plainview is destroying. But Eli's ambition and avarice match Plainview's stroke for stroke. The two clash at a number of critical junctures in the story. In an epic, lifelong struggle, each is out to best and humiliate the other.

The penultimate scene of the film takes place at Plainview's coastal mansion and a confrontation with H.W. (who is now adult and, still deaf, works with an interpreter).

H.W. wants to take his experience as an oilman to Mexico so that he can establish his own empire separate from his father's. Plainview curses his son as a bastard and disowns him. The next morning, Plainview is woken from a drunken stupor by his nemesis, Eli, who, in a stunningly cruel scene, he belittles and then kills. "I'm finished," he says as his butler enters and surveys the murder scene. The movie ends with Plainview's face framed in an expression that is simultaneously success and failure. Illness/success/remedy/failure coexist in *There Will Be Blood*, and connect us as viewers to an emotionally and politically rendered Western landscape that is excessive in its capacity to be consumed and tenuous in its push towards intimacy.

Conclusion

Bruno's tender mapping project finds form by collapsing disciplinary boundaries as well as the borders between genres. In working from Scudéry's seventeenth-century challenge to French imperialism, she provides a way of connecting motion and emotion to film.[33] In this essay, we have used this subversive, feminist cartography to address the relations among men, imperial power, spectacles, and a particular filmic representation. We worked from Bierstadt's paintings of the American West, suggesting the creation of a landscape that is part consumable empire and part unappeased craving. Following this, our tender mapping of *There Will Be Blood* recognizes the desire for connection, family, and community.

We are left at the end of *There Will Be Blood* with a disturbing depiction of the consequences of avarice and greed, which foreclose upon the main character's ability to engage in any kind of intimacy. This is part of the illness prescribed through patriarchal imperial capitalism. Given that the violence at the movie's end is also a form of intimacy – albeit cruel and savage – there is nonetheless an embodied and simultaneous connection to landscape and the emotional work of fathering that carries through *There Will Be Blood*. It is this connection – so ably constructed during the film's beginning – that points to a tender mapping. Its mapping between Plainview and all his relations (his brother, Eli, H.W., and the landscape of Southern California) is confused and misplaced, but always tender. Its tender mapping of illness and remedy is prescribed by an inchoate and slippery intertextuality.

A focus on *motion* and *emotions* is only beginning to make inroads in the spatiovisual disciplines. Overly prescribed by semiotics, psychoanalytic theory, and discourse analysis, these disciplines tend towards somewhat structured and mechanistic ways of knowing that are often static spatial frames, relatively easily incorporated into imperial strategies that foreclose political challenge. Emotion, always a central part of the humanities, until quite recently has been under-theorized in geography, architecture, and planning. Films, buildings, plans, and other geovisual data are always representations, but they are also more than representation. And it is with this in mind – from the affective – that subversive feminist cartographies enable an opening of the political that is *also* and *at the same time* intimate and personal.

With the movie over, we reflect on connections, rewind the moments that pulled us into new spaces of intimacy and appreciation. We discuss the anger, the fear, the love; we embrace the affective localities and reinvent the moving landscapes into something we use. We create a mapping of our own moving pictures: an intimate geography, a fragile landscape with political muscle.

Notes

1 *There Will Be Blood*, directed by P. Anderson, USA: Paramount Vantage International: Miramax Films, 2007.
2 *Citizen Kane*, directed by O. Welles, USA: RKO Pictures, 1941.
3 *Giant,* directed by G. Stevens, USA: Warner Bros., 1956.
4 G. Agamben, *Infancy and History: The Destruction of Experience*, London: Verso, 1993, p. 15.
5 M. Rogin, "'Make My Day!': Spectacle as Amnesia in Imperial Politics," *Representations*,1990, Winter:106, 29, 99–123.
6 For Critiques see T. Creswell and D. P. Dixon (eds), *Engaging Film: Geographies of Mobility and Identity*, Lanham, MD: Rowman and Littlefield, 2002, p. 13–31.
7 G. Bruno, *The Atlas of Emotion: Journeys in Art, Architecture, and Film*, London and New York: Verso, 2002.
8 b. hooks, *Reel To Real: Race, Sex, and Class at the Movies*, London and New York: Routledge, 1996, pp. 1–2.
9 D. P. Dixon and J. Grimes, "On Capitalism, Masculinity and Whiteness in a Dialectical Landscape: The Case of Tarzan and the Tycoon," *Geojournal*, 59(4), 2004, 265–75.
10 Rogin, *op. cit.*
11 Rogin, *op. cit.*, p. 106.
12 N. Thrift, *Non-Representational Theory: Space, Politics, Affect*, London and New York: Routledge, 2007.
13 S. C. Aitken, "Leading Men to Violence and Creating Spaces for their Emotions," *Gender, Place and Culture,* 13(5) 2006, 491–507.
14 M. Crang, "Rethinking the Observer: film, mobility and the construction of the observer," in T. Creswell and D. P. Dixon, (eds), *Engaging Film: Geographies of mobility and identity*, Lanham, MD: Rowman and Littlefield, 2002, p. 20.
15 G. Bruno, *Streewalking on a Ruined Map*, Princeton, NJ: Princeton University Press, 1992.
16 Bruno, *Atlas, op. cit.*
17 Bruno, *Atlas, op. cit.*, pp. 55–56.
18 *Ibid.*
19 Bruno, *Atlas, op. cit.*, p. xi.
20 *Ibid.*
21 T. Conley, *Cartographic Cinema*, Minneapolis, MN: University of Minnesota Press, 2007, p. 127.
22 Bruno, *Atlas, op. cit.*, p. 2.
23 A. Bierstadt, *The Rocky Mountains, Lander's Peak*, Oil on Canvas, 1863.
24 A. Bierstadt, *Mt. Whitney*, Oil on Canvas, 1875.

25 A. Bierstadt, *The Yosemite Valley*, Oil on Canvas, 1868.

26 H. Melville, *The Confidence Man* [1857], H. Parker (ed.), New York: W. W. Norton, 1971, p. 158.

27 L. C. Mitchell, *Westerns: Making the Man in Fiction and Film*, Chicago and London: The University of Chicago Press: 1996, p. 64.

28 B. Massumi, *A User's Guide to Capitalism and Schizophrenia: Deviations from Deleuze and Guattari,* Boston: MIT Press, 1992.

29 Mitchell, *op. cit.,* p. 66.

30 M. Walsh, "Review of *Atlas of Emotion* by Giuliana Bruno," in *Senses of Cinema*: *An Online Journal Devoted to the Serious and Eclectic Discussion of Cinema,* 2003. 3. http://archive.sensesofcinema.com/.

31 L. Williams, "Film Bodies: Gender, Genre and Access," *Film Quarterly* 44(4), 2–13, 1991, 11.

32 G. Bruno, *Atlas, op. cit.*, pp. 241–242.

22

Altered landscapes

Philip Govedare

The Pacific Northwest is a region of remarkable geographic diversity, scenic beauty, and varied land use, both commercial and recreational. With great forests, rivers, and lakes, and rich agriculture, it has an abundance of natural resources. East of the mountain ranges, the remote areas far from major population centers have been selected for a variety of purposes military and commercial, including nuclear weapons research and development (Hanford, Washington), disposal sites for chemical weapons (Umatilla, Oregon), and toxic waste disposal in southern Idaho. Logging, hydroelectric projects, warming water temperatures in lakes and rivers, and melting glaciers have altered ecosystems and reconfigured the landscape.

For contemporary artists, landscape presents a compelling challenge with many questions that are unique to our world today. For an informed person, it is no longer possible to experience landscape without some sense of loss and foreboding. Every aspect of our natural environment today is affected by development, technology, and modern industry. Global warming is a phenomenon that crosses all natural and political boundaries and has ominous implications for the future of life on earth. This gathering crisis of environmental degradation on a global scale influences how we respond to landscape as individuals and as artists. The meaning that lies beneath the surface of any response to our natural world in the form of art can generate discourse and awareness, and may ultimately shed light on our own condition and how we choose to live. Art can be transformative, making the commonplace extraordinary and allowing us to experience the familiar in ways that challenge conventional thought and reveal deeper meanings.

My paintings are derived from specific landscape sites that are both visually compelling and charged with implications of use, development, and ownership. In 2003, I began a series of paintings inspired by the highly industrialized Duwamish Waterway in south Seattle, a designated Superfund site. This subject was chosen for its visual interest and its history, but also because it represented a landscape that had been dramatically altered and defiled by human enterprise. My most recent paintings are inspired by a variety of sites in the vast western landscape that have been impacted or transformed through some form of development or human activity.

My work is both a response to and an interpretation of the world which imparts sentiment through projection that comes from a perspective of anxiety about the condition of landscape and nature in our world today. My paintings are a fiction based on an observed phenomenon, a metaphor that is infused with a blend of celebration,

Figure 22.1 Excavation series #2 by Philip Govedare. Oil on canvas, 63" x 55" (2009).

anxiety, and doubt about our place in the natural world. In this manner, painting can allude to the past and simultaneously project into the future. The visual terms of painting can allow us to experience the familiar in our world in ways that may be unexpected and revelatory. In this sense, my paintings reflect the landscape we inhabit with all its complexity and layered meanings.

4 SPATIAL HISTORIES
Geohistories

Douglas Richardson

From within the constellation of the humanities disciplines, geography has enjoyed its longest and deepest traditions of interaction with history. Historical Geography has been practiced with distinction by geographers for centuries,[1] and many historians have contributed texts with geographical grounding and dimension that have made them classics in the study of geography.[2] Work in landscape representation and analysis also has long generated fruitful zones of engagement between geographers and historians.[3]

While cognizant of previous collaborations, this book seeks to better understand new interactions between geography and the humanities at their current growing edges. During the past few years, significant new forms of interaction between geographers and historians have begun to emerge. The Geohistories section which follows explores multiple dimensions of recent efforts between geographers and historians to integrate historical/geographical research agendas using spatial analysis and new geographic technologies, and the potentially transformative implications of this work for both history and geography. This rapidly evolving cross-disciplinary zone of intellectual and technological creativity, referred to here as "Geohistory," relies heavily on historical geographic information systems (HGIS) to combine space and time for collaborative research and scholarship. Large-scale geohistory projects worldwide are now sparking substantial debate and fundamental re-evaluation of historical research methods and interpretation, and are revealing new possibilities for better understanding our past.

It may reflexively appear counter-intuitive, or at best peripheral, to portend a role for geographical technologies in relationship to the humanities, and especially to the practice of historical scholarship, as technology in recent years has been highly suspect in the humanities, and at various times in geography as well. Technologies are often considered inherently reductionist or as levers of power; they are approached warily and dismissively by many in the humanities whose training and methods of working traditionally have for the most part not encompassed technological approaches. They are also generally absent from the rich traditional methods of historical scholarship and research, which include narrative, metaphor, story-telling, presentation of multiple perspectives, and careful sourcing of extant historical texts or other artifacts to discern patterns and meaning from the bewildering complexity of bygone time, with all of its inherent ambiguity and uncertainty. Even the several much touted Digital Humanities initiatives have primarily been limited to simply scanning and digitizing historical texts and images, for use in traditional ways, with little thought given to how integrative digital technologies such as GIS might substantively or qualitatively impact scholarly research. The aversion to technology in scholarship by many in the humanities is understandable as a product of disciplinary tradition, style, and training, yet puzzling still in some ways as technology is an inseparable aspect of culture, and of history itself.

However, recent interactions between geographers and historians suggest that fundamental new approaches to historical scholarship may indeed now be underway. Historians are starting to understand key aspects of GIS, such as its ability to integrate, analyze, and visualize large amounts of both spatial and temporal data, from multiple disciplines and sources, and its ability to move across multiple scales, both spatially and temporally, or geographically and historically. This ability to combine time and space in one integrated system has profound implications for research in both history and geography.

Many historians are just now beginning to grasp the significance of incorporating a spatial dimension across multiple scales into historical research, despite the barriers of disciplinary tradition and training to its adoption in history. Yet several collaborations described in this book illustrate just how synergistic these early transdisciplinary collaborations have been, and hint at their future potential. One prominent historian has argued, for example, that "of all modern information technologies, GIS may have the most potential for breaching the wall of tradition in history," noting that, "its ability to integrate disparate information drawn from the same place at the same time allows scholars to simulate the complexity of history."[4]

Not surprisingly, however, there still remain significant obstacles to better understanding these new technologies and how they can build upon the traditional strengths and methods of both geography and history. The introduction of radically new technology or ideas into established disciplines is almost always perceived initially as challenging to or displacing of traditional disciplinary strengths. On the other hand, for many leading historians and geographers, the new geographic technologies instead have proven to be a means to revitalize, strengthen, and diversify these same disciplinary traditions. Just as technologies such as the microscope or DNA sequencing have each revolutionized research, education, and applications in biology, and in so doing made the work of Linnaeus and Darwin ever more important to modern science and to modern medical applications, so too the new geographic technologies such as GIS and interactive GPS/GIS have the potential to extend research horizons in traditional areas of geography and history. Seemingly disparate trends within disciplines often have a habit of eventually finding synergy and compatibility, as is increasingly the case of GIScience and critical theory in geography.[5] Similarly, history's traditional methods and strengths will not likely be threatened by historical GIS and geospatial technologies, but rather will find new intellectual terrain and extended research frontiers in which to operate.

These new engagements between geographers and historians hold collaborative interdisciplinary promise beyond the sharing of methods and technologies. By their very nature, the geographic technologies such as GIS also facilitate interdisciplinary research collaborations with other colleagues throughout the university. As Stanford University historian Richard White notes, "Recent advances in geographical information system technologies promise a way out of the problems that historians have faced in tackling the historical construction of space. These new techniques allow scholars to explore spatial variation without getting boxed in by a single cartographic representation. ... GIS creates the possibility of extending spatial analysis beyond the local scale ... We can tell more complex stories more clearly and coherently ... Spatial history not only creates the possibility of history becoming more collaborative, it virtually necessitates it."[6]

For geographers it is also clear that geographic technologies are integral to the intellectual core of our discipline, and that an understanding of their evolution and impact is essential to understanding the history and philosophy of geography as a discipline. Our ways of thinking and doing as geographers have been and will continue to be intertwined with advances in technologies which, while neither intrinsically good nor bad, in the best of hands help us to see beyond, to integrate the disparate, to visualize complexity, to communicate the remarkable commonplace as well as the merely extraordinary, to bridge continents and disciplines, and to create understanding of and in our world. The same might also be said to be relevant to the goals of history.

As historians are now discovering, historical GIS allows historians and related scholars to ask new questions from new perspectives, and to integrate and analyze large amounts of historical data that previously remained impervious to traditional historical methods alone. As Civil War historian Ed Ayers demonstrates in the following essay, *Mapping time*, which examines the location and frequency of lynchings in the American South over time, the use of geographic technologies such as GIS can have a profound effect on historical research, and create radically new understandings and interpretations of our past.

Often coupled with traditional methods of historical and geographical analysis, the number of major international historical GIS projects continues to expand. Examples of such projects include the China Historical Database; the Great Britain Historical GIS Database; the Tibetan and Himalayan Digital Libraries; the Hawai'i Island Digital Collaboratory for Humanities and Science; the U.S. National Historical Geographic Information System; the Holocaust Historical GIS Project; and many others. Although the field continues to grow, scholars still face many core theoretical and conceptual challenges. For example, in this section's concluding essay, *What do humanists want?...*, Peter Bol outlines the complex ontological and linguistic issues still to be resolved to understand the meaning and significance of geographic places and of place names over time and space, and their correlation with events of varying duration and geographic location across multiple temporal and spatial systems of measurement.

Other essays in this section explore the contours and emerging possibilities of these new interactions between geography and history, and illustrate the philosophical, cultural, methodological, and technical work underway, based on cutting edge research in the Geohistories and various specific project implementations.

Trevor Harris and colleagues, for example, note the fundamental tension between history's narrative traditions and geography's spatial and analytical approaches to understanding the world and events, and support approaches to interpreting experiential and personal histories into Historical GIS research in their essay *Humanities GIS: place, spatial storytelling, and immersive visualization in the humanities.*

In his perceptive essay, *Spatiality and the social web: resituating authoritative content*, Ian Johnson describes challenges to authoritative content in History and Geography, how these converge in Historical GIS research, and provides insight into the opportunities and obstacles which the rise of user-generated content present. Karen Kemp explores the representation of cross cultural knowledge in the context of geographic information science and historical GIS projects, through a geocollaboratory with indigenous peoples that is experimenting with ways to interact native Hawaiian epistemologies, cultural traditions, landscape concepts, and other ways of knowing and engaging the environment with concepts of Western science and, in particular with GIScience and systems. Amy Hillier shows how Historical GIS may be used in education and in interactive community outreach, based on her work in developing the W. E. B. Du Bois historical GIS education project in Philadelphia.

Authors von Lünen and Moschek illustrate the use of GIS for cultural-geographical and historical analysis of Limes, or Roman Empire fortifications along the Rhine Valley of Southern Germany, while Robert Schwartz, Ian Gregory, and Jordi Marti Henneberg provide a compelling example of the ways in which National HGIS projects can generate a font of historical research at the local and regional scales. In this case study, the Great Britain Historical GIS provides the foundation and context for detailed research on the historical development of Wales.

Many of the prominent scholars represented in the Geohistories section of this book were also present at a seminal Geography and Humanities Symposium organized in 2007 by the Association of American Geographers (AAG) and co-sponsored by the University of Virginia.[7] Discussions at the Symposium underscored the pressing need for a volume such as this, examining the experimental and growing edges within new zones of convergence between geography and the humanities. Also recognized at this Symposium was a need for the creation of ongoing interactive "places" and resources for those engaged in these transdiciplinary zones, including in particular an online forum for geographers and historical researchers who are exploring geographic technologies to address the significant challenges involved in creating GIS-based historical archives, or who are conducting historical research using GIS. Many established national historical GIS programs, as well as incipient HGIS research projects around the globe remain isolated with no common thread to pull them together, or to link comprehensively with other efforts underway. To address these needs, the AAG, with support from the National Endowment for the Humanities (NEH), has created a Historical GIS Clearinghouse and Forum[8] to provide a global inventory of existing HGIS projects and programs, and an interactive discussion forum for leading researchers and students in this area of research. The AAG HGIS Clearinghouse and Forum provides an exchange venue to facilitate standards development, to allow other interested researchers to draw on best practices or identify common pitfalls to be avoided, and to discuss topical and regional interests as well. Most importantly, it also encourages and fosters on-going critical thinking and debate among geographers and historians around the many key theoretical and conceptual issues still unresolved regarding the fusion of spatial and temporal methods and information in our research.

The Historical GIS Clearinghouse and Forum is maintained by the AAG, but available also via links from many other university and related websites, including those of the National Endowment for the Humanities (NEH) and the Library of Congress. By funding this project, the NEH has placed itself at the forefront of an important and growing trend at the nexus of historical and geographical research, and we are grateful for their support. Readers of this book who wish to continue to follow or participate in these collaborations between historians and geographers are invited to use this website as a portal to understanding new developments in this realm.

So what will be the future of our collective Geohistories interactions? The precise trajectory of such dynamic and creative processes as those in the Geohumanities is hard to project with any certainty, but the potential is clearly significant. Historian Bodenhamer and others see outcomes for history as both a means and as a medium. As a means, historical GIS would serve as a powerful new tool for analysis of historical evidence, giving geographical context and depth to their interpretation. As a medium, he suggests that, "historical GIS offers the potential for a unique postmodern scholarship, an alternate construction of the past that embraces multiplicity, simultaneity, complexity, and subjectivity." It may have its greatest impact "not as a positivist tool but a reflexive one: integrating the multiple voices and views of our past, allowing them to be seen and examined at various scales; creating the simultaneous context that historians accept as real but unobtainable by words alone ... In sum, historical GIS offers an alternate view of history through the dynamic representation of time and place within culture. This visual and experimental view fuses qualitative and quantitative data within real and conceptual space."[9]

Amidst much speculation and intellectual hard work, however, there is one certainty. Historical GIS is destined to be the common ground of a long marriage between the disciplines of geography and history, if for no reason other than that so many conceptual, research, and mutual interdisciplinary challenges remain to realizing its full potential. Creating massive national scale historical GISs, or even the equally intriguing tiny but highly textured neighborhood and personalized historical GIS projects, are engagements still in the embryonic stages of their development. The new historical methods, narratives, and stories they will generate are just now becoming apparent, and we can only imagine today what the ultimate outcome of this will be in the years ahead. The authors of this section address the evolving and highly creative new ways of interacting between these two old friends, geography and history, coexisting and in conversation for millennia as two of the oldest disciplines of study and research, as we embark on revitalizing and re-energized new collaborations, fresh with promise and laden with enormous new potential for understanding not only our past, but each other.

Notes

1 D. W. Meinig, *The Shaping of America: A Geographical Perspective on 500 Years of History*, 4 vols, New Haven: Yale University Press, 1986–2004; C. E. Colten and P. J. Hugill, T. Young, and K. Morin, "Historical Geography," in G. L. Gaile and C. J. Willmott (eds), *Geography in America: At the Dawn of the 21st Century*, New York: Oxford University Press, 2003.

2 F. Braudel, *Civilization and Capitalism, 15th–18th Centuries*, 3 vols, London: HarperCollins Publishers, 1979; F. J. Turner, *The Significance of the Frontier in American History*, 1893.

3 D. E. Cosgrove, *Social Formation and Symbolic Landscape*, Madison: The University of Wisconsin Press, 1998; S. Schama, *Landscape and Memory*, New York: Vintage Books, 1996; S. Daniels, D. DeLyser, N. Entrikin, and D. Richardson (eds), *Envisioning Landscapes, Making Worlds: Geography and the Humanities*, London: Taylor & Francis, 2011.

4 D. J. Bodenhamer, "History and GIS: Implications for the Discipline," in A. K. Knowles (ed.), *Placing History: How Maps, Spatial Data, and GIS Are Changing Historical Scholarship*, Redlands: ESRI Press, 2008, pp. 222, 225.

5 D. Richardson and P. Solis, "Confronted by Insurmountable Opportunities: Geography in Society at the AAG's Centennial," *The Professional Geographer* 56(1), 2004, 4–11; D. Richardson, "Foreword," in S. D. Brunn, S. L. Cutter, and J. W. Harrington, Jr., *Geography and Technology*, Dordrecht: Kluwer Academic Publishers, 2004, pp. xi–xii; M-P. Kwan and T. Schwanen, "Quantitative Revolution 2: The Critical (Re)Turn," *The Professional Geographer* 61(3), 2009, 283–91.

6 R. White, "Foreword," in A. K. Knowles (ed.), *Placing History: How Maps, Spatial Data, and GIS Are Changing Historical Scholarship*, Redlands: ESRI Press, 2008, pp. x–xi.

7 D. Richardson, "Geography and the Humanities," *AAG Newsletter* 41(3), 2006, 2, 4.

8 http://aag.org/historical_GIS/index.htm (accessed May 17, 2010).

9 D. J. Bodenhamer, *op. cit.*, pp. 230–31.

23

Mapping time

Edward L. Ayers

Our tools for dealing with terrestrial space are well-developed and becoming more refined and ubiquitous every day. GIS (Geographic Information System) has long established its dominion, Google permits us to range over the world and down to our very rooftops, and cars and cell phones locate us in space at every moment. It is hardly surprising that geography and mapping suddenly seem important in new ways.

Historians have always loved maps and have long felt a kinship with geographers. The very first atlases, compiled 600 years ago, were historical atlases. But space and time remain uncomfortable – if ever-present and ever-active – companions in the human imagination. Maps, even in the newest technologies, grant us freedom to move in space by fixing a moment in time.

Historians reciprocate: we hold space constant whenever we move people across time. Indeed, asked the great historian Hugh Trevor-Roper, "How can one *both* move *and* carry along with one the fermenting depths which are also, at every point, influenced by the pressure of events around them? And how can one possibly do this so that the result is readable? That is the problem." Modernist and postmodernist novelists routinely play with time and space, of course, and moviemakers jerk us all over the place temporally and geographically, but historians tend to tell our stories straight. We need our readers to know where they are in space and time and we need to keep the relationship between the two as clear as we can. That's our job, a responsibility not unlike that of geographers.[1]

It is possible that people simply do not have the neural bandwidth to deal with space and time simultaneously, in the same cognitive space, without the tricks of narrative or the aid of machinery. We tend to think of cause and effect in linear forms because that is how we get through life. We time travel constantly in our heads, telling ourselves stories from the past one more time to try to figure out what went wrong or what we might do differently next time. But we seem able only to tell ourselves one story at a time. We cannot sustain images of simultaneity or envision complex processes without at least writing things down or, better, drawing pictures – or much better yet, creating moving pictures. Scientists can do this no better than historians or geographers.

Scott Nesbit, Nathaniel Ayers, and I have been experimenting to see if new technologies might not permit us to approach this challenge in a new way. We began by trying to convey the unfolding patterns of the complex historical processes in the massive dislocations of the American Civil War and emancipation.[2]

The Civil War seems the least mysterious of subjects. Everyone thinks they know what caused the Civil War and what it means. Yet no one, abolitionist or secessionist, enslaved person or politician, expected a war that would kill the equivalent of 6 million people today and make the largest change in the history of this nation: the immediate emancipation of 4 million people who had been held for centuries in perpetual bondage. We have tamed too often that vast conflagration with a few stock images and easy explanations.

To throw us off balance a bit, to show the limitations of our formulaic understandings of the geography of the Civil War, we have focused on a boundary, a border, at the center of our work. The Shenandoah Valley was crucial to the entire Civil War, for it was the avenue that stretched from north to south, the route to and from Antietam and Gettysburg. We chose two places in the Valley, one on each side of the Mason-Dixon Line, and followed them through the war from John Brown's raid to the end of Reconstruction, a twinned microhistory of the entire Civil War. We created a vast digital archive that included massive evidence about all the people who lived in those two communities – black and white, male and female, soldier and civilian.

Putting the border in the middle of the story disrupts the easy stories we have been taught of a modern North against an agrarian South, of past against future. It forces us to confront just how weird this war was, how amazing it was that the South, a place larger than Continental Europe, could almost overnight forge a nation state and an army that could hold off the richest country in the world for four years.

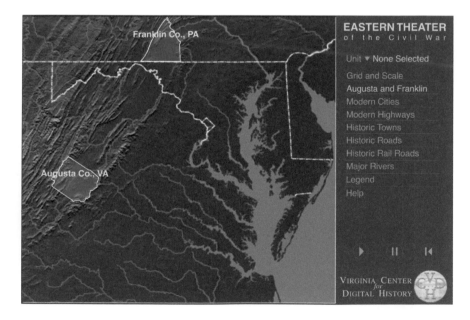

Figure 23.1 Eastern Theater of the Civil War.
Source: Virginia Center for Digital History.

To explain this process, I use the phrase "deep contingency." Only a process that reached throughout a society, deep into its ideology and psychology and even theology, could explain how millions of people could suddenly pivot into new identities, deep enough to kill for. Only contingency could explain how unexpected events, such as the *Dred Scott* decision and John Brown's raid, could lead to unforeseen consequences such as the crystallization the Republican Party and the election of Abraham Lincoln. Only depth and surprise could explain how two places so alike in every way but one – one had slavery, and one did not – redefined themselves so quickly and thoroughly. Deep contingency shows history moving tectonically, vast plates suddenly shifting, consequences connecting continents, people finding themselves standing on new landscapes of politics and culture and self-understanding.

Emancipation, the great and unlikely outcome of the war that began in 1861 with no mention – or hope – of ending slavery instantly and in place, embodied another deep contingency. Abraham Lincoln said he would leave no card unplayed to save the Union. He soon discovered, thanks to the bravery of escaping enslaved people, that undermining slavery in the Confederacy would be a powerful accompaniment to military action. A year and tens of thousands of deaths into the war, Lincoln proclaimed the Union effort a war to destroy slavery in the South, an act he could not have imagined only a year before.

Even as the war consumed a generation of young men, slavery's future remained uncertain, the consequences of emancipation undetermined. Indeed, while the coming of the Civil War was like a lens, focusing everything that came before in what we now call the "antebellum era," emancipation was like a shattered mirror. Every family, black and white, followed its own path through these years, picking its way through the broken images and sharp edges of history.

Emancipation might be imagined as something like the Big Bang. We have to follow the patterns of emancipation the way astronomers trace the expansion of the universe, extrapolating mass, size, speed, force, and dark matter from observable if faint points of evidence and perturbations of expected patterns. Just as we can no longer see the Big Bang we can no longer see emancipation, even though it occurred under our feet less than 150 years ago. We have only faint traces on pieces of paper, lost markings on the landscape. We have only scattered and incomplete testimony from the people making themselves free. Those 4 million people tend to dissolve into images of figures waiting for history to happen to them.

To capture the first decisions of freedom, we began with standard techniques of GIS to locate people on landscapes and put them down one layer after another: of race, of wealth, of literacy, of watercourses, of roads, of railroads, of soil type, of voting patterns, of family structure. We located newly freed people on the landscape, with greater detail than anyone else has ever attempted. We mapped churches, schools, and social networks. We mapped the relationships that newly freed people announced to the Freedmen's Bureau, showing how their marriages stretched far back into the darkness of slavery.

To set them in motion, we have begun to experiment with forms of mapping that are more fluid, dynamic, and cinematic. My colleague Cindy Bukach, a cognitive neuroscientist, tells us that "our perceptual system is not designed to perceive the

Augusta County

Race of household head, according to census:

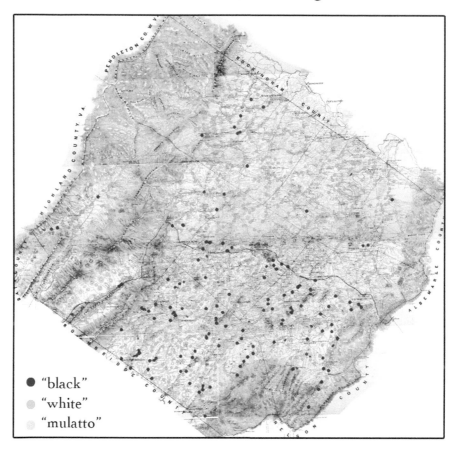

● "black"
○ "white"
○ "mulatto"

Figure 23.2 Augusta County, Race of Household Head.
Source: Census Bureau.

passage of time, but it is designed to see the movement of objects through space. By converting time to motion, we can visualize the passage of time (as one watches the hands of a clock move). This same principle can operate not only on the scale of seconds, minutes and hours, but also on the scale of years."

Our brains like seeing these patterns, it seems, because maps of time take advantage of our "multimodal cognitive system." Motion and temporal sequencing are key to our constant triangulation of causation. "These dynamic patterns can be simultaneous, allowing inferences of common causes, or they can be sequential, suggesting causal relationships," Bukach points out. "Motion captures attention. Displaying

Population of African Americans, 1880

Population of African Americans, 1900

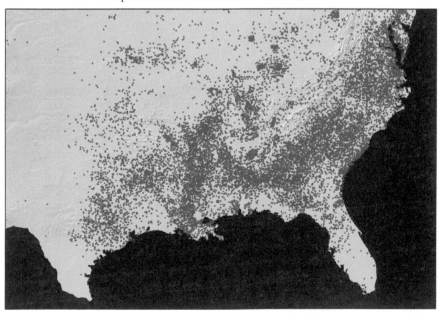

Figure 23.3 Population change of African Americans, 1880 to 1900.

historical information in a motion map guides the viewers' attention to changes in a somewhat automatic way, guiding even the most naïve observer to perceive the relevance of emerging trends and relationships."[12]

The techniques we have used thus far are simple – morphs and dissolves – but they represent something closer to the moving images of historical processes we imagine when we try to picture vast numbers of people enacting significant changes. They are something like time-lapse photography of plants opening, of leaves unfurling in particular shapes, of vines reaching to grasp a nearby structure, of diseased or thwarted processes. Or perhaps they are like models of streams and rivers, with currents folding back on themselves, of flows around submerged objects. They cannot move on the pages of a paper book, so the examples that follow need to be understood as stills from moving images that can only be seen live in electronic environments.

Let's look at a few stills that focus on the period between Reconstruction and the Great Migration. In most accounts of US history, those decades are lost in African American history. They are the time simply of sharecropping, of immobilization, of waiting for history to happen. But let's look at the pattern of population movement between 1880 and 1910.

Two static maps, from 1880 and 1900, for example, might suggest that nothing much happened in that time.

The great majority of black Americans remained black Southerners. And the great majority of them lived where their parents had lived in slavery, in a vast band from the largest slave state – Virginia – to the Mississippi River and beyond. But playing the film slowly, and moving over the same time with several passes, we see that as many black people moved during these years as they did during the Great Migration of World War I and following. The difference was that they moved *within* the South, to the very places we think of as being the Old South (the Delta, for example) but that were in fact new places for black people. Texas, Arkansas, Louisiana – these were places of promise. We see a dispersion and then a reconcentration, an escaping from the South into the West and the North. And we see a large population growth, as the maps of population density grow brighter and more intense.

We also see something that doesn't fit the usual stories: the emergence of cities. As it turns out, the New South period saw a growth of small towns and cities faster than that of the United States as a whole. There were more small towns in the South a hundred years ago than there are now. Look at this very different kind of map, one that looks more like what you might expect a historian or a social scientist to show (Figure 23.4).

Moving back and forth across time, we see patterns of great subtlety that would be hard to see in other ways. Entire regions of the South turn into places laced by small towns. We see the Carolina Piedmont, now the home of Charlotte, taking shape around textile mills. We see Florida and Texas change quite substantially. We see the cotton belt changing less rapidly than the areas to its north and south.

We can see the reasons for this change on this map (Figure 23.5).

In 1870, the South had many fewer rail lines than the North (even though the South was still the third most railroaded society in the world, after the United States and England, in 1860). But when the movie plays we see that the South is more

Percentage of Population in Small Towns, 1880

Percentage of Population in Small Towns, 1900

Figure 23.4 Percentages of People in Small Towns, 1880 and 1900.

Railroad Construction, 1870

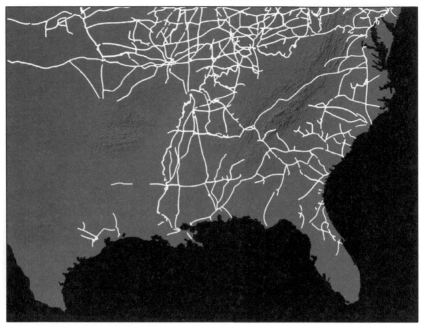

Railroad Construction, 1890

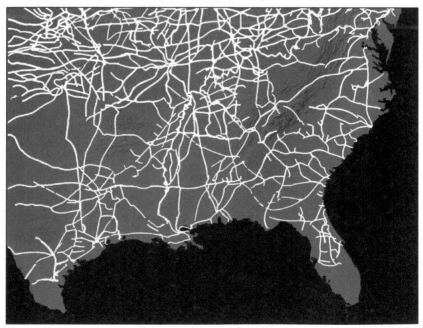

Figure 23.5 Growth in Railroad Construction, 1870 to 1890.

transformed than the North in these decades of the Gilded Age. During a time when supposedly not much was happening in the South, rail lines are racing through Texas, between the North and the South, through the coal fields of Appalachia, into the new citrus groves of Florida, up and down the Mississippi. By 1890, 9 out of every 10 Southerners lived in a county with a railroad. The scale, suddenness, and complexity of this bright lattice of rail lines is more compelling and its effects more comprehensible if we can see it unfold before us. If we overlay the small-town map on the railroad, we see a strong correlation between town growth and railroads.

Two other maps show that we discover things with dynamic mapping that we could not see otherwise. In the first (Figure 23.6, top), we have counted the number of reported lynchings by subregion.

This period was the heyday of this incredible brutality, in which black men were seized and murdered somewhere in the South virtually every day. This map shows some surprising patterns: lynching rates were not highest in the areas with the most black men, nor in the notoriously brutal cotton belt, but rather in the Gulf Coastal Plain, in the mountains of Appalachia, and in the newly settled plains of northern Louisiana.

In the second map (Figure 23.6, bottom), we show where the largest proportion of black Americans managed to acquire the most land.

Looking at the two maps in conjunction, we see a surprising juxtaposition: the areas with the most lynchings were also some of the areas with the greatest amount of black landholding. The areas of the greatest terrorism, in other words, were the areas where black people, despite all the odds against them, managed to save enough money, through heroic means, to buy small pieces of land.

So where does this point us in our understanding of geography and history and the other humanities? How might we use maps for discovery, not just the representation of what we already know? How might we combine the obvious strengths of geographic understanding with the traditional strengths of the humanities – the focus on the ineffable, the irreducible, the singular? How might we integrate structure, process, and event?

Perhaps we can return to the notion of deep contingency and use a metaphor from GIS, that of the "layer." In GIS, we imagine layers for topography, for rivers, for people. That metaphor is a fiction, of course, since the layers continually interact and the "top" layer of humans constantly changes the "bottom" layer of landscape. But it is a useful fiction, since it reminds us of the structural depth of time and experience. GIS is about patterns and structures; history is about motion. By integrating the two, we can see layers of events, layers of the consequences of unpredictability. Deep contingency is a contingency that penetrates all those layers.

The great historian Marc Bloch wrote that time is the "very plasma in which events are immersed, and the field within which they become intelligible."[3] Historians are obliged to deal with time. The beauty and utility of history is that it deals with the all-important fourth dimension in which we live, and of which we humans, alone of living things, are aware. With history, time can be mapped as it cannot be in our own lives – and history is the only tool we have to even guess at where our location in time might be.

Rates of Lynching in the New South

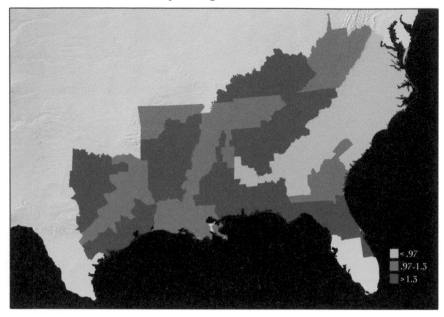

Black Landholding in the New South

Figure 23.6 Comparing Black Landholding to Rates of Lynching in the New South.

Despite – or perhaps because of – the sometimes uneasy relationship between space and time in our neural machinery, deepening our understanding of one dimension deepens our understanding of the other. Combining them, we might be able to glimpse the plasma of time in which we move and live.

Notes

1 Quoted in K. Thomas, "A Highly Paradoxical Historian," *New York Review of Books*, April 12, 2007, p. 56.
2 For electronic versions of the maps that follow, see www.vcdh.virginia.edu/emancipation/.
3 Personal communication from Bukach to Ayers, November 15, 2007.
4 J. L. Gaddis, *The Landscape of History: How Historians Map the Past*, New York: Oxford University Press, 2004, p. 29.

24

Humanities GIS

Place, spatial storytelling, and immersive
visualization in the humanities

Trevor M. Harris, Susan Bergeron, and L. Jesse Rouse

Introduction

There have been longstanding linkages between geography and the humanities that
go back at least to the early writings of Carl Sauer and his focus on cultural landscapes, history, and archaeology.[1] Perhaps most prominent of these traditions has been
in the subfield of historical geography and its exploration of geographies of the past
through the conjunction of history and human geography.[2] Despite the changing
centrality of historical geography within geography since the 1970s, the historic links
to the humanities have remained important to the work of historical geographers. In
addition to the lynchpin role of historical geography, the traditions, methodologies,
and domain areas of geography have also touched the humanities in many other ways
and especially in archaeology, literature, history, and religion.[3] These encounters
reflect more the tendency of geography and geographers to cross disciplinary boundaries rather than any underlying symbiotic melding of the disciplines. Thus mapping
applications *per se* have provided a focal point of exchange and dialog between geography and related disciplines, but these interfaces have not necessarily provided cor-
ollary substantive theoretical developments.

In recent years in the United States there has been a reawakening of interest in the
role that geography could play in humanities disciplines; witness the 2007 AAG
Geography and Humanities Symposium and this volume for example. A significant
part of this resurgence and curiosity in linking the humanities and geography is due
to the impetus provided by the availability of Geographic Information Systems (GIS)
and Internet-based mapping systems. These geographic technologies have piqued the
interest of a growing number of humanities scholars,[4] though whether this rekindled
interest in mapping would have come about without the technological legitimation
of GIS is more difficult to discern. Some geographers have equally been attracted to
the challenges that arise from applying GIS in humanities disciplines. In this essay we
suggest that the interface of geospatial technologies, geography, and the humanities
provides unique challenges but also opportunities. Specifically, we argue that in order
to address some of the significant data and epistemological issues associated with

interfacing a predominantly positivist science with disciplines that have predominantly humanistic traditions, scholars might draw upon the power of geovisualization and in particular GIS-informed virtual reality and gaming engines to address core issues of place, immersion, and spatial storytelling. Together these approaches could contribute to a uniquely humanities GIS.

Geography, space, and place in the humanities

The humanities are undergoing change through the impact of technology. The use of terms such as digital humanities, visual history, humanities GIS, and historical GIS are prevalent. In humanities disciplines that have explored GIS, this digital revolution has been accompanied by what has come to be called the "spatial turn." The spatial turn represents awareness not only of GIS but of the important role that space plays in human actions and events. Much of the interest in GIS has largely revolved around a (re)discovery of the power of the map. The humanities have long been at risk of treating space, the backdrop to all human behavior and events, as being neutral – a spatial vacuum – an isotropic backdrop to human affairs. Indeed, a perusal of many maps incorporated in humanities texts would imply that events take place in landscapes seemingly devoid of any terrain, hydrology, infrastructure, human culture, or other geography. Recently, however, some humanities scholars have found that mapping phenomena and cultural objects provides additional insights not previously known. Although the Electronic Cultural Atlas Initiative (ECAI)[5] is not the only example of this interest in mapping humanities and cultural data, its evangelizing work as certainly helped stimulate a global awareness of the value and potential contributions of GIS and Internet mapping in the humanities. Other forums such as the Social Science History Association and several independent initiatives such as the Virtual Center for Spatial Humanities have likewise documented the adoption of GIS by a number of humanities scholars as a tool in their analytical arsenal.

While GIS has had demonstrable success in the physical sciences and in spatially integrated social science,[6] its usage in the humanities has lagged these early adopters. GIS uptake in the humanities has been disparate, uncoordinated, and largely project and application driven. GIS has been used most effectively in applications that have drawn on the traditional strengths of the technology: inventorying, mapping, and spatial data management. As a result, there has been an early heavy emphasis on boundary demarcation, gazetteers, mapping, historical atlases, and the creation of important foundational coverages such as the historical census geographies that support the greater use of major databases.[7] Contemporaneous with humanities scholars cautiously embracing and utilizing GIS, some geographers have been active in exploring and adapting GIS to the unique needs of humanities disciplines. The primary focus, however, on GIS usage as a digital mapping system has left much of the analytical functionality of the systems largely untested in the humanities arena.

This lag in GIS uptake in the humanities is not without reason, for the humanities raise fundamental epistemological and ontological issues for GIS applications and these issues are challenging to GIS technology and to practitioners alike. One important underlying issue is that the geospatial analytical power of GIS does not easily lend itself to the ways in which humanities scholars explore or interpret phenomena in their research. As a result, using GIS as a lens to understand how space influences human behavior is challenging in the humanities and arguably more so than in the physical and social sciences. A second concern is that whereas the strength of GIS lies in analyzing space through digital mapping, database management, and spatial analysis, the humanities are predominantly focused on issues of place, sense of place, and the uniqueness of particular places. These differing perspectives, reflecting positivist and humanistic approaches to the geography of space, have significant implications when dealing with the application of GIS in the humanities.

Third, the humanities are dominated by a heavy reliance on qualitative data and yet GIS is based on a quantitative representation of the world involving spatial primitives and the topological relations between them. Incorporating and accessing qualitative information such as photographs, paintings, oral history, moving images, text, and sketches are less easily accomplished in GIS. Fourth, scholars in the humanities tend to express ideas in a linear format as characterized for example by an emphasis on linear text and the narrative format of the book. Work in digital environments, however, predominantly emphasizes non-linear modes of expression and analysis of which the World Wide Web is an archetypal example. Thus, merging these two traditions is challenging. Fifth, the humanities place heavy emphasis on narrative and essentially a storytelling form. This is one reason why the general public will take a history book for vacation reading and yet shy away from other academic texts for leisure reading. This storytelling form is again rather at odds with the digital spatial analytical nature of GIS. Finally, the humanities seek to pursue multiple and equivocal perspectives on issues and retain the ambiguity and pre-emptive resolution to issues that characterizes the sciences and social sciences. This ambiguity is again not easily represented or addressed in GIS and, when coupled with the issues outlined above, demonstrates why GIS to date has found more rapid adherence in science and social science disciplines than it has in humanities research.

The lag in the uptake of geospatial technologies in the humanities reflects these ontological and epistemological challenges related to how disciplines view, explore, and understand the world in different ways. These are critical issues that permeate the development of meaningful synergies between humanistic disciplines and geospatial technologies. For these reasons we suggest that while the early use of GIS in the humanities has focused almost exclusively on GIS as a *system*, this emphasis must also be accompanied by a consideration of GIS as a *science*. It is in this latter area that the greatest opportunities for meaningful reciprocity between the use of geospatial technologies in the humanities and the development of a Humanities GIS resides. The GIS and Society and Critical GIS discourses have certainly provided a valuable context for exploring many of the issues associated with the epistemology, assumptions, and biases of GIS.[8] The socio-theoretic critique of GIS explored the assumptions that were implicitly or explicitly involved in the way in which GIS is used to represent nature and society. GIS operates on a digital abstraction of reality obtained through the observation and testing of phenomena using the scientific method. In so doing, particular types of data, information, representation, logic systems, and ways of knowing are privileged over others. Whereas much of the focus of the GIS and Society discourse has centered on Participatory GIS, these substantive issues are equally relevant in the application of GIS in the humanities.

In the context of these issues we suggest that the early adoption of GIS in the humanities may have been at the loss of exploring other spatially informed technologies that could equally contribute to humanities analysis and, in conjunction with GIS, may more closely align the technology to the analytical and methodological traditions of the humanities. In this respect we are reminded of the story of a man who, on returning home late at night, came across another man on his hands and knees under a street light searching for something. When asked what he was searching for, the man replied that he was looking for his car keys. When then asked if he had dropped them in the area in which he was searching his reply was, "I don't know, but this is where the light is." This metaphor reflects the early applications and inroads of GIS into the humanities. GIS has shed light on some aspects of the spatial humanities and has enabled scholars to pursue the spatial turn in the humanities in ways not previously possible. However, these are very early days and we suggest that other geospatial technologies such as geovisualization, immersive and experiential visualization, and serious gaming among others, allied with GIS may have equal potential to contribute to the humanities. The emphasis in the humanities on qualitative data, linearity, narrative, storytelling, and multiple perspectives does not merge easily with the heavily positivist and scientific emphasis in GIS. Because of the conceptualization of space in the humanities as humanized place or occupied space, we suggest that the pursuit of GIS-enabled spatial storytelling, immersive visualization, and experientialism provides a potentially powerful analytical environment that goes beyond traditional uses of GIS, and offers a contextual and spatialized understanding of place.

Space, place, and humanities GIS

How, then, might geographic technologies emulate and capture the nature and essence of place that are so powerfully evoked in literary and artistic works? Take a description by Rebecca Harding Davis, a leading American literary figure of the nineteenth century, who lived in Wheeling, then Virginia and now West Virginia, who wrote evocatively about the industrializing nature of the antebellum town:

> A cloudy day: do you know what that is like in a town of iron works? The sky sank down before dawn, muddy, flat, immovable. The air is thick, clammy with the breath of crowded human beings. It stifles me. I open the window and looking out, can scarcely see through the rain the grocer's shop opposite, where a crowd of drunken Irishmen are puffing Lynchburg tobacco in their pipes. I can detect the scent through all the foul smells ranging loose in the air. ... Stop a moment. I am going to be honest. This is what I want you to do. I want you to hide your disgust, take no heed to your clean clothes, and come right down with me, here, into the thickest of the fog and mud and foul effluvia. I want you to hear this story.

There is a secret down here, in this nightmare fog, that has lain dumb for centuries: I want to make it a real thing for you.[9]

How would one capture the vividness of this textual description of place within the vector and raster representations of a traditional GIS? If one of the defining characteristics of place is the notion of meaning through experience, then exploring methods that can somehow represent these qualitative elements without destroying the essence of that meaning would be a valuable complement to historical and literary research.

To draw on another metaphor, consider the map of Middle-earth in Tolkien's epic *The Lord of the Rings* (www.middle-earth-map.com/). The map displays the well-known locations so central to the story and to the travels of Frodo Baggins and provides a graphical rendering of the space and topography of Middle-earth. As a spatial representation, however, the map misses much of the richness of place portrayed in Tolkien's writing and as captured so well in the film trilogy of the same name. The storytelling, the journey pilgrimage, the significance of the narrative as the story unfolds, the symbolism of place and structure, the stunning qualities of the landscapes through which Frodo traveled, the oral histories and personal experiences that dominate the story, the landscapes of memory, and the contextuality of image, text, and sound all contribute to a rich and enthralling sense of place which the film captured so well. The sense of "being there" and of being immersed in the place as the story unfolds was central to the success of the book trilogy and the films. We suggest, then, that in pursuing Humanities GIS we must go beyond the map and spatial analysis of conventional GIS applications, and explore a sensuous and reflexive system that is aware of, and sensitive to, the richness and needs of the humanities.

The role of space and place, as intertwined phenomena, are central to the works of geographers[10] and historians[11] and others who have focused on the relationship between humans and their environment through experience and perception. For Tuan, the individual's relationship with the livable world, their being-in-the-world, is at the heart of understanding place and the larger networks of places that is space. Tuan argues that aspects of human experience, while highly individualistic, are also based on biological and cognitive traits that are common to everyone. As people mature from infancy to adulthood, so do human perceptions of space and place also evolve. And yet, although we experience myriad environments throughout the world, we share innate traits that allow us to also share our experiences of space and place. In the context of prehistoric archaeology, Bender[12] argues that landscapes are not static, passive entities, but are constructed, mediated, and contested places. There is thus the "need to mesh an understanding of embodied landscapes with a political landscape of unequal power relations."

Any experience or perception of place carries with it the observer's own worldview, reflexivity, as well as social, cultural, and political meanings that have been layered onto the landscape. While the works of Tuan, Jackson, Bender, and others are compelling in their focus on the lived experience of space, place, and landscape, these ideas are difficult at best to translate into the medium of GIS and geospatial technologies. The success of GIS as a tool for synthesizing, displaying, and analyzing geographic information is clear, but the limitations imposed by the current data models and methods of representation within GIS make it extremely difficult to develop a compelling digital framework for humanistic and experiential aspects of space and place. The individualistic nature of perception and experience presents a challenging task for those who would generate digital representations of spaces and places defined by culture, behavior, or experience. If, for example, a landscape is experienced by each person in his or her own unique way, how then might another person truly share that experience? Even if, as Tuan argues, our shared human heritage can create experiences that are similar to those of others who have inhabited or experienced a landscape, how can we even know that and how might these experiences be represented through the medium of GIS? One possible answer may lie in the application of a phenomenological approach to exploring and interpreting landscapes and places that are relative, mediated, and socially produced. Phenomenology's focus on perception and experience and being in the world offers one possible framework for understanding the unique characteristics of place.[13]

A phenomenological approach to cultural landscapes through the medium of digital representation and reconstruction is fraught with challenges. These difficulties become more pronounced when the focus shifts to historic or prehistoric landscapes. Not only do the cultures and peoples that created and peopled the land no longer exist, but the realities and experiences of the inhabited landscapes have also changed and been altered by subsequent physical, cultural, economic, and social processes. It is impossible to accurately reconstruct and experience previous cultural landscapes in the same way that earlier cultures did. Not only can we never know with certainty that we have replicated the physical landscape features, we cannot access the accumulated and embedded social and cultural experiences and symbolic meanings that earlier peoples brought to their experience of space and place.

The representation of space and place within GIS is thus of central importance to the use of geospatial technologies within the humanities: "To understand place in a manner that captures its sense of totality and contextuality is to occupy a position that is between the objective role of scientific theorizing and the subjective role of empathetic understanding."[14] Place encompasses not just the physical environment but also the symbolic, the emotional, and the many meanings that are associated with place. The quantitative nature of GIS favors representations of space that are absolute, such as the familiar two-dimensional Cartesian plane of x, y coordinates. Data that cannot be easily translated into such a conception of space, such as relative space or place, become problematic. Incorporating and drawing upon qualitative representations of place such as paintings or drawings, text, photographs, audio, or video into a digital and spatially precise GIS is challenging. A number of historical GIS projects have explored the use of multimedia embedded within traditional GIS platforms as a method for incorporating and spatializing qualitative data in GIS.[15] Spatial multimedia applications can also aid the exploration of space and place in a historical context as can be seen in the *Valley of the Shadow Project* (http://valley.vcdh.virginia.edu/choosepart.html) that explored the comparative experiences of two communities in the Shenandoah Valley of Virginia at the time of the Civil War. Through spatial multimedia, the embedded source materials enable users to explore the historical data and to construct their own narrative path through the written record. Addressing issues of place requires a sensuous and reflexive GIS that is sensitive to the traditions and epistemology of the humanities. This is a significant challenge to GIS and we suggest that GIS-based geovisualization and immersive approaches might provide one avenue of inquiry worthy of pursuit in this context.

GIS, geovisualization, and immersive technologies

Within the last few years, a number of developments within the fields of computer science, Internet and web technologies, virtual reality, geovisualization, and geospatial technologies have broadened the possibilities for generating virtual representations that we suggest more intuitively address the qualitative aspects of space and place encountered in the humanities. Furthermore, affordable and powerful desktop computing, combined with advances in graphics hardware, now provide opportunities for the development of impressive rendered immersive representations that more closely emulate the world around us.[16] The development of Second Life (http://secondlife.com/) and other collaborative virtual worlds also demonstrate the power of virtual environments for the generation, exploration, and interpretation of digitally created spaces and places. In linking phenomenology to these GIS-enabled technologies we suggest that the Humanities are well placed to explore the immersive and experiential elements of place through the sense of "being there."[17]

Geovisualization is a visual-cognitive process that builds on highly interactive and dynamic graphical displays to create mental models that transform the user from a passive observer into an active participant controlling the way in which complex information is displayed.[18] The fusion of GIS, virtual reality, immersive visualization, and serious gaming technologies now allow researchers to move far beyond the static two-dimensional paper map and enable multiple viewpoints in the exploration of data through these dynamic, interactive, and multidimensional representations. These geovisual and immersive opportunities to advance understanding in the Humanities are relatively untapped to date and yet they have the potential to address some of the epistemological challenges to GIS that arise from humanistic disciplines, and indeed in science and social science.[19] Gillings and Goodrick[20] offered what was perhaps the first example of virtual reality integration with GIS, within the field of archaeology. While this approach has been utilized and subsequently expanded upon elsewhere,[21] Gillings and Goodrick proposed a theoretical basis for the integration of virtual reality into archaeology while criticizing the drive toward hyper-realism and the reconstructions of famous structures because the models are "seen to stand in an inferior position to an original referent."[22]

There have also been parallel developments outside of virtual reality that contribute to the creation of immersive environments offering differing levels of immersion. Video game hardware and software, for example, have placed into the hands of everyday users incredibly powerful graphics engines capable of generating impressive visual digital constructions of landscapes and virtual worlds. Ironically the graphical quality and power of these systems far exceed what is available to many university students within classroom and even research settings. Although video games have mainly focused on the field of entertainment, an important and growing research area that focuses on cutting-edge efforts to utilize the concepts and technology of the gaming industry to develop "serious" applications for training, simulation, and education across a wide range of industries and fields is gaining ground.[23] Commonly known as "serious gaming engines," these technologies have brought the latest computer graphics hardware and software, including virtual reality, to a variety of users in government and industry, and to simulation and training projects involving potentially hazardous activities.[24] We propose that education and the humanities fields represent additional rich settings for exploration of these technologies.

Within the last few years, a rapidly broadening range of scientific, educational, and private sector projects have leveraged serious gaming technologies and the advanced graphics and physics of game engines without incorporating the play-based aspects of games. The term "immersive learning simulation" has been suggested to distinguish these recent developments from "game-based" virtual technologies. Such immersive learning simulations provide users with the advanced simulated environments and interactive navigability that create a sense of presence within the virtual environment. As a new generation of digital natives matures, so their expectations and willingness to explore the unfamiliar with familiar technologies such as digital gaming environments opens up possibilities not readily embraced by previous generations of students and scholars. A critical, self-aware approach to these technologies is important for it is difficult to separate the layers of meaning that are embedded in any landscape, space, or place, and recombine them in ways that are meaningful within the context of experiencing landscapes in digital environments. The utilization of advanced digital technologies to explore virtual representations of space and place must critically examine the social significance of such representations. Bender[25] argues that as people interact with inhabited spaces and places we are assigning meaning and "engaging and re-engaging, appropriating and contesting the sedimented pasts that make up the landscape." Likewise Tuan[26] suggests that, "Space and place are basic components of the lived world; we take them for granted. When we think about them, however, they may assume unexpected meanings and raise questions we had not thought to ask." These arguments suggest that any representation of physical space is mediated by our own experiences and worldviews, both in the creation and in the experiencing of such digital landscapes. Consequently, while exciting possibilities exist for exploring the past, it is important to be aware of the reflexive role that users play in forming and modifying the pasts that they depict and examine.

Geovisualization and immersive technologies: a case study

To explore the fusion of GIS, geovisualization, and immersive simulation as a means to understand place and spatial storytelling we have created a historical case study based on an area well known to us in Morgantown, West Virginia. *Virtual Morgantown* is a project focused on a period of rapid industrialization in this area at the turn of the nineteenth century and provides a vignette of how GIS and 3D graphics, modeling, and serious gaming technologies can be used to create an immersive and interactive environment for data exploration and the study of place. Although the term 3D is used here to denote visualization in three dimensions, in reality it is a 2.5D representation of the world in that the third dimension is not an independent variable. Now in its fourth iteration, *Virtual Morgantown* began as a 2D web-mapping application that focused on the industrial heritage of Morgantown and the rapidly gentrifying Wharf District. The interactive web-mapping application utilized a multimedia enhanced GIS platform to provide spatialized access to embedded qualitative historical media such as text, audio, narrative, photographs, and images through the map portal of the GIS. Features on the map were hot linked to URL addresses that allowed these media to be easily retrieved. The base geographic information was derived from the Sanborn Fire Insurance Maps of the early 1900s, from which building footprints, heights, and materials were identified, along with streets and sidewalks. A second version migrated to a desktop computer and expanded beyond the initial areal extent to include historic districts on the National Register as well as embedded historical census information (Figure 24.1). A third generation system migrated to 3D visualization and again expanded the study area to include the downtown area of Morgantown, as it may have appeared between 1898 and 1906. This representation was based on a combination of the Sanborn maps and photographic evidence.

The well known ESRI ArcScene module within the ArcGIS desktop software package provided the base software platform for early development work and enabled direct access from the 3D scene to the GIS database (Figure 24.2). However, the system struggled when rendering the base terrain and surface data and over 400 structures, trees, and street furniture. This rendering issue, along with the need for a more robust and dynamic connection to multimedia, led to the current iteration of *Virtual Morgantown* that is based on a pseudo 3D gaming engine. Game engines offer significant technological and graphical advantages over GIS visualization software currently offered through vendors and these contribute to superior navigation, interaction, immersion, customization, and visual representations. Software was developed to support "collision" detection whereby movement through a scene could generate access to qualitative data resources through proximity to a building or a street or an area. At a collision point, multimedia pop-up resources become available for viewing (Figure 24.3). A growing number of researchers are recognizing the utility of serious gaming engines in scientific research. Indeed, the rapidly broadening range of projects in these game engines, especially those that support visualization and navigation based on a first person perspective, provide a sophisticated user interface to the virtual world that, importantly for the Humanities, directly connects the user to the visual experience of the scene. Navigation tools in these engines tend to conform to a standardized model so that users may easily migrate between games. *Virtual Morgantown* draws upon these navigation models to enable system interaction and movement within the immersive learning simulation through keyboard, mouse, and game controller. When navigating the system, users gain considerable insight interacting with the virtual image and experience the place as it might have been over a century ago.

Much of the current effort on *Virtual Morgantown* has been expended on establishing the system, generating the virtual world, and attaching historical information such as images, texts, narrative, audio, and video to the scene objects. We have experimented

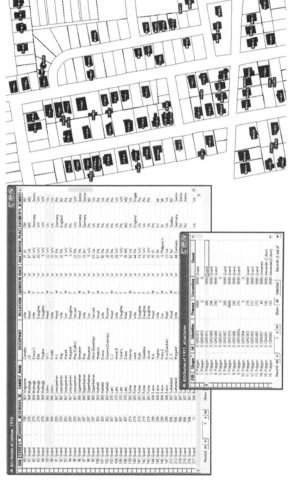

Figure 24.1 Spatially embedded multimedia in a GIS.

Figure 24.2 Virtual Morgantown in ESRI's ArcScene.

with extending the sensual nature of the system through the use of ambient sound and smell. For example, a beating heart has been used to provide a sense of fear and emotion in the historical portrayal. As a user moves through a scene, the sound of a beating heart quickens to indicate proximity to areas that have been identified by writers of the time as dangerous or places to be feared. In other instances we have experimented with smell as a way of triggering other senses that are well recognized as being powerful in evoking memory and a sense of place.

Figure 24.3 XNA image and GIS generated media, in this case a census record, triggered by spatial proximity to the building.

Other forms of interaction with the system and its data content are being explored such as connections with avatars within the environment. These avatars provide guidance and information about the place and can be interrogated in a variety of ways, including conversational interaction. The user is thus provided some structure in the spatial storytelling world for narrative linearity can be embedded through a variety of means including the use of signposts, avatars, and pathways. All of these are forms of interaction in a game environment that enable the "reader" to become author and to construct his or her own interpretation of the scene. These interactions with objects and avatars also contribute to levels of user immersiveness and connectedness to the virtual world itself.

Finally, the images have been projected within a Cave Automatic Virtual Environment (CAVE) to provide a full immersive experience (Figure 24.4). The CAVE is a specially designed room, including walls and a floor, onto each of which are projected images that are computationally made to appear seamless. A user or users stand in the CAVE space wearing 3D glasses and with their peripheral vision captured such that they are immersed in a lifelike sensory environment. The experience of such immersion is to be disengaged from reality and immersed within the image rather than being exogenous to the image.[27] The user can walk virtually within the displayed scene, have fly-through navigation, and interact with the imagery in real-time. The sense of immersion is a powerful psychophysical experience that renders the

Figure 24.4 Exploring the virtual landscape of Morgantown through immersion in a CAVE.

user present in the virtual environment.[28] This immersive experience is powerful in providing the user a sense of "being there," as suggested by Tuan, by drawing on the creative power of the mind to move seamlessly between the physical, virtual, and imaginary environments. In this way an analogy may be drawn to dreamscapes where the dreamer has spatial awareness and freely interacts with others in that imaginary (but seemingly real) environment. The result is to create a medium for exploring humanities phenomena and seeking a greater understanding of cultural space and place.

Humanities GIS: challenges and opportunities

In this essay we identify some of the underlying issues associated with the conjoining of GIS, geography, and the humanities. We suggest that GIS has been an important catalyst contributing to the spatial turn in the humanities and to a renewed interest in mapping cultural and historical phenomena. The somewhat late embrace of GIS by the humanities reflects underlying ontological and epistemological issues associated with integrating a predominantly positivist science with humanistic disciplines. The current focus of the humanities on GIS mapping and spatial analysis draws upon the obvious strengths of the technology, and historical gazetteers, mapping systems, historical boundary digitization, and spatial database development are an important start. But these areas represent the low-hanging fruit and there is considerably more that geography and geographers can contribute to such interdisciplinary work. Indeed if geography and GIS are to move beyond these early contact areas then GIS has to become more sensitized to the ways in which humanities scholars practice their craft. To date the exposure of the humanities to the principles, concepts, and potential contributions of geography to these disciplines has been subsumed within a focus on method and a GIS perspective of space. In this respect the focus on method and tool and lack of engagement with the concepts of geography is analogous to asking historians for a list of dates without peering into the richer ways in which historians seek to understand human behavior and events. The importance of the spatial turn in the humanities is to not only build on the methods that are central to geography but to also shift focus toward geographical concepts and spatial thinking in order to gain spatial insights not previously seen.

We also suggest that while GIS has become an important focal point in the spatial turn in the humanities it is necessary to think more broadly in terms of Geographic Information Science and the concepts that surround that term rather than such a heavy focus on GIS as system. It is in the arena of GISc that the more substantive intellectual engagement and reciprocity between geography, GIS and the humanities will emerge. We similarly argue for a broader perspective in the humanities that extends beyond a sole focus on off-the-shelf GIS toward a fusion of geospatial and related technologies including virtual reality, geovisualization, immersion, and serious gaming. Drawing upon these approaches and technologies will assist in addressing substantive epistemological issues in the humanities and contribute to a Humanities GIS that is sensuous, reflexive, and sensitive to issues of humanities place as well as geographical space.

Notes

1 C. Sauer, "The Morphology of Landscape," [1925] reprinted in J. Leighly (ed.), *Land and Life: Selections from the Writings of Carl Ortwin Sauer*, Berkeley, CA: University of California Press, 1963, pp. 315–50; C. Tilley, *A Phenomenology of Landscape: Places, Paths, and Monuments*, Providence, RI: Berg Publishing, 1994.

2 A. R. H. Baker, *Geography and History: Bridging the Divide*, Cambridge, UK: Cambridge University Press, 2003.

3 T. M. Harris, "GIS in Archaeology," in A. K. Knowles, (ed.), *Past Time, Past Place: GIS for History*, Redlands, CA: ESRI Press, 2002; B. M. MacDonald and F. A. Black, "Using GIS for Spatial and Temporal Analysis in Print Culture Studies: Some Opportunities and Challenges," *Social Science History*, 24, 2000, 505–36; I. N. Gregory and R. G. Healey, "Historical GIS: Structuring, Mapping and Analyzing Geographies of the Past," *Progress in Human Geography* 31, 2007, 638–53; ARDA, The Association of Religion Data Archives, www.thearda.com/, 2008 (last accessed May 6, 2010).

4 I. N. Gregory and P. S. Ell, *Historical GIS: Technologies, Methodologies and Scholarship*, Cambridge, UK: Cambridge University Press, 2008; Knowles, *Past Time, op. cit.*; A. K. Knowles, (ed.). *Placing History: How Maps, Spatial Data, and GIS are Changing Historical Scholarship*, Redlands, CA: ESRI Press, 2007; J. B. Owen, "What Historians Want from GIS," *ArcNews* 29 (2), 2007, 4–6.

5 ECAI, Electronic Cultural Atlas Initiative, www.ecai.org (Last accessed May 6, 2010).

6 M. F. Goodchild and D. G. Janelle (eds), *Spatially Integrated Social Science*, New York: Oxford University Press, 2004.

7 I. N. Gregory, "The Great Britain Historical GIS," *Historical Geography* 33, 2005, 132–34.

8 S. C. Aitken and S. Michel, "Who Contrives the 'Real' in GIS?: Geographic Information, Planning and Critical Theory," in *Cartography and Geographic Information Systems* 22 (1), 1995, 17–29; D. Weiner and T. M. Harris, "Reflections on Participatory Geographic Information Systems," in *Handbook of Geographic Information Science*, 2007, J. Wilson, and A. S. Fotheringham (eds).

9 R. H. Davis, *Life in the Iron Mills*, Google Books, 1861, pp. 1–2.

10 Y. F. Tuan, *Topophilia: A Study of Environmental Perception, Attitudes, and Values*, Englewood Cliff, NJ: Prentice-Hall, 1974; Y. F. Tuan, *Space and Place: the Perspective of Experience* (reprint), Univ. of Minnesota Press, 2001.

11 J. B. Jackson, *A Sense of Place, a Sense of Time*, New Haven: Yale University Press, 1994.

12 B. Bender, *Stonehenge: Making Space*, Oxford: Berg Publishers, 1999, p. 38.

13 C. Tilley, *op. cit.*

14 J. N. Entrikin, *The Betweeness of Place*, Baltimore, MD: Johns Hopkins University Press, 1991.

15 T. M. Harris and L. J. Rouse. 2001. Grave Creek Mound: an Internet GIS project. Electronic Cultural Atlas Initiative. http://ark.geo.wvu.edu/grave_creek/default.html (last accessed March 1, 2008).

16 P. Fisher and D. Unwin (eds), *Virtual Reality in Geography*, London: Taylor and Francis, 2002.

17 T. M. Harris and V. Baker, "Immersive Visualization System Promotes Sense of 'Being There'," *ArcNews*, Winter 2006/2007.

18 J. A. Dykes, M. MacEachren, and M. J. Kraak, *Exploring Geovisualization*, Amsterdam: Elsevier, 2005; P. Hodza, "Developing and Evaluating a GIS-Supported Immersive Visualization System for Soil Resource Mapping," unpublished PhD, West Virginia University, 2007.

19 J. Raper, *Multidimensional Geographic Information Science*, Boca Raton, FL: CRC Press, 2000.

20 M. Gillings and G. Goodrick, "Sensuous and Reflexive GIS: Exploring Visualization and VRML," *Internet Archaeology* 1, 1996, http://intarch.ac.uk/journal/issue1/gillings_toc.html (last accessed May 6, 2010).

21 Harris and Rouse, *op. cit.*

22 M. Gillings, "Virtual Archaeologies and the Hyper-Real: Or, What Does it Mean to Describe Something as Virtually-Real?" in P. Fisher and D. Unwin (eds), *Virtual Reality in Geography*, New York: Taylor & Francis, 2002, p. 20.

23 M. Lewis and J. Jacobson, "Game Engines in Scientific Research," *Communications of the ACM* 45 (1), 2002, 27–31; J. Dovey, and H. Kennedy, *Game Cultures: Computer Games as New Media*, Open University Press, 2006.

24 B. Bergeron, *Developing Serious Games*, Boston, MA: Charles River Media, 2006; I. Bogost, *Persuasive Games: The Expressive Power of Video Games*, Cambridge, MA: MIT Press, 2007.

25 Bender, *op. cit.*, p. 25.

26 Tuan, *Space and Place*, *op. cit.*, p. 3.

27 Harris and Baker, *op. cit.*,

28 Hodza, *op. cit.*

25

Without limits

Ancient history and GIS

Alexander von Lünen and Wolfgang Moschek

Introduction

At first sight, the Roman Limes appears to be a border in the modern sense, like the former inner-German border, the border between North and South Korea or the US border with Mexico. Over the longest time in Roman history, however, there were no such border installations (as the Limes) at the boundaries of the Roman Empire. Throughout the main periods of the expansion of the *Imperium Romanum* we find a kind of frontier that is in a way similar to that in American history with single forts or settlements facing a "terra incognita" – at least in the mind of the settlers. The question remains why the Romans all of a sudden began to build the Limes (i.e., the walls, palisades, and trenches), although there were no other states, kingdoms, bigger tribes, or threatening military forces on the other side of it? Can it be regarded as some kind of displacement activity for the soldiers, as for example J. C. Mann argued?[1] Other historians, like E. Luttwack,[2] claim to have detected a long-term "Grand Strategy" of a deep defense of the Imperial Roman borders in it. The military function of the Roman Limes has always been the main point in modern interpretation, from the beginning of the *Limesforschung* (Limes Research) in the nineteenth century until today. Newer publications started to merge this viewpoint with the interpretation of the Roman Limes as having been a controlled economic borderline.[3]

Delving into these questions, one immediately encounters the main problem of the Limes: no ancient author or inscription gives any reason whatsoever for building a wall, palisade, and trench from the north of England to North Africa from the middle of the second to the late third century AD. Nevertheless, there are few written sources available and many if ambiguous, archaeological findings – does this mean that the work of the historian or the archaeologist becomes futile?

Geography and borders

Without new sources it seems much more promising to look at the roots of historical writing (e.g., Herodotus) for inspiration, written in a time when history was enriched by geography and examined the interdependence of human actions in time and space.

To us, it appeared to be of much greater value to look into older concepts of political geography. By doing so, we do not want to encourage a backlash in historiography, but to support the call for a *geographical* turn opposed to the abstract *spatial* turn so often discussed in today's humanities. Such an approach is fitted to analyze the nature of borders, boundaries, and frontiers much better as it uses a mixture of historical and geographic methods.

A look at late nineteenth-century historiography might furthermore help to explain the misinterpretation of historians of that period. Albeit a great interest of historians in geography in that time was widely spread, most could not let go of their contemporary conceptions of geographical entities. That the Roman Limes was researched while bearing in mind the context and mentality of modern state boundaries comes as no surprise when taking into account the general obsession of late nineteenth-century European *Zeitgeist* with monolithic state boundaries and competing and quarreling nations. To the average Western European, fixed national borders as lines of demarcation against competitors appeared to be something natural. One of the few notable exceptions was the German geographer Friedrich Ratzel (1844–1904), whose geographical survey of Northern America inspired him to think of the *open* nature of *frontiers* at a time when no rivaling political entities (in the modern sense) claimed the same space.

For the Roman campaigns into the north, most of Frederick Turner's definition of the *frontier* as being a "barren land open for the taking" corresponds to Roman mentality. During their conquests north of the Alps, no large power, no closed political entity or system was encountered, but open land that was barely settled by diverted tribes, some of them eager to cooperate with the occupiers, some of them not. For the main period of the Roman expansion – from the earliest time of the Roman Republic up to Emperor Trajan (98–117 AD) – there where no artificial linear barriers at the geographical ends of the Roman Empire towards the "non-Roman" rest of the world.

The way of Roman spatial thinking[4] and how they conceptualized different spaces (psychologically and sociologically) resembles our modern view. On the other hand, the Romans had no "scientific" maps like ours[5] and used natural (rivers, coasts, mountains) and artificial (watchtowers, forts, or colonies) landmarks to define their territory according to the will of their gods. Only by means of the sanctification and protection of their gods, could Roman borders (*termini*) become effective to religious, legal, and mental perceptions of space and place.

From this point of view, it becomes evident that both, frontiers and borders, are closely related to the society, to the politics, to the economics, and to the culture which defined them. In a nutshell: by analyzing the type of border, one is able to analyze the society which established it.[6] The ancient Roman Empire is no exception to this rule. On the contrary, one cannot grasp the nature of the artificial border (the Limes) it fabricated without looking at the mentality of the border's creator.

By the time of Emperor Hadrian's reign (117–138 AD), the Roman idea of expansion without limits came to an end. Emperor Trajan, Hadrian's predecessor, nearly extended the Empire beyond proportions. Hadrian put the focus back to Rome again, and instigated a policy and culture of consolidation in the provinces.[7] Hadrian knew he might lose the whole Empire if he would not change Roman colonization policy decisively, and was willing to sacrifice a few sandy areas in the East to stabilize the Empire.

But then again, what made these palisades and walls, the Limes, so attractive for this kind of spirit? Was it just the technical superiority or *majestas imperii* of the Roman culture to be showcased in a palisade reaching out from one horizon to another?[8] Yet again we have to focus on the internal historical processes or the cultural development of the Roman Empire. The age of Hadrian and Antoninus Pius was called, the "Golden Age" of mankind (like others before),[9] because of its prosperity and fewer wars than in other decades. This age was also a time of cultural and political "harvest" of the previous century where political and economic power, together with the Romanization of the world, accumulated in the will to construct and to strengthen the Empire from the inside. This goal could only be met, in the view of the Romans, by strong bounds between the several cultures and regions under the primacy of the Roman culture. As a first step, a common culture with clear structures was defined, followed by a step to incorporate all the different cultures within Roman territory. The latter step is dedicated to establishing a policy of exclusion of non-Roman life from Roman territory, in order to sustain the forces that support the Empire (at least in the eyes of Hadrian), and those who are supposedly apt to decompose it.

The best of both worlds

From this point of departure, we deliberated a strategy to utilize GIS software for an historical inquiry that, as will be pointed out below, could not rely on empirical data alone. The empirical data, chiefly from archaeological excavations, would give rather little insight into the mentality of the Romans and their motivation for erecting long stretches of ramparts – not only in the German Provinces, but all along at the edges of their Empire. We therefore sought methods to incorporate cultural studies into data-driven analyses. We found that geography, and specifically Geographic Information Systems (GIS), could greatly contribute to such a research project, although GIS software as such is not particularly well suited to meet this approach.

Most of the systems implemented and presented as "Historical GIS" are actually not about *history*, but rather about *Historical Geography*, and so is their methodology. In the majority of cases, the focus is on building such "Historical" GIS for constructing a repository of historical-geographical data.[10] Historical Geography, however, is not History, and the two disciplines have distinct methodological approaches. Because of its nature as a geographical subdiscipline, the average Historical GIS uses techniques established in that field, like geostatistics, to evaluate the data collected. Such an empiricist approach is fully legitimate in the context of Geography, and not much contested in that discipline.

Among many historians, however, such quantitative approaches are not always easy to establish. Many researchers and research fields have turned to hermeneutics and do not favor statistics too much anymore, whereas fields such as Economic or Social History have traditionally clung to quantitative methods and are also embracing GIS for their analyses. Other historical subdisciplines have a somewhat different viewpoint on such methods, and proponents of digital or quantitative techniques are often met with suspicion.[11]

The issue for us, however, was not *quantitative* versus *hermeneutic* approaches. As a matter of fact, our investigation makes good use of statistical evaluations. For our project, in contrast, we had to find something in between textual and geographical analyses. For a purely GIS-based study, the data available were too sparse and too unreliable. Nearly all of our datasets were taken from archaeological excavations, and either provided to us by the regional archaeological agency, or entered from the yearbooks or reports from archaeological societies. Most of the findings were not dated, rendering a precise historical analysis futile. However, for our study most of the data in question were not of utmost relevance anyway, as we were interested in an overall settlement strategy of the Roman Empire, rather than in individual sites. For such purpose, we wanted to use GIS as a tool for the detection of structural patterns, rather than for data mining. This detection was designed so that we could gain pointers for further researches within the written sources, and to verify the plausibility of our theses.

It thus became evident that we could easily ignore the greatest amount of archaeological data we were able to track down and focus on buildings, that is, building types. These had the advantage of being comparably big and immobile, and consequently the data of their location are secure, whereas smaller items like pottery or horse shoes could have been moved hundreds of miles away from their original location through the centuries. Also, for many buildings the dates were fairly easy to narrow down through inscriptions in the walls or other sources. Problems with uncertain dates therefore imposed little or no issue for our investigation.

The general problem in regard to a GIS-based analysis we had to tackle was chiefly the lack of boundary data. For modern history, these regional boundaries are usually given through the historical boundaries of administrative units, so that statistics on certain aspects (like unemployment rate, population density, etc.) can be handled and compared by these regions. This, however, is not the case with the ancient Roman Empire. Although the Romans had their territory well organized, everything below the provincial level is not well documented, and little to no knowledge of their administrative policy in this respect is available. In short, there were no small-scale administrative districts for our area of investigation that could be sensibly exploited. Simply put, we could only differentiate between the Roman versus the non-Roman side of the Limes. Our major interest was directed towards the Roman side, as the non-Roman side was largely uninhabited – although this circumstance alone is a very important cornerstone of our investigation, as we will point out below.

It thus became clear that we had to partition the area of interest by algorithm, and Voronoi diagrams appeared to be the natural choice.[12] With those virtual partitions, we were able to analyze the infrastructure of the Roman province *Germania Superior* and combine it with other analytical techniques into a survey of possible scenarios of the Roman settlement strategies.

Analysis of the Roman infrastructure in the German provinces

For our inquiry we focused on the area north of the River Main (the region nowadays called Wetterau in Germany, see Figure 25.1), for we could conclude from preliminary investigations that this area seemed to be of utmost interest to the Romans, as the principal infrastructure suggests. Furthermore, the course of the Limes in that area indicated that its military value seemed not to be the only criterion for erecting such a structure.

As can be seen in Figure 25.1, there was only one Germanic village in the direct vicinity of the Limes. There was one other village *c*. 80 km (50 miles) north-north-east of this one, and a small group of Celts are believed to have settled some miles west. All in all, the non-Roman presence in the area north of the river Main was relatively minor, and the strong military fortifications in the area do not appear to have been ultimately necessary. Also, from a mere military point of view, the slope the Limes takes in the Wetterau region does not seem to make much sense. Looking at Figure 25.3, however, also reveals that there were a great number of farms located in the Wetterau, whereas there were next to none in the area south of the River Main. A map of the soil quality in the area provides a quick resolve: the soil in the Wetterau region was extremely fertile, whereas the soil south of the River Main was not. Furthermore, the ground there was not very suitable for construction because of its swampyness. Consequently, most farms were situated north of the River Main (with some further south, not shown in Figure 25.4, where the soil became richer again), and the area south of the Main along the Rhine, largely avoided by the Romans. The Wetterau region thus became the "granary" of the Roman provincial capital Mogontiacum (today's Mainz).

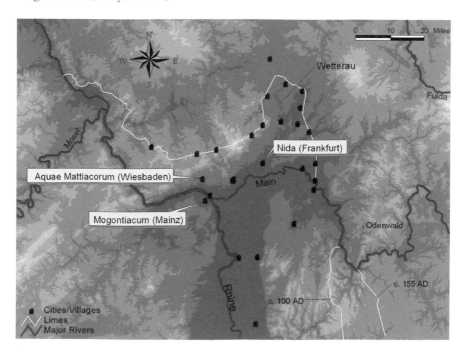

Figure 25.1 The area of interest for our research: the Rhine-Main area in the first century AD, with Mogontiacum as its provincial capital. Our research focused on the Wetterau region.

That this important supply of food was valuable for the Romans is obvious, but does not yet exclusively explain the efforts being undertaken to secure the area. The next step in our investigation included the creation of a Voronoi diagram over the Roman forts (see Figure 25.3). As Figure 25.2 shows, the infrastructure of the Roman presence in the Germanic territories was well deliberated, with small forts at the border and larger castles in the center, all well connected by roads. Figure 25.3, in turn, displays the locations of farms and what we called "productive sites," that is, potteries, brickyards, blacksmiths, and so on. Comparing Figures 25.1 and 25.2 also shows that the villages in the Wetterau were usually close to forts. As a matter of fact, the major object of the villages was to support the forts, and also to function as trading points, by having the granaries for the farms in the area in their vicinity. Furthermore, the larger forts were almost always placed at important road junctions, supposedly protecting the trade traffic in the area. As we knew from the sources, although the German tribes were too few to start a military invasion into the region, raids by comparably small groups of marauders were frequent, for the region hosted a great number of goods, as explained above.[13]

This conclusion became even more apparent after we did some geostatistical analyses over the Voronoi diagram, relating the size of a fort to the building types and numbers in the respective cell.[14] We could establish that there was a correlation between the number of economically important buildings and the size of a fort. This rule also held true in the proximity of the Limes. Usually, there were only rather

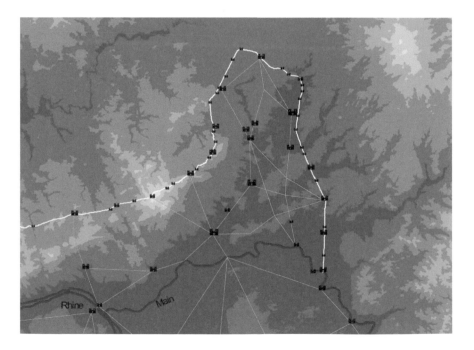

Figure 25.2 The Roman forts in the area, scaled by their size, also showing the principal road network.

small forts directly at the Limes, with the large forts in the interior area to protect trade and traffic, except when the "productive site" was also located near the border. As these sites first and foremost functioned as trading hubs, the Romans obviously had a vital interest to secure them against raiders.

We therefore came to the conclusion that the Romans, rather than being concerned with a military invasion into their territory, adapted their general infrastructure to protect their economy and to establish Roman life in their areas of influence. However, this still does not sufficiently explain the efforts being put into the erection and securing of the Limes. As we pointed out above, ramparts were constructed all around the Empire, in regions with little or no economic interest (like the Sahara region in the Roman provinces in northern Africa), or little or no enemy presence (like Hadrian's Wall in northern England). Looking at the Voronoi diagram in Figure 25.3 again, the two distinct spatial set-ups become apparent: a fan-like structure with a rather small lateral distance between each fort, but large longitudinal dimensions to make the territory virtually impenetrable,[15] compared to an *inner space*, as we would come to call the area enclosed by the Limes (chiefly the Wetterau area), that was guarded by rather large forts with lower spatial density. Leaving the economic reasons we detailed above aside, the structure appears as if the Romans were paranoid that any barbarians might come into their territory without notice, whereas they were more relaxed when it came to the level of surveillance of the interior regions. Besides the more mundane reasons for their construction efforts (like protecting the economy), the Romans

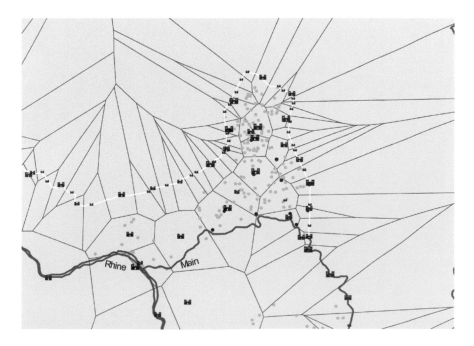

Figure 25.3 A Voronoi diagram created over the forts, also showing the farms (grey dots) and the "productive sites" (black dots) in the area.

seemed to be obsessed by some kind of *territorial imperative*. As mentioned, the area on the other side of the Limes was quite deserted, so there were no immediate reasons for them to be afraid. Yet, as our maps indicate, the Limes was not only a mere *economic border*, but also a *cultural border*, since the clear distinction between the "civilized" Roman territory versus the "barbarians" appears to be of utmost importance to the Romans.

As a matter of fact, the Limes in Germany had been erected *after* Roman campaigns into northern Germany had failed, and the Romans had turned their back in disgust (and awe) at most of Germany, regarding the lot of German tribes as inapt to get "civilized," for they resisted assimilation by the Romans. Most contemporary Roman chroniclers and poets like Tacitus depict those unoccupied German territories as rather gloomy and sinister places, best to be avoided.[16] The Roman Limes was therefore not simply a military-utilitarian building, but also an expression and signpost of Roman culture.

Only after they encountered resistance did they come to think of demarcations to separate themselves from the rest of the world. The Limes was therefore not erected to protect the Roman Empire from the invading barbarians, but to signify the line where the Roman (a.k.a. "civilized") world ends.

Conclusion

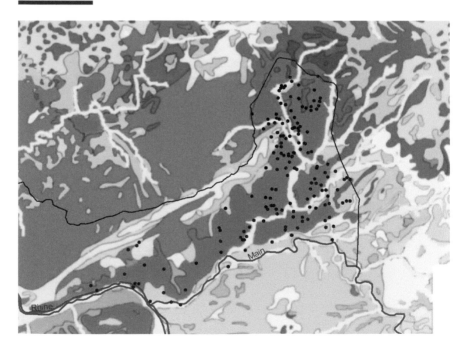

Figure 25.4 The Limes and the farms lay over a map showing the soil quality, ranging from orange and red (very fertile soil) down to grey and ochre (poor soil). The map of the soil quality was taken from the Hessian Historical GIS, LAGIS, at the University of Marburg, Germany, (available at) http://web.uni-marburg.de/hlgl/lagis/projekt.html. The map was slightly edited to fit into our dataset.

In our investigation, we attempted to combine quantitative techniques with hermeneutics. The methods of geographical analyses were not new, and we never had the incentive to create methods specifically fostered to support historical research, albeit we acknowledge the need for those special techniques.[17] Frankly speaking, most historians struggle to cope with geographical thinking in general and the technology of GIS in particular. Our stance, on the other hand, is that historical investigations can greatly benefit from both GIS and Geography, as German philosopher Immanuel Kant (1724–1804) once elaborated: "There is nothing that trains the mind like geography."[18]

Our use of geography and GIS, consequently, is one of *epistemology* rather than mere empirics. The maps produced by us in the course of our research deliberately represent *intermediate* stages in that investigation. We regard the use of GIS as an *active* component in our work, and not as a tool to visualize its outcome. We therefore are not focused on producing appealing maps for publication, but rather to exploit the data and create variate spatial configurations that help us to see historic constellations from a different angle, and to detect possible alternative scenarios. In short, we employ GIS as a *geographically induced heuristics*[19] in historical research. Geography and GIS can become important cornerstones in a dynamic historical research agenda. It bridges geography and history, as it gives credit to both geographical and hermeneutic methods, thereby reconciling quantitative with nonquantitative methods, which today often imposes a dividing line between historians of different schools.

Notes

1 J. C. Mann, "The Frontiers of the Principate," in W. Haase and H. Temporini (eds), *Aufstieg und Niedergang der römischen Welt: Geschichte und Kultur Roms im Spiegel der neueren Forschung*, Volume II, part 1, Berlin: De Gruyter, 1974, p. 532.

2 E. Luttwak, *The Grand Strategy of the Roman Empire from the first Century A.D. to the third*, Baltimore: John Hopkins University Press, 1976.

3 G. Klose and A. Nünnerich-Asmus (eds), *Grenzen des römischen Imperiums*, Mainz: Zabern, 2006.

4 K. Brodersen, *Terra cognita: Studien zur römischen Raumerfassung*, Hildesheim: Olms, 2003.

5 R. Talbert, "Cartography and Taste in Peutinger's Roman Map," in R. Talbert and K. Brodersen (eds), *Space in the Roman World. Its Perception and Presentation*, Münster: Lit-Verlag, 2004, pp. 113–41.

6 G. Simmel, "Soziologie des Raumes," in *Gesamtausgabe*, Volume 7: Aufsätze und Abhandlungen 1901–1908, O. Rammstedt (ed.), Frankfurt: Suhrkamp, 1995, 132–83.

7 M. T. Boatwright, D. J. Gargola, and R. J. A. Talbert, *The Romans: From Village to Empire*, Oxford: Oxford University Press, 2004, p. 373.

8 Cf. Alföldy, Geza. 2004. "Die lineare Grenzziehung des vorderen Limes in Obergermanien und die Statthalterschaft des Gaius Popilius Carus Pedo," ed.

E. Schallmeyer, *Limes imperii romani: Beiträge zum Fachkolloquium"Weltkulturerbe Limes", Lich, November 2001*, Saalburg-Schriften 6. Bad Hoburg v.d.H.: Saalburgmuseum, p. 10.

9 E. Gibbon, *Decline and Fall of the Roman Empire*, Great Books of the Western World, Chicago: Encyclopedia Britannica, 1952, p. 32.

10 A. K. Knowles (ed.), *Past Time, Past Place. GIS for History*, Redlands: ESRI Press, 2002. Also: A. Okabe (ed.), *GIS-based Studies in the Humanities and Social Sciences*, Boca Raton: CRC Press, 2006.

11 For example, the editorial comment on L. J. McCrank, "Historical information science: history in information science; information science in history," in C. Barros and L. J. McCrank (eds), *History Under Debate: International Reflection on the Discipline*, Binghamton: Haworth Press, 2004, pp. 177–98.

12 Also called "Thiessen-Polygons" in many GIS-packages. We won't detail the nature of Voronoi diagrams here, for details see M. Berg, O. Cheong, M. V. Kreveld, and M. Overmars, *Computational Geometry: Algorithms and Applications* 3rd edn, Berlin: Springer, 2008.

13 D. Baatz, "Zur Funktion der Kleinkastelle," in A. Thiel (ed.) *Forschungen zur Funktion des Limes* Volume 2, Stuttgart: Theiss, 2007, p. 24.

14 The Romans had a standardized system for their forts, in which a certain footprint size corresponded to a certain category of fort. Each category had a standardized number of soldiers, so that the number of soldiers could be derived from the footprint of a fort, something readily available from archaeological excavations. A. Johnson, *Roman forts of the 1st and 2nd centuries AD in Britain and the German provinces*, London: Black, 1983.

15 There was also a watchtower at the Limes approximately every 700–900 meters (*c.*1/2 mile), manned with two to five soldiers.

16 H. Jahnkuhn, "Terra ... silvis horrida," *Archaeologia Geographica* 10/11, 1961/63, 19–38.

17 J. B. Owens, "What historians want from GIS," *ArcNews* 29(2), 2007, 4–6.

18 I. Kant, *Physische Geographie*, 2nd edn, Mainz/Hamburg: Vollmer, 1816, p. 17. Translation ours.

19 Or "geographically integrated history," as J. B. Owens (fn 17) calls it.

26

History and GIS
Railways, population change, and agricultural development in late nineteenth-century Wales

Robert M. Schwartz, Ian N. Gregory, and Jordi Marti-Henneberg

Introduction: historical GIS and national historical GIS databases

Historical Geographical Information Systems, or historical GIS, has become a rapidly growing field within historical research.[1] Historical GIS is an interdisciplinary approach that involves applying GIS technology to the study of history. Using a conventional database of only statistical information, a researcher can search for *aspatial* patterns of variation and change. Using historical GIS, the researcher is now equipped to identify patterns of change that occur simultaneously over time and across geographic space. Additionally, because all of the data are located in space, they can be integrated with any other data that are also located in space. With historical GIS we can get closer to understanding the complexity of change and historical reality.

A major impetus behind the growth of historical GIS has been the significant investments made by a number of countries in National Historical GIS (NHGIS). Among the countries that have built or are building NHGIS based on census data, the best developed include Great Britain,[2] the United States,[3] and Belgium.[4] Such systems are costly because they require not only that census information be entered, but also that administrative boundary changes through time be researched.

Increasingly NHGIS hold more diverse data sources than censuses and other statistics. Gazetteers, information on settlement patterns, historical maps, dynastic information, travel accounts, and literature are all examples of the types of material that are becoming increasingly common. In some cases, such as in the China Historical GIS[5] and the German Historical GIS,[6] diverse data sources have been the main emphasis from the start. In other cases, a census-based NHGIS has been extended to a more diverse system, such as the expansion of the British NHGIS to include the Vision of Britain through Time website.[7]

Is all the effort and money expended worth the cost? This essay offers a strong affirmative answer using the Great Britain Historical GIS (GBHGIS) as an example of a highly developed NHGIS. It brings together census data with (1) a database on the development of the rail network derived from Cobb,[8] and (2) a collection of agricultural statistics[9] to look at how the development of the rail network influenced population and agriculture in a small but diverse part of Britain, namely the principality of Wales. The essay illustrates the type of substantive results that historical GIS enables by bringing together disparate sources and exploring how the relationships within and between them change over space and time. GIS is not the philosopher's stone of course, but it is capable of contributing to our knowledge of the past, all the more so when combined with other tools in the historian's kit.

The development of railways in nineteenth-century Wales

In nineteenth-century Wales, of what importance were railways in agricultural development? This question is interesting because of a gap in the literature. Specialists in railway history rarely give more than passing attention to agricultural developments, while those in Welsh and agrarian history usually give scant attention to railways.[10] In an effort to bridge this gap, D. W. Howell revealed interesting connections between railways and agricultural change. He offered telling examples of the great cost savings of rail transport over horse-drawn carting, and suggested that cost savings stimulated trade and benefited farmers. There were also negative consequences, he believed. A case in point was the demise of traditional markets as railway agents and dealers purchased directly from farmers. He concluded that railways reduced isolation and benefited Welsh farmers who raised cattle by cutting the cost of marketing livestock and increasing the volume of trade with English graziers (merchant-farmers who fattened store animals prior to their sale for meat). Railways, he added, also facilitated the migration of workers from rural Wales, improving the lot of those who stayed on the land.[11]

The use of GIS and newly available data on railways and agriculture makes it possible to explore the role of geographic variation in describing the nature and timing of agricultural change. As we shall see, railways played a substantial role in helping Welsh farmers adjust to changing market conditions during the agrarian depression of the 1880s and 1890s. The effects of rail transport also varied geographically, benefiting upland livestock farmers somewhat differently than their lowland counterparts.

The dawning of the railway age in Wales began in 1848 with the opening of a line from Chester, England, to the coastal town of Bangor in Caernarvonshire County. Built for a train known as the *Irish Mail*, its purpose was to quicken the speed of communications between the capital and Ireland.[12] The line was extended to Holyhead in Anglesey and south to Caernarvon in 1852. In the same year, the South Wales Railway reached the agricultural county of Carmarthenshire.[13] As the network continued to grow in the 1850s and 1860s, new lines served the northern lead and slate mining districts in Caernarvonshire and Merionethshire, and others opened in the lowlands of Cardiganshire and along the Merioneth coast. On the southern peninsula, lines reached the main agrarian districts of Pembrokeshire in 1862, an advance that would prove a boon for the county's dairy and livestock farmers.[14] This development is summarized in Figure 26.1. The advent of rail transport in central Wales proceeded more

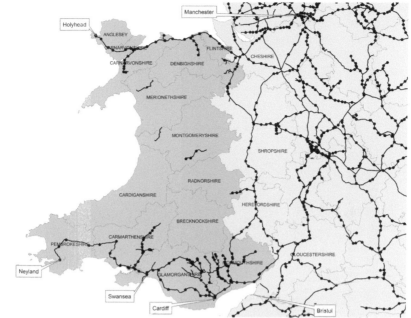

A. 1850s

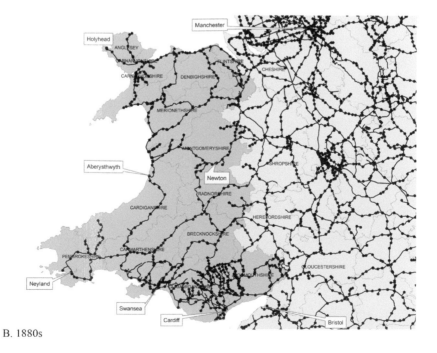

B. 1880s

Figure 26.1 The development of the railway network.

slowly. Railway contractors and private finance faced the most challenging terrain yet to master, as is demonstrated in Figure 26.2.

In 1859, the first lines in Montgomeryshire opened, connecting Newton and Llannidloes. Further expansion in the 1860s and early 1870s linked these and other upland agrarian districts with coastal towns such as Aberystwyth and with the industrial

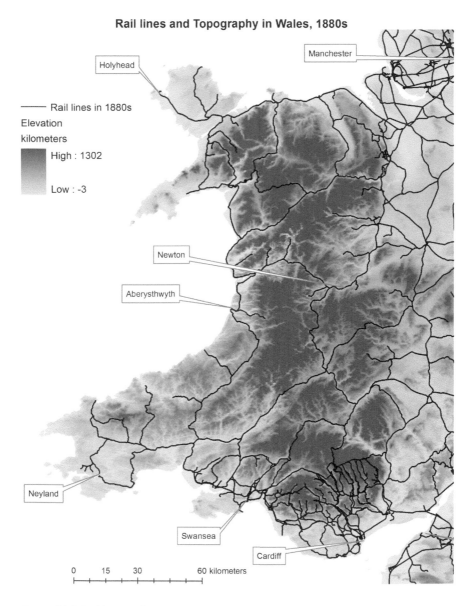

Figure 26.2 Rail lines and terrain.

areas of south Wales. By the end of the 1870s, with the partial exception of Cardiganshire, the favored agrarian districts of southern, central, and northern Wales, all with proximate rail service, enjoyed a level and pace of communication impossible to imagine only three decades earlier.

Agriculture in nineteenth-century Wales

In the 1840s, before the railways arrived, most Welsh farmers engaged themselves in whatever mixed farming their climate, terrain, tradition, and communications permitted. Wales was generally a grazing region, and its rugged terrain and poor roads made transportation slow and costly. Isolation characterized rural life. Under these conditions, a considerable portion of the land was devoted to arable farming and cereal production to provide sufficient food for local household consumption and livestock over the winter. Trade and commerce were predominantly local in character and centered on market towns. Stock raised for sale were known as store cattle and were taken on foot to feeding or "fattening" regions by drovers. Even as the growing market for food in the industrial southeast increased opportunities for Welsh farmers, transportation costs limited agricultural trade between producing areas and the shops in Swansea, Cardiff, Newport, and Merthyr Tidfyl.

Railways fundamentally changed this situation. By reducing transport costs and improving access to distant markets, rail transport helped alter farming practice and agricultural land use, accelerating a shift from cereal production to dairy and livestock farming throughout Britain and Western Europe. Within this broad pattern, specific developments in Wales reflected its distinctive culture, landscape, climate, and transportation facilities. We focus here on landscape, rail transportation, agrarian practice, and the changing market conditions to which Welsh farmers had to respond.

The biggest challenge to agricultural livelihoods came during the agricultural depression in the 1880s and 1890s. Vast increases in agricultural production in North American and other settler countries[15] – brought to world markets via steamships, railways, and the telegraph – intensified international competition and depressed farm prices. In Britain, as growing shipments of imported foodstuffs were dispatched by rail from the docks of Liverpool and London, the price of wheat fell steeply, that of livestock, butter, and meat somewhat less. In France and Germany, tariffs were introduced in the 1880s to support farm incomes against foreign competition. In Britain, the government held firmly to its policy of free trade. With little hope of reversing that policy, English and Welsh farmers pressed the government to reduce rail freight charges.[16] Although rail companies made some adjustments, British farmers remained convinced that freight rates were too high.[17]

During the depression, Welsh farmers probably suffered somewhat less than their English counterparts.[18] The grazing economy of the principality was both less vulnerable to the effects of severe depression in cereal farming and suited to the expansion of livestock. Large markets for Welsh beef and mutton were comparatively close at hand in the industrial southeast of the country and in the urban and industrial centers of Lancashire and the Midlands, all areas in which the demand for fresh milk, butter, and meat was growing. The expansion of railways would help farmers meet that demand. The key to success was proximity to a rail station, where store cattle and sheep as well as dairy products were gathered for shipping. There, important agricultural supplies were received – lime for treating soil, agricultural tools, mail, market information, and stone and slate used to construct better cow sheds and barns.[19]

Such proximity was a reality for many if not most Welsh farmers in the 1880s. The increased accessibility of rail service in that decade can be seen in Figure 26.3 showing the average distance from rural parishes to the closest station over four decades from the 1850s to the 1880s. By the 1880s, there were still clusters of parishes where journeys of 15 kilometers or more were required to reach the nearest station, a distance easily requiring a full day (or more) to move cattle to a shipping point. Nonetheless, for the majority of rural districts isolation was reduced and the distances from farm gates to rail stations manageable for most farmers.

Based on available agrarian statistics, a retreat from wheat cultivation, coupled with a gradual expansion of the pastoral economy, got under way in the mid-1870s,[20] and the pace of the expansion increased during the agrarian crisis of the 1880s and 1890s.

The expansion of the pastoral economy was both more marked and geographically more uniform in cattle raising than in dairy or sheep farming. For example, an upward trend in cattle raising was dramatic in Anglesey. A small, lowland county, it had benefited as early as the 1850s from its rail connections to markets in Lancashire and Cheshire. From the early 1870s to the early 1880s, in a single decade, fields of wheat disappeared in favor of new pasture and expanding herds of beef cattle. In Pembrokeshire, a somewhat slower rate of growth, associated with much larger herds, produced the largest numerical increase of cattle in the principality. Thanks to lush, improved pasture land, good rail transportation, and astute management, Pembrokeshire became the principal region in Wales to which store cattle and sheep were sent for fattening. Elsewhere in Carmarthenshire and Montgomeryshire, modest rates of growth in the large herds of store cattle provided the fattening areas with a good supply of young animals.

From the mid-1870s to the mid-1880s, when struggling farmers generally looked first to rearing more cattle, sheep flocks were reduced by about 15 percent on average. Then, in the following decades, and despite the continued fall in the price of wool, sheep farming expanded in response to relatively favorable prices of mutton and lamb.[21]

In comparison to cattle rearing, the development of dairy farming was generally less extensive. In the rural counties, relatively small herds and growth rates of about 10 percent per decade characterized the situation in Anglesey and Flintshire, for example, whereas large herds and similar growth rates were the norm in Pembrokeshire and Caernarvonshire. The highest rates of growth occurred in the industrial counties of Glamorgan and Monmouth, where the urban population in 1882 amounted to more than 60 percent of the two counties' totals.

Increased Parish-level Rail Accessibility from the 1850s to the 1880s

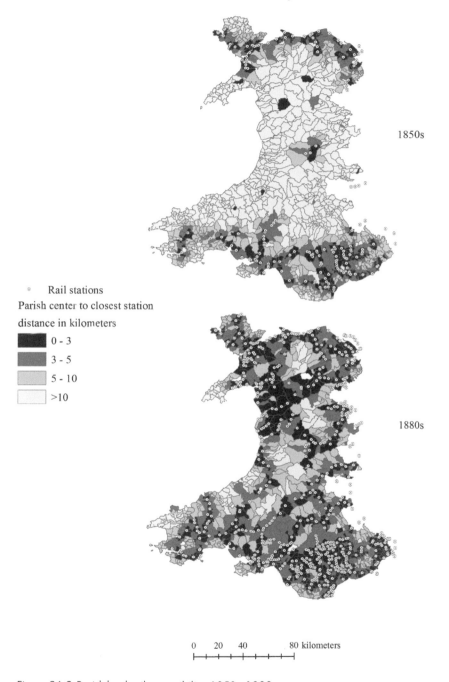

Figure 26.3 Parish-level rail accessibility, 1850s–1880s.

Agriculture and the railways

What was the relationship between these varied and shifting patterns of livestock farming and the development of the railways? Ideally we would use data at the level of the parish or registration district to identify spatially varying relationships across the counties and the principality as a whole. Although geo-referenced data on railways is at hand, agricultural data are currently available only at the county level. At that scale of resolution, statistically significant spatial variation in the agriculture data is masked by the aggregate data for the 13 Welsh counties. In the absence of spatial autocorrelation, a good tool to explore the relationship between railways and agricultural developments is ordinary least-squares (OLS) regression.

Regression results prove both interesting and complex. Several simple versions of a multivariate regression were tested to gauge the effect of rail transport accessibility on the proportion of cattle, dairy cows, and sheep in the counties in 1882 and 1892, near the beginning and the end of the agricultural depression. Each analysis examined different indicators of rail accessibility together with measures of terrain elevation, the density of the population, and its urban versus rural proportion. Neither of the population variables were significant and were put aside.

In the best-fitting regressions for the two periods, from 80 to 95 percent of the variation in the proportion of one or other of the three kinds livestock (cattle, dairy cows, and sheep) was accounted for by the combined effects of rail station density and mean terrain elevation. In the case of cattle and milk cows, the relationship between the dependent variable (proportion of livestock) and independent variables (rail station density and train elevation) was inverse, and the effect of terrain elevation was consistently greater than that of rail accessibility. As for sheep, elevation also had a greater influence on the proportion of sheep than did rail density, but the effects of elevation and rail density were, nonetheless, both positive.

The effect of rail transport on livestock farming was clearly important, but it differed depending upon the type of stock and the average elevation and character of the terrain in a given county. As Figure 26.4 shows, in counties with higher mean elevations the proportion of sheep tended to be greater, especially in areas where the availability of rail service was relatively good. This suggests that even though lowland sheep farmers were apt to have better access to rail, they used it less because their flocks were small. One reason for that was that lowland breeds were more vulnerable to disease. Other varieties thrived in upland and mountainous areas.[22] Hence, the expansion of sheep farming begun in the late 1880s was primarily a highland development, one that was substantially facilitated by the upland railways that came into operation from the late 1860s through the 1880s.

A different pattern emerges with regard to cattle raising and dairy farming. Here, our results show that the proportions of cattle and of milk cows were typically greater at lower elevations and where there were fewer stations per aerial unit than elsewhere. Pembrokeshire and Carmarthenshire are examples of the inter-relationships in question. Both were lowland counties with relatively moderate rail station densities (1.9 and 2.1 stations/100 km^2 respectively) and relatively large proportions of cattle (31 percent and 20 percent of livestock respectively). In absolute numbers, these

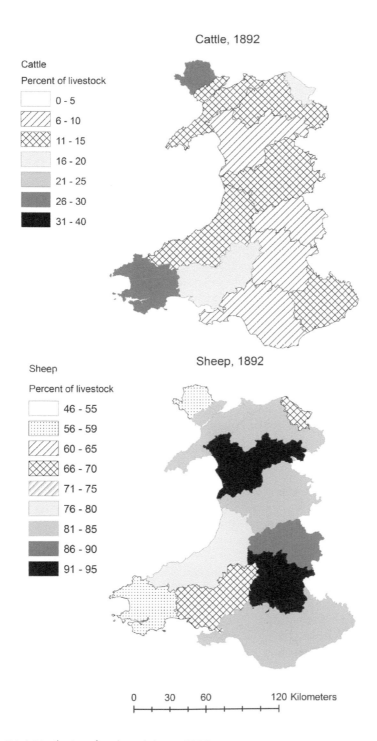

Figure 26.4 Distribution of cattle and sheep, 1892.

counties had the largest stocks of cattle in Wales, and between 1872 and 1892, stocks grew by 43 percent and 33 percent respectively. During the same period, the increase in milk cows was comparatively modest, at 17 and 12 percent respectively. In comparison to cattle and sheep, their proportions declined by about 3 percent. Stability seems an apt characterization of dairy farming in these counties.

Signs of a weakening relationship between railways and dairy farming in the 1890s probably attest to the little success that most Welsh farmers, and those in upland counties in particular, had in exploiting the growing market for fresh milk in urban areas. In upland areas, lower quality milk, greater distances from urban markets, and slower speeds of transportation all combined to encourage farmers to rely on butter as the major dairy product. Unlike fresh milk, foreign competition in the butter trade was intense by the 1870s and 1880s, so profits were increasingly under pressure. Lowland farmers nearer the urban markets in Swansea, Cardiff, and Newport benefited from faster rail transport than that available to highland farmers and expanded their herds of dairy cows accordingly.

Railways clearly stimulated the expansion of livestock farming in Wales in ways that differed between lowland and highland areas. In responding to market conditions and agricultural environments, farmers in upland Montgomeryshire, Radnor, Merioneth, Breconshire, and the higher plateaus of Glamorganshire, relied on railways to substantially expand their flocks of sheep. In the lowlands, farmers in Anglesey, Pembrokeshire, and Carmarthenshire, for example, could exploit their lush pasture and growing herds of cattle and dairy cows due to their speedier rail connection.

Railways, agricultural development, and population change at county level

Whether these agrarian developments improved the lot of farmers and agrarian communities is another important question that we can explore indirectly by turning to related population movements. Howell's claim that railways improved the situations of Welsh farmers and rural communities by facilitating rural out-migration is open to revision. The arrival of accessible rail services likely created new opportunities for commercial agriculture that would have stemmed the pace of out-migration, given different regions, varied market conditions, and other factors. Even as drovers were being displaced by rail shipments of stock, an increased demand for horse-drawn carting of goods from farm gates to stations would employ more carters and provide additional work for wheelwrights. On the other hand, the decline of arable farming in favor of livestock raising usually diminished the demand for agricultural laborers. In dairy farming, however, expansion typically increased the need for boys and young women to handle the milking and related tasks. A further element that accompanied intensified price competition and declining incomes for both tenant farmers and landowners was the increased viability of small-scale, family farming. The Settled Land Act of 1882 permitted landowners to sell off farmsteads without entail restrictions, and the number of small farms in Wales rose dramatically by the mid-1880s.[23] Small farms were viable because they depended upon family labor to work the land and raise stock.[24]

With these agrarian processes in mind, we can use OLS regression to describe the temporally varying relationships between livestock rearing, arable farming, rail transport, and rural population change in Welsh counties. Once again the available data on agriculture limit the analysis to aggregate patterns at the scale of the county effects completely masked by aggregation. At the scale of the country, the dominant geographic and temporal trend was rural depopulation in all the counties except for Flintshire, Glamorganshire, and Monmouthshire over the period 1840–1910. In the majority of counties, median rates of depopulation were substantial in the 1860s, 1870s, and 1880s and then became relatively moderate in the 1890s and 1900s. As noted above concerning railways, the opening of new lines and stations had begun to open up rural districts by the end of the 1860s; nevertheless, the influence of rail service on agriculture and rural populations seems to have taken hold widely only in the 1870s.

Regression analysis showed that across the 13 counties from the 1860s to the 1900s rural population change was strongly associated with the decline of arable farming and the expansion of livestock rearing. In the 1870s and after, population change was also strongly influenced by the accessibility of rail service. With regard to arable farming, the relative extent of land under tillage proved significant and positively related to rural population change in the 1860s, 1880s, and 1890s. In these decades the decline of tillage per 100 acres of county land helped account for rural depopulation: as fewer hands were needed in arable farming, the rural population in agrarian districts tended to decline. In the 1860s, 1870s, and 1880s additional downward pressure on agricultural employment came from expanded cattle raising. In contrast, dairy farming required comparatively more labor than cattle and sheep farming. Hence, in the 1870s and 1900s the expansion of dairy farming stemmed depopulation by creating more jobs for service workers in the industry. Likewise the growth of rail service, beginning in the 1870s, had a positive impact on rural demographic change: in counties with relatively greater access to stations, rural populations more likely sustained themselves or increased. Except in industrializing Glamorganshire and agrarian Flintshire, where the rural population grew substantially, the general context of rural demographic change, to underline the point, was that of decline or, in the case of Monmouthshire, stability. In the majority of the counties, then, the combination of greater rail transport and the expansion of dairy farming slowed the decline of rural population associated with the shift from mixed farming and arable cultivation to cattle and sheep raising.

The railways and population change at parish level

While agricultural data are only currently available at county level, population data are available with far more spatial detail at the level of the parish. There were 1,257 parishes in Wales in 1911 of which 1,163 can be classed as rural if we define "rural" as having a population of less than 2,500.[25] This gives us a potentially very detailed picture of the rural population; however, parish-level census data suffer from two major problems: first the census only publishes total populations for these units, and second the boundaries of parishes changed frequently between censuses, making changes over time very difficult to explore. The second of these problems can be resolved using a GIS-based technique called areal interpolation. As the GIS can encode the boundaries used in the different censuses, we can calculate the degree to which parishes at one date intersect with those at a different date. This information provides us with ways of estimating what the populations for one set of parishes, in this case we will use the 1911 arrangement, would have been at earlier dates.[26] This in turn allows us to calculate the intercensal population changes. It must be noted however that population change has two components: migration and natural increase.

We compared the median population change of parishes with the distance from their centroid[27] to the nearest station. If Howell's assertion is correct, parishes with easy access to the railways would likely have higher rates of population loss than those further away. We found that in the 1850s there was little relationship between a rural parish's population change and its distance from a station. However, in the 1860s, 1870s, and 1880s, the population change is highest (above average) among parishes whose centroid was within 2 kilometers of a station, and this pattern declined with distance. This distance decay became more pronounced over time. In the 1860s and 1870s population growth dropped below average by 4 to 6 kilometers from the nearest station. By the 1880s all population centroids nearer than 10 kilometers from a station had above average growth while all further away had below average.

In contrast to Howell's assertion, these results suggest that being close to a station actually encouraged population to stay in a parish, likely because of the economic advantages that rail service provided. At a time when high rates of rural out-migration rates were the rule, the greater the rail density in any given rural district, the greater the likelihood that the rate of out-migration was less than elsewhere in the countryside.[28]

Two major conclusions can be drawn: first, accessibility to the rail network appears to have had a major impact on the development of agriculture in Wales. Second, Howell's assertion that the railways led to out-migration needs revision. As in the rest of Britain – indeed in most of Western Europe – rural out-migration was the general trend in Wales in this period. And yet, areas with better accessibility to the rail network seem to have had lower rates of population loss than those further away.

With additional data, this research can be extended. We are currently digitizing parish-level data from the British Agricultural Returns, which will allow us to explore the relationship between population change and agricultural change in detail. The number of parishes and the volume of data for each parish – there are usually over 50 columns of statistics for each parish – makes this a slow and laborious process. Once done, however, the spatial detail that these data provide will give us a far better and nuanced understanding of the relationship between population, railways, and agriculture by allowing us to conduct the entire analysis at the parish level.

Conclusions: historical GIS and the humanities

This discussion of Wales points to some of the strengths and limits of a study based in part on a census-based National Historical GIS. On the positive side, a National

Historical GIS provides the data needed for detailed investigations of long-term historical change over extensive geographic space. It serves as an effective means of integrating data from census returns, railway atlases, agricultural statistics, digital elevation models, and potentially many other sources, making it possible to pursue research questions that used to be beyond our capacity. With initial results in hand, additional questions can be addressed by including additional data. Although preparing geo-referenced data requires a large commitment of money and time, the resulting databases not only lead to new discoveries by the researcher but can be made available to others for their use.

Like any approach, one based on historical GIS has its weaknesses. A notable weakness is limiting research questions to fit the available data in GIS form. Among humanities scholars, this kind of data driven research is the target, rightly so, of much criticism. The findings may offer less than a desirable degree of explanation. Results that identify only what is where and when inevitably raise "the why questions," and if they are not pursued, the findings remain both incomplete and less satisfying. In the case at hand, we have made good progress toward a more nuanced understanding of change over time and space, but we readily admit that our study is incomplete and needs additional work to realize the promise of the subject and interdisciplinary methods.

In sum, GIS is not this era's philosopher's stone but a new and very useful tool for conducting historical research in an interdisciplinary context. It allows researchers to structure, integrate, explore, and describe geographical and temporal patterns in ways that are far beyond the capability of any other approach. Historical GIS does not replace the more traditional approaches to history. It complements and enriches them.

Acknowledgments

This essay benefited from support from the National Endowment for the Humanities under grant RZ-50577-06, and from the European Science Foundation under Eurocores – Inventing Europe grant FP-005 (Water, Road and Rail). The railway data from M. H. Cobb's (2006) *The Railways of Great Britain: A historical atlas* were digitized at the University of Lleida. We would like to thank Antonia Valentín, Laura Ortiz, Eloy del Río, and J. Martí-Domínguez for their work on this. Partial funding for their work was provided by the Spanish Ministry of Education (SEJ2007-64812 – SEJ2007–29474-E/SOCI), the Government of Catalonia (AGAUR), and the Jean Monnet Programme (141000, 2008). For full details of staffing and funding of the Great Britain Historical GIS project see: www.gbhgis.org. Shading schemes used on some of the maps in this essay are based on the Colorbrewer website (www.colorbrewer.org) produced by M. Harrower and C. Brewer.

Notes

1 A. K. Knowles, "Emerging trends in Historical GIS," *Historical Geography* 33, 2005, 7–13; I. N. Gregory and P. S. Ell, *Historical GIS: Technologies, methodologies*

and scholarship. Cambridge: Cambridge University Press, 2007; I. N. Gregory and R. G. Healey, "Historical GIS: Structuring, mapping and analysing geographies of the past," *Progress in Human Geography* 31, 2007, 638–53.

2 I. N. Gregory, C. Bennett, V. L. Gilham, and H. R. Southall, "The Great Britain Historical GIS: From maps to changing human geography," *The Cartographic Journal* 39, 2002, 37–49.

3 C. A. Fitch and S. Ruggles, "Building the National Historical Geographic Information System," *Historical Methods* 36, 2003, 41–51.

4 M. De Moor and T. Wiedemann, "Reconstructing Belgian territorial units and hierarchies: An example from Belgium," *History and Computing* 13, 2003, 71–97. For general reviews of some of these systems see: I. N. Gregory, "Time variant databases of changing historical administrative boundaries: A European comparison," *Transactions in GIS* 6, 2002, 161–178; and A. K. Knowles (ed.), "Reports on National Historical GIS projects," *Historical Geography* 33, 2005, 293–314.

5 M. L. Berman, "Boundaries or networks in historical GIS: Concepts of measuring space and administrative geography in Chinese history," *Historical Geography* 33, 2005, 118–33. P. Bol and J. Ge, "China Historical GIS," *Historical Geography* 33, 2005, 150–52.

6 A. Kunz, "Fusing Time and Space: The Historical Information System HGIS Germany," *International Journal of Humanities and Arts Computing* 1, 2007, 111–122.

7 See www.visionofbritain.org.uk. Accessed August 18, 2009. H. R. Southall, "A Vision of Britain through Time: Making sense of 200 years of census reports," *Local Population Studies* 76, 2006, 76–84.

8 M. H. Cobb, *The Railways of Great Britain: A historical atlas, 2 editions.* Shepperton: Ian Allen, 2003, 2006.

9 J. Williams, *et al.,* July 2001. Digest of Welsh Historical Statistics : Agriculture, 1811–1975. AHDS History, Colchester, Essex [UK].

10 D. S. Barrie, *South Wales, A Regional History of the Railways of Great Britain. Vol. 10 Regional History of the Railways of Great Britain,* Newton Abbot; North Pomfret, VT: David & Charles. 1980; P. E. Baughan, *North and Mid Wales, A Regional History of the Railways of Great Britain. V. 11.* Newton Abbot; North Pomfret, VT: David & Charles. 1980; E. J. T. Collins, "The Great Depression, 1875–1896," in E. J. T. Collins (ed.) *The Agrarian History of England and Wales. Volume VII 1850–1914,* Cambridge: Cambridge University Press, 2000; R. J. Moore-Colyer, "Wales (Farming Regions)," in T. Collins Edward John (ed.), *The Agrarian History of England and Wales,* Cambridge: Cambridge University Press, 2000; B. Reay, *Microhistories: Demography, Society, and Culture in Rural England, 1800–1930, Cambridge Studies in Population, Economy, and Society in Past Time,* Cambridge: Cambridge University Press, 1996; J. Simmons, *The Railway in Town and Country, 1830–1914,* North Pomfret, VT: David & Charles. 1986; M. E. Turner, "Agricultural Output, Income, and Productivity," in E. J. T. Collins (ed.), *The Agrarian History of England and Wales. Volume VII 1850–1914,* Cambridge: Cambridge University Press, 2000; D. Turnock, *An Historical Geography of Railways in Great Britain and Ireland,* Brookfield, VT: Ashgate, 1998.

11 D. W. Howell, "The Impact of Railways on Agricultural Development in Nineteenth-Century Wales," *Welsh History Review* 7, 1974–5, 40–62.

12 R. Millward, "Railways and the Evolution of Welch Holiday Resorts," in A. K. B. Evans and J. V. Gough. (eds) *The Impact of the Railway on Society in Britain. Essays in Honour of Jack Simmons*, Brookfield, VT: Ashgate, 2003, 211.

13 Howell, *op. cit.*, 46–47.

14 British Parliamentary Papers, "Royal Commission on Land in Wales and Monmouthshire Second Report," *Reports of Commissioners*, 1896, pp. 373, 417, 438.

15 Countries colonized by European states, including Australia, New Zealand, Argentina, Chile, Brazil, and British India.

16 E. J. T. Collins, "The Great Depression, 1875–1896," in E. J. T. Collins (ed.), *The Agrarian History of England and Wales. Volume VII 1850–1914*, Cambridge: Cambridge University Press, 2000; N. Koning, *The Failure of Agrarian Capitalism: Agrarian Politics in the UK, Germany, the Netherlands and the USA, 1846–1919*, London: Routledge, 1994; K. H. O'Rourke, "The European Grain Invasion,1870–1913," *The Journal of Economic History* 57, 1997, 775–801; R. Perren, *Agriculture in Depression, 1870-1940, New Studies in Economic and Social History*, Cambridge: Cambridge University Press, 1995; R. Price, *The Modernization of Rural France Communications Networks and Agricultural Market Structures in Nineteenth-Century France*, New York: St. Martin's Press, 1983.

17 British Parliamentary Papers, "Royal Commission on Agricultural Depression Final Report," *Reports of Commissioners,* 1897, pp. 132–33, 222–23, 347–49.

18 British Parliamentary Papers "Royal Commission on Land in Wales and Monmouthshire Second Report," *Reports of Commissioners*,1896, pp. 834–35. ibid. "Royal Commission on Agricultural Depression Final Report," *Reports of Commissioners,* 1897, p. 19.

19 British Parliamentary Papers, "Royal Commission on Land in Wales and Monmouthshire Second Report," *Reports of Commissioners*,1896, p. 763.

20 Williams, *et al., op. cit.*

21 British Parliamentary Papers, Alphabetical Digest, 1897: p. 194; A. K. Copus, "Changing Markets and the Development of Sheep Breeds in Southern England," *Agricultural History Review* 37, 1989, 36–51.

22 Moore-Colyer, *op. cit.,* p. 447.

23 British Parliamentary Papers. "Royal Commission on Land in Wales, and Monmouthshire Second Report, 1896, *Reports of Commissioners*; pp. 344, 575. Moore-Colyer, *op. cit.,* pp. 447–48. Williams, *et al., op. cit.*

24 D. B. Grigg, "Farm Size in England and from Early Victorian Times to the Present," *Agricultural History Review* 35, 1987, 179–89; N. Koning, *op. cit,* pp. 26–36.

25 C. M. Law, "The growth of urban population in England and Wales, 1801–1911," *Transactions of the Institute of British Geographers* 41, 1967, 125–43.

26 I. N. Gregory and P. S. Ell, "Breaking the boundaries: Integrating 200 years of the Census using GIS," *Journal of the Royal Statistical Society, Series A* 168, 2005, 419–37.

27 A centroid is the geographical centre of a polygon. It provides a useful way of measuring the distance from a polygon to another feature.

28 R. M. Schwartz, *New Tools for Clio: GIS, Railways, and Change over Time and Space in France and Great Britain, 1840–1914*, University of Nebraska at Lincoln, 2007; R. M. Schwartz and I. N. Gregory, "Railways, Uneven Geographical Development, and Globalization in France and Great Britain, 1830–1914," http://www.mtholyoke.edu/courses/rschwart/railways/. South Hadley: Mount Holyoke College, 2007.

R. M. Schwartz, I. N. Gregory, and T. Thevenin, "Spatial History: Railways, Uneven Development, and Population Change in France and Great Britain, 1830–1914," *Journal of Inter disciplinary History* (Forthcoming).

27

Spatiality and the social web

Resituating authoritative content

Ian Johnson

Introduction

Digital recording methods, user-generated content and global connectivity – even if still unevenly distributed – are fundamentally changing the ways that information is created and distributed, not least in the broad genre of cultural and historical atlases. Static repositories of general knowledge such as printed atlases, dictionaries, encyclopedias, and other 'reference books', are being rapidly supplanted by sources on the web, sweeping with them the long-established structures for establishing the authority of content such as publisher's imprint, author reputation, and book reviews. Community-based history projects and personal academic blogs nibble at the edges of academic writing, while the online publication of scholarly source data forces us to review existing methods of valuing academic output in the Humanities which has been based more on works of synthesis than on works of compilation. New terms have slipped into the vernacular – social networking, crowd-sourcing, folksonomies, tags, geotags, blogs, wikis, bookmarks, googling, and mashups – to describe new modes of information creation, discovery, and sharing which scarcely existed at the turn of the twenty-first century.

The first generation of the web brought information on almost any subject – not necessarily coherent, accurate, or the best available – within instant reach, although a modicum of computing skill was needed to publish to this medium. Dale Dougherty and Tim O'Reilly[1] coined the term Web 2.0 in 2004 to characterize a set of related shifts, not so much in the technology itself, but in the way technologies are integrated. While there is still debate on whether the Web 2.0 term is just marketing hype, the landscape of the web continues to shift, democratizing content creation through collaborative web tools such as wikis, blogs, image albums, and social networking sites. One no longer needs the support of a recognized organization, a peer review process or a commercial publisher to contribute to the most used general reference works such as Wikipedia.

Spatially located content and map data creation is also becoming democratized through consumer access to GPS, location-enabled cameras, and mobile phones; through geotagging and web-based services such as Google Earth, WikiMapia, and Mappr; and through collaborative map creation projects such as OpenStreetMap[2] or

Urban Tapestries.[3] In a sign of the increasing recognition of the *long tail* of spatial data collection by members of the public, Google piloted public reporting of corrections for its map data in 2007, as well as tracking mobile phones running Google Maps to generate live traffic flow information.

These new modes of content creation present a challenge to traditional top-down modes of scholarly content creation in the Humanities based on authorial expertise and peer recognition, and create considerable unease if not resistance among museum curators, academic historians, and others. Yet they provide new opportunities to collect content which might not (yet) be the subject of scholarly interest and to give interested amateurs and members of the community a voice which they would not have in traditional scholarly practice. This essay reviews the genre and some key components of online digital historical atlases, and addresses the ways in which community participation can be leveraged while still providing mechanisms for professional historians to engage in peer-reviewed content production.

Databases and digital media

Digitally published historical content – in contrast with more traditional textual modes of publication such as books and academic articles – is often characterized by the scale and detail of the information which can be represented; that is, the ability to deliver large amounts of *granular* information (broken down into individual observations of entities, their attributes, their relationships and their spatial and temporal locations) and to expose original documents to the reader's examination. These modes of delivery provide extensive factual detail rather than a coherent narrative or interpretation. The reader – if "reader" is still the correct term – is faced with making his or her own sense of the material, with the task of tying this information together into a coherent view.

This shift of the onus from producer to consumer is of course not a problem unique to historical data. Traditional modes of validation have been overtaken by the exponential rise and fluidity of web resources. New ways of taming this monster are developing, from review sites embedded in a community of users (such as the UK's Intute academic web site review service)[4] through heuristic methods of promoting "relevant" resources in search engines (such as Google's cross-referencing algorithms), to social bookmarking and rating of web resources (such as Diigo[5] and CiteULike[6]).

Digital historical content is most easily delivered through a conventional top-down approach, in which experts construct a narrative which delivers interpretation, supported by referencing and in some cases qualitative or quantitative appendices, in the same way as they would for a published volume. In this model (e.g. the *Encyclopaedia of Chicago*[7]), the online historical encyclopedia genre is an electronically delivered book, liberated from the cost constraints of size, color images, and multimedia resources. The book serves as narrative scaffold which can be enriched with interactive maps, timelines, and granular data, providing rich links, background sources, and searchability.

As producers shift towards born-digital historical databases, there is a developing genre of digital atlases/encyclopedias in which the user's experience is structured through higher level overviews, leading to more detailed entries and eventually to databases of individuals, places, events, documents and, in some cases, census and statistical information. Projects such as the Dictionary of Sydney,[8] the Prosopography of Anglo-Saxon England[9] or the VICODI concept browser[10] prototyped in EuroHistory. Net[11] are moving towards more structured methods of searching and organizing content, including directed and facetted browsing, fully integrated use of maps and timelines, and display and navigation of relationships between resources. While development of data is still generally top-down by a group of experts, the extensive use of a content management approach rather than static resources invites the development of interfaces which manage content creation beyond the confines of the project (either through an editorial workflow, which is likely to be favored by most historians for accountability, or through a wiki style process of community checking) and allow users to actively remix content for their own use.

Collaboration and public content generation

The Electronic Cultural Atlas Initiative[12] attempted a form of cultural (mostly historical) content collaboration over 10 years ago, using an open access clearinghouse to which content developers could contribute spatial datasets, which could then be mapped together or downloaded for offline use. This initiative failed to develop much momentum because the size and composition of the target (academic) audience did not lead to a critical mass of content and there was no adequate reward structure (primarily academic recognition) for the costs of participating (the technical expertise required to prepare spatial data and the effort of metadata creation with somewhat primitive software and no immediate benefits). More recent crowd-sourcing initiatives depend on simple, highly intuitive interfaces and a much larger population of contributors, leveraging the enthusiasm and spare time of the long tail, rather than the more focused interest and time poverty of a finite group of academic participants.

The rise of user-generated content on the web has been a significant development over the last few years, whether one espouses the label Web 2.0 or not. The idea that the general public can comment on and tag historical databases, let alone contribute content to them, challenges the top-down authority-based model of historical content creation and validation used in most historical research and public information delivery projects. Yet the success of general reference resources such as Wikipedia – including the surprising, if unknowable, quality of much of its historical content – and the empowering of the holders of recent historical knowledge through community-based history projects, such as Brisbane Stories[13] or Urban Tapestries,[14] point out the fallacy of ignoring both the resources and enthusiasm of the general public and the rich sources of material which can be available within the community.

An intermediate step towards the incorporation of user-generated content into historical encyclopedia projects can be seen with the use of a tightly controlled volunteer group by the Dictionary of Sydney to seek out and identify historical sources in libraries, archives, and local collections, and build databases of granular historical information which are entered in the same database and subject to the same editorial process as the articles written by academic and local historians. A Web 2.0 implementation would simply take this one step further by allowing an open-ended group to collect and contribute content and provide their own interpretations. Such an approach would allow members of the public to contribute information resources such as old family photographs, personal accounts, and articles on topics which might not be of interest to academic historians, or to identify locations, dates, or people in the images collected; mechanisms can easily be put in place to allow vetting of new material either by volunteer members of the community or by an editorial committee.

Rather than being afraid of this content, historians should welcome it as a potential source of new material and a vehicle for engagement with communities of interest. Even if the core data are tightly controlled, it does not preclude a two-layer system in which users can add new information which is clearly separate from the core, tag and rate information according to their interests, create connections between resources, engage in discussion and annotation of resources, follow other users' tags to identify interesting resources, and last but not least construct their own narratives, save them within the system, and publish them for others to read.

Spatio-temporal context

No historical observation or interpretation can be separated from either its spatial or temporal context, although these are often implicit rather than recognized. Spatial and temporal data provide context through the location of relevant information by spatial and temporal searching, through analysis of spatial, demographic, topographic, and environmental relationships, and through the visualization of these relationships using maps, timelines, and other forms of graphical representation.[15]

History commonly deals with events, trends, social and economic structures, and so forth which have clear spatial and temporal patterning, yet spatial and temporal information are often used only to provide maps or timelines for broad contextualization or for the illustration of particular points of argument; they are rarely used as a primary structuring metaphor, a search mechanism, or a basis for spatial or temporal analysis.

When spatial and temporal data are used analytically, they can bring significant new insights to existing historical questions. Cunfer,[16] for example, uses GIS to reassess the origins of the dustbowl and challenge the assumptions that it was due to overintensive agriculture. Ell and Gregory[17] use census data to re-examine the spatial demography of the Irish potato famine. The application of GIS methods to historical data is becoming more common, exemplified in a series of publications edited by Anne Knowles,[18] but still remains a specialist domain; traditional historians are generally not comfortable with disaggregating textual argument into databases and with the acquisition of the necessary technical skills to deal with them.

Collection

Within the last few years, simple mashups (web applications which generate pages connecting with multiple data sources) involving web services such as Google Maps allow contributors of content to easily add geographic data by onscreen digitizing. While this will often be on a modern map or satellite image, it can also be set up fairly easily to use historic map backgrounds specific to a particular application. User-contributed content will typically be identified by point locations (owing to ease of implementation), or through named places (towns, suburbs, administrative boundaries, street addresses) which can be used to geocode location. Historical text may also be susceptible to automatic geocoding through text analysis aimed at identifying and disambiguating references to places which can be identified in a gazetteer – see for example work of the Perseus project.[19]

With the increasing availability of consumer level GPS and location-enabled mobile phones and cameras, it is becoming increasingly easy for community participants to provide locations for their images and other observations, either directly or though matching camera datestamps with a GPS tracklog. Recent developments in automatic feature matching between digital photographs to generate 3D models and interpolate the camera position – Microsoft's PhotoSynth[20] and the University of Leuven's ARC 3D modeller[21] – provide as yet untapped possibilities for community input on extant tangible heritage.

Searching

Spatial and temporal searching has great potential as a means of organizing access to historical information in publicly accessible systems. It has significant advantages over textual metadata searching which is dependent on agreed (if implicit) vocabularies. Unlike classificatory schemes, for example administrative units (countries, states, counties, cities) or named time periods,[22] the spatial and temporal "dimensions" are not language dependent and are readily translated between alternative measures. Library catalog systems typically have provision for location and date applied to the content (as opposed to publication information), yet this is rarely recorded, let alone exposed to search mechanisms; the development of spatial and temporal search within digital libraries is a current topic in library science research.[23]

While spatial searching of point data (single x,y or latitude/longitude pairs) and temporal searching of specifically dated events is relatively straightforward, the identification and rating of relevance in spatially and temporally extensive data is more problematic. For example, the Electronic Cultural Atlas Initiative spatial data clearinghouse sorts results by spatial relevance based on proportional overlap between the target search and resources in the database, and provides visualization of the density distribution of map resources in order to identify hotspots of spatial data in unevenly distributed historical data. However, no equivalent methodology is applied to the temporal dimension. Significant research is needed to develop appropriate ways of combining and weighting spatial and temporal data in searching historical databases – perhaps emphasizing the spatial dimension for longitudinal questions, and the temporal dimension for study of historical snapshots – and to provide intuitive methods of describing searches in terms which reflect the understanding of users rather than the computational methods. Furthermore, historical events often have uncertain and/or diffuse spatial and temporal boundaries,[24] and to date formal methods for defining spatial and temporal uncertainty in historical data have not been widely adopted.

Visualization

Spatial and temporal data allow visualization of context through maps and timelines. Maps and timelines are powerful yet under-utilized modes of communicating history, but neither is adequate on its own. Static maps can be used to support specific points of argument, but fail to represent processes and change. Timelines are a useful way of organizing sequences of events, but conflate events spatially (it is worth noting that many historical timelines divide the timeline into bands representing geography). The spatial and temporal dimensions are rarely used in concert, although some attempt is made to combine them in printed historical and archaeological atlases.[25] Digital methods, on the other hand, open up the possibility of actively linking maps and timelines drawn from rich backend databases, allowing exploratory interaction on the part of the reader. There are numerous examples on the web of Google Maps being linked with a SIMILE timeline to display contemporary and historical events, from dinosaurs and wars to conferences and seismic events.

The HEML project[26] breaks new ground in explicit definition of temporal and spatial mark-up of historical events, the integration of mark-up with text and the potential to build databases from multiple marked-up sources, but timelines and maps are offered as alternative visualizations rather than linked into a single coordinated view. The interactive animated maps of the Salem Witch Trials and the Valley of the Shadow battle theater, developed at the University of Virginia,[27] go further towards illustrating the potential of combining the spatial and temporal dimensions to provide a better understanding of a set of events, although interactivity is limited. On a longer timescale, SahulTime[28] uses the relationship between time, sea level, and coastline to contextualize the history of Aboriginal sites and migration to Australia. However, unlike the HEML visualizations, these examples are not generic visualization tools as they are individually constructed from pre-built animation components rather than live query of a backend database.

GapMinder,[29] InstantAtlas,[30] and GeoVISTA Studio from Pennsylvania State University[31] are among the best examples of interactive data-driven visualizations of spatio-temporal data, offering multiple coordinated views of tabular data, statistics, maps, and graphs, filtered by time and other attributes. They allow users to interact with and visually analyze the spatio-temporal structure of the data, but do not address the inter-relatedness of historical information, dealing only with time-stamped observations. In order to model history effectively for geographic analysis and presentation, we need to move beyond simple time-stamped observations to modelling the inter-relatedness of historical events.

Modelling historical events and relationships

The modelling of historical events as spatial and temporal entities in a database provides a generic structure for representing history which has been surprizingly neglected. There is an extensive literature on spatio-temporal modelling in geography dating back to the 1990s[32] leading to an awareness of events as the shaping force behind a wide variety of non-static geographic entities. In the museum world there is increasing recognition that object descriptions are not static through time (see ABC Harmony Data Model v. 2 Section 2.2[33]) and collection management systems are moving from simple object description towards event-based models to track the history of objects, their conservation, display, storage, and publication.

Since their adoption by Europeans in the eighteenth century, some form of gazetteer has provided the spatial infrastructure for most geographic and historical description, yet they have largely been structured as atemporal geographic databases to which temporal information, in the form of date-stamped alternative names, is added as an afterthought, if at all. Gazetteers – in the sense of indexes of named places existing and changing through time – are a fundamental historical framework, increasingly provided by extensive infrastructure projects such as the Great Britain Historical GIS[34] and the China Historical GIS.[35] Unlike geographically inspired gazetteer projects, such as GeoNames[36] or the Getty Thesaurus of Geographic Names,[37] these historical GIS projects are built around administrative boundary changes and demographic data, providing essential infrastructure for historical analysis.

Mostern and Johnson[38] propose that, within historical disciplines, gazetteer entries should be treated as historical naming events rather than geographical entities and that these events should be represented in a Web 2.0 backend database allowing collaborative editing and construction of user-defined maps and timelines. By changing the emphasis in this way from geography to place – from physical location to human construction of space – the historical dimension is foregrounded over the purely physical. The system proposed – and developed in prototype using the Heurist collaborative database developed by the author[39] – emphasizes the *relationships* between events and provides explicit methods for recording and for navigating through a web of relationships. We argue that events linked into a web of relationships offer a better reflection of history than a series of snapshot views through time, an aspatial timeline, or rich inventory databases of separate items (people, physical objects, sites, named events etc.).

It is this character of connectedness which is critical to historical databases, in the same way that the linking of data is critical to the World Wide Web. As with the web, this freeform structure brings new challenges in providing effective methods of access, including visualizations capable of making sense of an extended web of complex relationships, and representing this geographically and through time. While projects such as the Temporal Modelling Project,[40] the SemTime project,[41] and the Continuum timeline project[42] attempt a more sophisticated view of relationships within the time dimension, none of them have attempted the visualization of space, time, and relationships together. It is only through embracing this complexity and visualising these three aspects together, rather than reducing them to simplistic aspatial or atemporal snapshots, that we can effectively represent the structure of event-based historical data.

Conclusion

The rapidly shifting domains of social computing and Web 2.0 are fundamentally changing the way that "readers" (consumers/users/students/public) interact – indeed expect to interact – with historical resources on the web, and in museums and visitor centres. Rather than simply delivering top-down authoritative content which can be passively absorbed, the onus is now on information providers to allow the consumer of information to contribute new material. This taps into the creative urge that sites like Wikipedia and Flickr have encouraged, and promotes community involvement and ownership of the resource.

Rather than worrying that the voice of the expert may be swamped by the unruly clamor of the crowd, historians need to find ways of validating information through peer review which recognize expertise rather than qualifications and the cumulative value of small contributions from personal experience. Content should allow remixing, annotation, extension, and saving for reuse, including the possibility of publishing this enriched content for others to view. It should emulate the basic reader needs of paper-based sources – marking pages, highlighting, marginal annotation, note-taking, clipping, and copying. None of this is incompatible with curated editorial content and these functions may be layered to allow ownership and sharing within community groups or student classes.

The way forward is represented by a hybrid genre that we can see in a new generation of historical encyclopedias. In this genre, authoritative historical writing and databases of content assembled by knowledgeable individuals (professional or otherwise) are integrated with tools for remixing and community contributions. Without this genre, serious scholarship will be increasingly swamped by the sheer volume of popular history, with limited attribution and therefore of unknowable quality, in community-generated resources such as Wikipedia.

Digital historical resources which manage editorial control of contributed data including spatial data, layered Web 2.0 functions, extensive cross-linking of resources, and database-driven visualizations of geography, time, and relationships are complex projects requiring close collaboration between historians, information technologists, interface designers, and communities of stakeholders. The end product, however, is likely to be far more than the sum of its parts.

Notes

1 T. O'Reilly, *What Is Web 2.0: Design Patterns and Business Models for the Next Generation of Software*, Available online at: www.oreillynet.com/pub/a/oreilly/tim/news/2005/09/30/what-is-web-20.html (accessed October 22, 2009).

2 http://openstreetmap.org (accessed October 22, 2009).

3 G. Lane and S. Thelwall, *Urban Tapestries: public authoring, place and mobility,* London: Proboscis, 2006.

4 www.intute.ac.uk/ (accessed October 22, 2009).

5 www.diigo.com/ (accessed October 22, 2009).

6 www.citeulike.org/ (accessed October 22, 2009).

7 www.encyclopedia.chicagohistory.org/ (accessed October 22, 2009).

8 http://dictionaryofsydney.org/ (accessed October 22, 2009).

9 www.pase.ac.uk (accessed October 22, 2009).

10 G. Nagypál, R. Deswarte and J. Oosthoek, "Applying the Semantic Web: The VICODI Experience in Creating Visual Contextualization for History," in *Literary and Linguistic Computing* 20(3), 2005, 327–49.

11 www.eurohistory.net/ (accessed October 22, 2009).

12 http://ecai.org (accessed October 22, 2009).

13 http://brisbane-stories.powerup.com.au (accessed October 22, 2009).

14 http://urbantapestries.net (accessed October 22, 2009).

15 I. Johnson, "Contextualizing archaeological information through interactive maps," *Internet Archaeology* 12, 2002.

16 G. Cunfer, "Causes of the Dust Bowl," A. K. Knowles (ed.), *Past Time, Past Place: GIS for History*, Redlands: ESRI Press, 2002, pp. 93–104.

17 P. S. Ell and I. N. Gregory, "Demography, depopulation and devastation: Exploring the Geography of the Irish Potato Famine," in *Historical Geography* 33, 2005, 54–75.

18 A. Knowles (ed.), *Past Time, Past Place: GIS for History*, Redlands: ESRI Press, 2002; A. Knowles (ed.), "Emerging Trends in Historical GIS", *Historical Geography* 33, 2005; A. Knowles and A. Hillier (eds), *Placing History: How Maps, Spatial Data, and GIS are Changing Historical Scholarship*, Redlands: ESRI Press, 2008.

19 G. Crane, "Georeferencing in historical collections," *D-Lib Magazine* 10(5), 2004.

20 http://livelabs.com/ (accessed October 22, 2009).

21 www.arc3d.be/ (accessed October 22, 2009).

22 M. Feinberg, R. Mostern, S. Stone and M. Buckland, "Application of digital gazetteer standards to named time period directories," 2003, available online at: http://ecai.org/imls2002/time_period_directories.pdf (accessed October 22, 2009).

23 M. Buckland, A. Chen, F. C. Gey, R. R. Larson, R. Mostern, and V. Petras, "Geographic Search: Catalogs, Gazetteers, and Maps," *College and Research Libraries Journal* 68(5), 2007, 376–87.

24 I. Johnson, "Mapping the fourth dimension: the *Time* Map project," in *CAA97: Computer Applications in Archaeology 1997* BAR International Series 750, 1999.

25 See for an example J. Black (ed.), *Atlas of World History: Mapping the Human Journey*, New York: Dorling Kindersley, 2005.

26 B. Robinson, *Introduction to HEML*, 2005, Available online at: http://heml.mta.ca/heml-cocoon/description (accessed October 22, 2009).

27 www.iath.virginia.edu (accessed October 22, 2009).

28 http://sahultime.monash.edu.au/ (accessed October 22, 2009).

29 www.gapminder.org (accessed October 22, 2009).

30 www.instantatlas.com (accessed October 22, 2009).

31 www.geovista.psu.edu (accessed October 22, 2009).

32 For example, G. Langren, *Time in Geographic Information Systems*, London: Taylor and Francis, 1992; A. M. MacEachren, *et al.,* "Constructing Knowledge from Multivariate Spatio-Temporal Data: Integrating Geographical Visualization with Knowledge Discovery in Database Methods", *International Journal of Geographical Information Science* 13(4), 1999, 311–34.

33 http://metadata.net/harmony/ABCV2.htm accessed (October 22, 2009).

34 www.gbhgis.org, (accessed October 22, 2009).

35 www.fas.harvard.edu/~chgis (accessed October 22, 2009).

36 www.geonames.org (accessed October 22, 2009).

37 www.getty.edu/research/conducting_research/vocabularies/tgn/ (accessed October 22, 2009).

38 R. Mostern and I. Johnson, "From Named Place to Naming Event: Creating Gazetteers for History," *International Journal of Geographical Information Science* 22(10), 2008, 1091–108.

39 http://HeuristScholar.org (accessed October 22, 2009).

40 J. Drucker and B. Nowviskie, "Temporal modeling: conceptualization and visualization of temporal relations for humanities scholarship," available online at www3.iath.virginia.edu/time/reports/ (accessed October 22, 2009).

41 M. Jensen, "Semantic Timeline Tools for History and Criticism," in *Proceedings Association for Computing in the Humanities/Association for Literary and Linguistic Computing conference, Paris July 2006*, 2006.

42 P. André, M. L. Wilson, A. Russell, D. A. Smith, A. Owens, and M. C. Schraefel, "Continuum: designing timelines for hierarchies, relationships and scale," in *UIST2007: ACM Symposium on User Interface Software and Technology*, Newport, Rhode Island, 2007.

28

Teaching race and history with historical GIS

Lessons from mapping the Du Bois Philadelphia Negro

Amy Hillier

Introduction

The Philadelphia Negro was the path-breaking work of a young W. E. B. Du Bois, optimistically embarking on an academic career with the hope of bringing about social change through research. Du Bois moved to Philadelphia with his new bride in 1896, when he was just 28 years old, to conduct a study about "Negroes" for the University of Pennsylvania and College Settlement Association. The book he produced, now claimed as a classic by the fields of sociology and urban history, was as noteworthy for its social scientific research methods as it was for its findings. Du Bois was committed to empirical research, using observation, key informant interviews, door-to-door surveys, and archival research, in an era dominated by racist-driven theories about why African Americans struggled to get ahead.[1]

For all its value to the then-emerging social sciences, *The Philadelphia Negro* belongs as much to the humanities. Du Bois conceptualized the book as one about the humanity of Philadelphia's black community. He accepted the invitation to conduct the study of Philadelphia's Seventh Ward, a stretch of downtown where 100 years ago 40 percent of residents were black, even though he understood that his white benefactors thought they already knew what was wrong with the people living there.[2] His major achievement was reframing the study from one about "the Negro problem" to one about the problems – primarily "race prejudice" – that blacks were facing. While he criticized blacks throughout the book for contributing to their own problems, he reserved the harshest judgments for whites. The "vastest of the Negro problems," Du Bois concluded, was that the "civilized world" denied the full humanity of "the Negroes of Africa" (386–87). This translated into limited housing, education, and employment opportunities for blacks in Philadelphia.

Despite its significance to both the social sciences and the humanities, *The Philadelphia Negro* is not read widely among college and graduate students today, much less high school students. "Few persons ever read that fat volume ...," Du Bois, himself, wrote, "but they treat it with respect, and that consoles me."[3] Historical

Geographic Information Systems (HGIS) provide a tool and a framework for reviving this respected – but largely ignored – text for teaching and making this dense text more accessible. By translating it into the language of the Internet, students can make connections between nineteenth-century racial discrimination and patterns of racial disparity today through the familiar clicking, linking, networking, and searching. HGIS also serve as a new tool for developing skills in critical thinking, spatial analysis, and working with primary sources. An HGIS of *The Philadelphia Negro* also provides an opportunity to teach students about Du Bois, a complicated intellectual and political figure whose impact on the American civil rights movement has been underappreciated in part because he was so controversial. Finally, teaching *The Philadelphia Negro* with mapping technology can tell us something about the possibilities and limitations of HGIS and provide a chance to assess how far the field has come and where it needs to go.

Mapping the Du Bois *Philadelphia Negro*

Mapping the Du Bois *Philadelphia Negro* is a web-based teaching, research, and public outreach project at the University of Pennsylvania.[4] Its centerpiece is the development and use of a HGIS of the Seventh Ward where Du Bois conducted a door-to-door survey. The Seventh Ward GIS includes information derived from fire insurance maps and the 1900 US census about the age, race, birthplace, occupation, and schooling for the 28,000 residents of the Seventh Ward. Census data also describe the size and make-up of households, indicating who had servants and boarders and who owned and rented their properties. Birth and death records from the Philadelphia City Archives provide additional information about health and healthcare at the time, noting the cause of death, name of physicians, and location of burials.

The building of the Seventh Ward GIS benefited from recent advancements in GIS software and online historical archives. Undergraduate and graduate students from history, social work, urban studies, and city planning digitized manuscript census data from microfilm and scanned records on Ancestry.com. Our team created a base map of the Seventh Ward parcels by tracing georeferenced historical fire insurance maps using on-screen GIS digitizing tools. We balanced this technical and often tedious work with discussions of the racial categories and social class standards used at the time, how Du Bois managed to accomplish so much research alone and without the aid of technology, and the relevance of Du Bois's work to the count of murders – mostly of young black men – updated daily by *The Philadelphia Inquirer*. The resulting Seventh Ward GIS, accessible through the Mapping Du Bois website, allows visitors to identify who lived at a specific property by typing in the address or clicking on a map, query the database of individuals and households to find people matching a particular profile, and create thematic maps.[5] Once we completed the HGIS, our challenge transitioned from one of collecting and integrating large amounts of data to trying to retell the story of Du Bois and *The Philadelphia Negro* in order to generate dialog among students about how race and racism shape our lives, today.

Maps and *The Philadelphia Negro*

GIS is an appropriate tool for teaching Du Bois' humanities lessons, in part because Du Bois used maps to do this himself. Three maps appear in *The Philadelphia Negro*. The first, a simple ward map with statistics along the side, shows the concentration of African Americans within the Seventh Ward and underscores Du Bois' narrative description of the area (see Figure 28.1). The map shows how small and centrally located the Seventh Ward was relative to the 140-square mile city of Philadelphia. The ward-level statistics along the side indicate that the Seventh Ward had more than twice as many black residents as any other area. While neither the map nor summary statistics can communicate the experience of living in a crowded area, it does provide some sense of the relative density. Another of his maps indicates the migration patterns of blacks, Italians, Jews, and Irish within Philadelphia between 1790 and 1890 (see Figure 28.2). Again, while this does not provide insight into the experience of

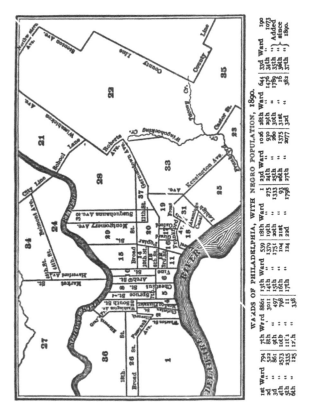

Figure 28.1 Black Population in Philadelphia by Ward, 1890. This map from *The Philadelphia Negro* shows how small the Seventh Ward was relative to the whole city while the statistics at right show that the Seventh Ward had more blacks than any other ward.

Source: The Philadelphia Negro, p. 59.

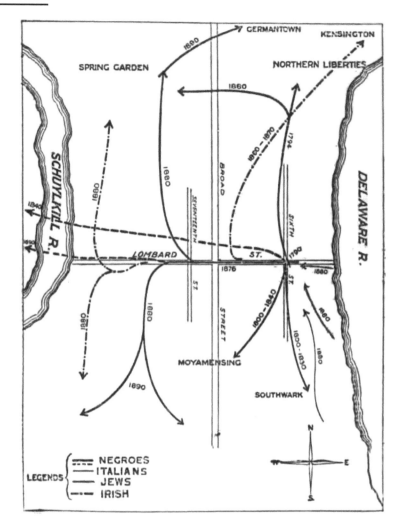

Figure 28.2 Black Migration, 1790–1890. This map diagram shows that Blacks, like Jews, Italians, and Irish, migrated to new parts of the city over time.

Source: The Philadelphia Negro, p. 306.

migration, it does communicate how movement characterized the experiences of all four of these groups, emphasizing the dynamic nature of urban life and residential patterns.

The premier map in *The Philadelphia Negro* was the pull-out color-coded parcel map showing social class for households at each property Du Bois surveyed in the Seventh Ward. Du Bois used the map to underscore his argument that the black community was not homogeneous, showing the class structure within even this small area. Using the categories of "middle class," "working people," "poor," and "vicious and criminal," he showed how blacks of all classes lived together, frequently along small back streets and alleys. In color-coding individual properties, Du Bois provided rich spatial detail and reminded us that behind summary statistics were thousands of individuals, each with their own stories and struggles.

GIS and *The Philadelphia Negro*

The Seventh Ward GIS can help teach lessons about racial discrimination by introducing students to real African Americans who struggled to make ends meet at the turn of the nineteenth century. In *The Philadelphia Negro*, Du Bois provides brief profiles of individuals whose experience of employment discrimination he documented in his section on "color prejudice." He used the social science convention of hiding names to protect his research subjects, referring, for example, to "A⁻" who had been a "porter at a great locomotive works for ten years" (343). But how can we expect students to empathize with nameless people or the summary statistics that were even more prevalent in *The Philadelphia Negro*?

Through the manuscript census data, made accessible in the online HGIS through an address search, name search, or simply by clicking on properties across the ward, we can meet Oscar and Katie Steward of 407 Iseminger Street. Oscar, a black waiter born in Alabama, and Katie, a white housekeeper from Ireland, had been married for 23 years. Du Bois provided a variety of statistics about such couples in his section on "intermarriage of the races," but he never provided names or addresses. Similarly, the Drapers living at 1113 Rodman Street can help illustrate the concern Du Bois had about families with daughters renting rooms to young adult men. "In such ways the privacy and intimacy of the home life is destroyed," Du Bois warned, "and elements of danger and demoralization admitted" (194). Seeing that daughters Lucille (13 years old), Hellen (6 years old), and Edner (8 years old) lived in the same home with lodgers Paul (21 years old) and Henry (26 years old) may make online visitors more sympathetic to Du Bois's Victorian moralizing.

The Seventh Ward GIS relies on census survey data, so it provides none of the details that a diary or extensive family records do. But the simple fact of who lived together reveals much about family structure, social class, and conditions of overcrowding. The Saunders family at 412 South Camac Street, for example, included nine people – six adults and three children – living in a small row house typical of the Seventh Ward's side streets. The adults held common occupations for blacks – laborer, housekeeper, and porter – and were all able to read and write, as were most black residents of the Seventh Ward. At age 45, Harriet Saunders had given birth to 14 children, only five of whom were still living.

A non-spatial database of individual-level census data could provide these same kinds of teaching examples; the added value of GIS comes in using geography and spatial relationships to make discoveries. Du Bois described in detail certain blocks within the Seventh Ward as though he were giving a tour. "Passing up Lombard, beyond Eighth, the atmosphere suddenly changes, because these next two blocks have few alleys and the residences are good-sized and pleasant," he wrote. "Here some of the best Negro families in the ward live. Some are wealthy in a small way, nearly all are Philadelphia born, and they represent an early wave of emigration from the old slum section" (60). By using the GIS map to identify the households along these blocks, we find Sarah Coker, a 39-year-old dress-maker born in Pennsylvania, David Tinitt, a 44-year-old storekeeper from Maryland, and Alexander Coats, a 62-year-old caterer born in Pennsylvania, all of whom likely qualified as "middle class" by Du Bois' standards.

The HGIS also presents students with thematic maps to read for spatial patterns. For example, they can create a map showing the dominant race/nationality of residents which reveals a great mix of black and white Philadelphia-born, black southern migrants, Russian and Polish Jewish immigrants, and other European immigrants when the entire ward is viewed. But upon closer inspection at the scale of individual blocks, it becomes clear that blacks frequently lived along the side streets in smaller houses while Philadelphia-born whites lived in the large row houses along the main streets of Spruce and Pine. They can also correlate patterns of race, homeowners, and overcrowding by viewing thematic maps of each one at a time and seeing that blocks with blacks were frequently more crowded and had few or no homeowners.

GIS also provides a way to investigate what happened to the neighborhood Du Bois studied. Animation tools hold promise for making GIS as effective at showing change over time as over space, but the simple method of using different map layers to represent different time periods can also be effective. At the end of the nineteenth century, the Seventh Ward was the heart of black Philadelphia, home to the highest concentration of blacks in the city, which was, in turn, home to more blacks than any other northern city. At the beginning of the twenty-first century, it is one of the most expensive residential areas in the Philadelphia region with a population only 7 percent black in

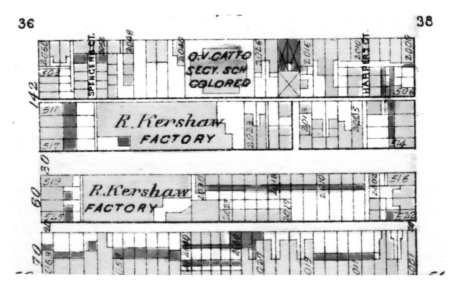

Figure 28.3 Land Use in the Old Seventh Ward, 1895. At the time of Du Bois' study, the Seventh Ward was full of factories, like this one at 21st and Naudain Streets, and small residential alleys like Spencer and Harpers Courts.

Source: Bromley Fire Insurance Map, courtesy of the Philadelphia Athenaeum.

a city that is now more than 40 percent black. Using georectified historical fire insurance and land use maps and current aerial photographs from GoogleMaps, we can see some of the specific changes. For example, we can see that the Kershaw Factory on 21st Street became a lamp factory in the 1960s and was replaced more recently by new housing. Of even greater symbolic value is the location of the elite Lombard Swim Club at the site where the Octavius Catto Secondary School for Colored once stood (see Figures 28.3, 28.4). Students can discover these kinds of connections on their own, leading them to question how such changes came about and their meaning for the racial justice for which Du Bois called.

The story of Du Bois

One cannot teach *The Philadelphia Negro* without teaching about Du Bois. Despite his protestations that science should be objective, he voiced concerns about racism based on his own experience, not just that of the Philadelphia residents he interviewed. "There is always a strong tendency on the part of the community to consider the Negroes as composing one practically homogenous mass," he lamented. "Nothing more exasperates the better class of Negroes than this tendency to ignore utterly their existence" (309–10). Such was his experience as a member of the black elite who had little in common with the large number of poor blacks whom he studied in Philadelphia.

Students familiar with Du Bois may be surprised too by the Du Bois who emerged in the pages of *The Philadelphia Negro*, with its fairly conservative analysis of social and economic inequality and reflection of the Victorian sensibilities of a New England upbringing. Du Bois' later political activism, including his long-time service as the editor of the NAACP's magazine, *The Crisis*, and his leadership in the international Pan-African Movement, came about in part as a response to his frustrating experience with academia. At the outset of his work in Philadelphia, Du Bois was optimistic – even idealistic – about the ability of research to bring about social change. "The world was thinking wrong about race, because it did not know. The ultimate evil was stupidity," he wrote in his autobiography, *Dusk to Dawn*. "The cure for it was knowledge based on scientific investigation" (58–59). But by the time he completed *The Philadelphia Negro*, he had abandoned his belief that empirical data would sway the public and lead a "benevolent despot" to address the conditions in which blacks were living. He also had to accept that, despite his respected scholarship, the University of Pennsylvania had no intention of hiring a black man for its permanent faculty. As Du Bois explained in *The Philadelphia Negro*, white men started life knowing that their success depended primarily upon their talent and effort. "The young Negro," on the other hand, "starts knowing that on all sides his advance is made doubly difficult if not wholly shut off by his color" (327). Even a Massachusetts-born, Harvard-trained PhD could not escape that fact.

The meaning of all this

Du Bois transitioned from the detailed, statistics-laden text of the first 384 pages of *The Philadelphia Negro* to his conclusions in a section titled, "The Meaning of All This."

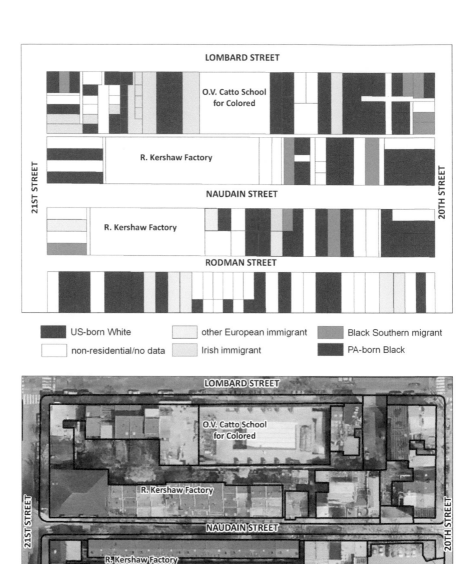

Figure 28.4 Race/ethnicity and migration status by household in the old Seventh Ward, 1900. Census data in the Seventh Ward GIS provide the basis for this thematic map showing race/ethnicity. Throughout the Seventh Ward, rich and poor, black and white, immigrant and Philadelphia natives lived in close proximity.

Source: 1900 U.S. Census.

He suggested that the original question he posed – "What is the Negro Problem?" – was no different from the questions about human welfare, poverty, righteousness, and identity "over which the world has grown gray" (385). While *The Philadelphia Negro* is remarkable for the unparalleled insight it provided into the daily lives and struggles of black Americans, its value ultimately comes from the timelessness and universality of the issues he raised relating to human experience. Beyond transporting students to a specific time and place to appreciate a different culture, *The Philadelphia Negro* challenges them to question why their world looks as it does today.

The Philadelphia that residents and visitors see today, as they stroll along the southern edge of the Old Seventh Ward, which is now the hippest destination in the city, or as they move guardedly through low-income areas challenged by drug activity and violence, has grown out of the industrial city that Du Bois studied. Students of other post-industrial cities will find the story familiar, and those living in suburban communities are encouraged to place themselves and their families within the larger context of metropolitan growth, white flight, and sprawl. *The Philadelphia Negro* challenges all of them to consider the relevance of Du Bois' main concern, that whites were denying the humanity of their black neighbors. Does this still ring true?

GIS provides a way of transforming *The Philadelphia Negro* into a web-based interactive experience for high school students and others who might otherwise never approach it. Mapping Du Bois is one of a growing number of HGIS projects that applies new technology to old stories, digitizing historical maps and creating databases from primary sources. It was one of four such projects featured in a session at the symposium on Geography and the Humanities organized by the Association of American Geographers at the University of Virginia in June 2007. In addition to congratulating the many HGIS project creators for their accomplishments, discussant Rickie Sanders, a Temple University geographer, challenged us to consider the value of these contributions for the humanities. What, she asked, can GIS teach us about the human condition?

The use of GIS in the humanities holds great promise for teaching us about what it means to be human – about suffering and discrimination, about hope and achievement – but this promise has not yet been realized. *What can GIS teach us about the human condition?* Dr. Sanders seems to caution us: Do not get caught up in the novelty of the technology or color maps. GIS only has a place in humanities scholarship if it can contribute to critical and imaginative thinking, understanding, and appreciating the experience of others, and making sense of our own human experience.

Our work with Mapping Du Bois had taught us that an HGIS version of the Seventh Ward offers a complement – not a substitute – to *The Philadelphia Negro*. Other representations of the history of the Seventh Ward, such as Diane McKinney-Whetstone's historical fiction work *Tumbling*, which tells the story of the displacement of black residents and institutions by the planned-but-never-built cross-town expressway, should be studied, as well. The Seventh Ward GIS is only one component of the Mapping Du Bois project; we have looked to other formats to more fully capture the humanities themes that Du Bois introduced. We have produced a documentary, filmed by two high school students, that fits interviews with scholars around the story of the chance meeting of a current white resident of the Seventh Ward and a black woman whose grandmother was born in his house and whose great-great grandmother, born a slave, died in his house. In our board game, students choose a character based on a real person who lived in the Seventh Ward and then try to navigate the everyday challenges of nineteenth-century life in a quest to improve their social class. Advancing through the map-based board game requires them to act out and draw topics from *The Philadelphia Negro*, answer fact-based questions, or relate quotations from the book to their lives, today. Finally, we worked with the Philadelphia Mural Arts Program to design and paint a mural honoring Du Bois and the residents of the Seventh Ward whom he studied.

HGIS has helped us to teach some of the important humanities themes in *The Philadelphia Negro*, and the Mapping Du Bois project has, in turn, taught us about the promise and limitations of HGIS. The data and tools to map demographic and land use data for historical communities are readily available, but we struggle to map some the concepts that matter most. What data and spatial data formals are available to represent the sting of racial discrimination shared by Seventh Ward residents? Or the hope of European immigrants and southern black migrants upon moving to Philadelphia and seeking a better life? What GIS tools allow us to show the network of complicated relationships between black and white residents of the Seventh Ward, perhaps living in close proximity but likely sharing little of their social lives? Research on historical GIS has already identified the need for documenting data sources in ways more familiar to historians, developing methods for representing uncertainty, and allowing for analysis of change over time. If the "H" in HGIS is used to represent the humanities and not just history, we will need the field and the tools upon which it relies to move in the direction that Dr. Sanders points.

Notes

1 See T. Zuberi, "W.E.B. Du Bois's Sociology: *The Philadelphia Negro* and Social Science," *Annals of the American Academy of Political and Social Science* 595, 2004, 146–156; M. B. Katz and T. J. Sugrue, "Introduction: The Context of The Philadelphia Negro," in Katz and Sugrue (eds), *W.E.B. Du Bois, Race, and the City: The Philadelphia Negro and its Legacy*, Philadelphia: University of Pennsylvania Press, pp. 1–37; E. Anderson, "Introduction," *The Philadelphia Negro*, Philadelphia: University of Pennsylvania Press, 1996, pp. ix–xxxvi.

2 Katz and Sugrue, "Introduction: The Context of The Philadelphia Negro."

3 W. E. B. Du Bois, *The Autobiography of W.E.B. Du Bois: A Soliloquy on Viewing My Life from the Last Decade of Its First Century*, New York: International Publishers, 1968, p. 198.

4 Funding for this project has been provided by the National Endowment for Humanities, University of Pennsylvania Research Foundation, Robert Wood Johnson Health & Society Scholars Research & Education Fund, Samuel S. Fels Fund, Hartford Humanities Center, and Penn Institute for Urban Research.

5 The Seventh Ward GIS can be accessed through the project homepage, www.mappingdubois.org, or directly from http://venus.cml.upenn.edu/UPennSD_PhilaNegro/ The online mapping application as created by Azavea, Inc.

29

Ha'ahonua

Using GIScience to link Hawaiian and Western knowledge about the environment

Karen K. Kemp, Kekuhi Keali'ikanaka'oleohaililani, and Matthews M. Hamabata

> not science, not religion, not culture ... just a way to converse about what needs to be done with the deepest respect and no compromise.
>
> Kekuhi Keali'ikanaka'oleohaililani

In 2007, a cross-cultural team of scholars, scientists, technologists, and indigenous practitioners gathered to build the foundation for a "geocollaboratory" that would bring together indigenous Hawaiian knowledge with Western research science on Hawai'i Island. Immersion learning in indigenous knowledge systems enabled Western scientists to see how Hawaiian knowledge systems could be synthesized with their own disciplines. Hawai'i knowledge practitioners saw how Geographic Information Science could provide a common basis for interpreting the essential relationships with the land that drive Hawaiian life and practice. True to the transformative power of Hawai'i, the team immediately outgrew their initial concepts and committed to *Ha'ahonua*, a revised collaborative vision to weave a unifying theme among a variety of community programs and research projects on Hawai'i Island. Led by The Kohala Center (TKC), the Edith Kanaka'ole Foundation (EKF), and the Redlands Institute, Ha'ahonua is now actively bringing together indigenous Hawaiian and Western knowledge; building GIS tools for conservation and management; and working to demonstrate practical alternatives to facilitate the revitalization of Hawai'i Island as a self-reliant, sustainable system.

This essay begins by exploring how geography provides a basis for the integration of Hawaiian knowledge and Western science. How this can be implemented in the context of GIS is discussed next. The essay ends with a brief overview of current Ha'ahonua activities.

Hawaiian knowledge as humanities and science

Hawaiian epistemology and cultural traditions are holistic and systemic. They encompass what Westerners label variously humanities, arts, social and natural sciences, including language, history, dance, religion, indigenous arts, crafts and skills, politics, economics, agriculture, forestry, and ecology. Hawaiian native traditions and practices reflect a deep connection with the *'aina* (land). The centrality of place pervades Hawaiian life and culture, not only in terms of geographic entities (e.g., sacred structures or natural features) but also cultural identity.

A notable example is the concept of the *ahupua'a*, a traditional division of land that still operates in contemporary island society. The *ahupua'a* is both a geographic marker and a cultural system. In Native Hawaiian epistemology, ocean and terrestrial ecosystems are integrated, and so the *ahupua'a* extended from the mountains into the sea. True to the multidimensional understanding of the *ahupua'a*, the sky (the heavens, the atmosphere) is also included. Corresponding with watersheds, the *ahupua'a* provided communities with access to the source of fresh water and to the ocean. Behavioral and cultural prescriptions obligated lineal descent groups to sustain each particular ecosystem in the *ahupua'a*, as goods moved vertically through the *ahupua'a* and horizontally across the island.

Prior to the arrival of Westerners, Hawaiian society had developed a complex and highly successful set of cultural practices that demonstrated comprehensive understanding of natural processes and environmental sustainability. As a community that depended on verbal and other non-written forms of communication, the record of these practices was embedded in language, myths, place names, dance, rituals, and other elements of a powerful cultural heritage. With the arrival of Europeans (usually marked by Captain James Cook's arrival in 1778), the overthrow of a sovereign government by American planters and missionaries (a series of events in the 1880s and 90s), and the forced annexation to the United States (1898), Hawaiian society experienced colonization and hybridization. The vital link between land and people was nearly severed were it not for the perseverance of lineal descent groups. These indigenous knowledge-holders kept Hawaiians' rich heritage, their sacred sense of ecology alive, often out of public view.

In their outward-looking society, Hawaiians, particularly the *ali'i*, the royal ruling class, quickly adopted writing once it was developed by the missionaries in the early 1800s. A recognition of the threat to their cultural traditions posed by the rapidly growing Western culture propelled a number of native Hawaiian scholars to record legends, stories, songs, and other cultural knowledge in articles in a series of Hawaiian language newspapers published in the Islands between 1834 and 1948. Fortunately, these have been preserved and are now freely available online in digital form, mostly still in the original Hawaiian. They serve as a major resource for anyone involved in cultural studies in Hawai'i.

In recent years, renewed interest in Hawaiian Studies has fostered projects that attempt to preserve, document, archive, and communicate indigenous knowledge and practices before they disappear. Educational institutions like Hawai'i Community College support Hawaiian Studies departments offering degrees and coursework focused on the Hawaiian lifestyle – past, present, and future. Native Hawaiian Charter Schools have emerged to bolster the language and cultural base of Hawaiian society. And Hawaiian cultural organizations are stepping forward to share their knowledge and sense of ecology.

At the same time, there is increasing interest in understanding the scientific basis of Hawaiian indigenous knowledge, which, like Western science, is empirical, based on centuries of careful observation and experimentation. While not published or peer reviewed in the Western sense, this empirical knowledge was recorded and transmitted through proverbs, legends, and other cultural practices. Environmental experimentation within the *ahupua'a* was demonstrated by the careful tending of the landscape and construction of various landscape "improvements" (in Western parlance) such as fish ponds and stone walls. As Sam 'Ohukani'ōhi'a Gon III, the former Director of Science for the Nature Conservancy of Hawai'i, wrote in a now archived webpage "Any knowledge that could not be practically replicated, or gave inconsistent results, would likely not be further promulgated. ... in a very practical sense, Hawaiian knowledge depended on reliability of results, another hallmark of the scientific method" (informal web communication, 2005).

For the Western scientific community, the Island of Hawai'i is a microcosm of the planet. The Island has long been noted as a world treasure by scientists because of the variety of its terrestrial and ocean ecosystems and unique geologic activity. The island's scale (4,000 square miles; 170,000 residents), its bounded system, and the intense pressure at the intersection of human and natural systems make the Island ideal for the development of systemic approaches to global environmental challenges. Thus, the Island of Hawaii serves as a model *of* and *for* the world: island solutions are global solutions.

However, in the same way that the tourist economy has had an impact on local lifestyle, this academic attention has sometimes been problematic. Island communities and their Native and non-Native members, environmentalists, and Western scientists have come into conflict over the use, access, and management of resources with both natural and cultural significance. Perhaps the most visible of these struggles revolves around the proposed expansion of the astronomical observatories, considered the world's best, on Mauna Kea, a mountain of significant sacred importance to the indigenous culture.[1]

Fortunately, awareness of the value of Hawaiian native knowledge to many aspects of life in Hawai'i today is growing. A leading proponent of this awareness is The Kohala Center in Waimea on the Island of Hawai'i. The Center requires that all Western research and teaching programs that it sponsors are oriented to the island's cultural and spiritual, and natural landscape. Working with Native Hawaiian partners such as the Edith Kanaka'ole Foundation, The Kohala Center's programs embrace Hawaiian culture and knowledge as part of the contemporary world, knowledge that contributes to the world's future with regard to issues of natural resource management and the alignment of human and natural systems. The collaborative work between indigenous knowledge holders and The Kohala Center affirm that such knowledge should not to be relegated to a romantic past but is a part of a dynamic future.

Linking Hawaiian knowledge and GIScience

Geographic Information Science (GIScience) is a subdomain of information science that focuses on the collection, transmission, storage, manipulation, and retrieval of *geographic* information. GIScience addresses the nature of geographic information and applications of geospatial technologies (including geographic information systems (GIS) and related technologies such as remote sensing and global positioning systems). The recently published first edition of the *Geographic Information Science and Technology Body of Knowledge* provides a means of articulating this diverse field.[2]

GIScience provides an intellectual framework for formalizing place-based concepts in such a way that they can be digitized and stored in the computer. Much of the core knowledge of GIScience involves concepts and methods for representing the physical world in the computer – which presents obvious challenges when dealing with complex natural and human systems. In an article musing about the impact of cultural differences in the development of indigenous GIS, Mark[3] noted "[s]patial data infrastructures encode entity types or feature codes in order to enhance the semantics of geospatial data. But such codings might not add value from an indigenous perspective, unless data are also encoded according to the categorical systems of the indigenous people." One of the challenges in linking Hawaiian knowledge and Western science is to find a means of formalizing Hawaiian knowledge about the landscape in a manner that represents their understanding of the *'aina* and all of its inherent relationships while remaining cognizant of how a technological representation may affect that knowledge.

Linking Hawaiian knowledge and GIScience is an effort that falls within the domain of Digital Humanities, an area that is currently undergoing intensive expansion, both scholarly and technological. The US National Endowment for the Humanities has been a strong supporter of this effort and in 2008 established an Office of Digital Humanities. In many ways, definitions of digital humanities correspond to those of GIScience:[4]

> It works at the intersection of computing with the arts and humanities, focusing both on the pragmatic issues of how computing assists scholarship and teaching in the disciplines and on the theoretical problems of shift in perspective brought about by computing ... Its object of knowledge is all the source material of the arts and humanities viewed as data.

John Unsworth defines *humanities computing* as "a tool for modeling humanities data and our understanding of it" and said:

> We are by now well into a phase of civilization when the terrain to be mapped, explored, and annexed is information space, and what's mapped is not continents, regions, or acres but disciplines, ontologies, and concepts. We need representations in order to navigate this new world, and those representations need to be computable, because the computer mediates our access to this world ... We should not refuse to engage in representation simply because we feel no representation can do justice to all that we know or feel about our territory. That's too fastidious. We ought to understand that maps are always schematic and simplified, but those qualities are what make them useful.[5]

The link to information science through the need to represent digitally, and in the case of integrating Hawaiian knowledge with Western science specifically with GIScience, moves beyond technology and into issues of epistemology and perhaps into the realm of ontology.

Unsworth goes on:

> Much of this map-making will be social work, consensus-building, compromise. … Consensus-based ontologies … will be necessary, in a computational medium, if we hope to be able to travel across the borders of particular collections, institutions, languages, nations, in order to exchange ideas.

Unsworth's call for collaboration is key. Through the principles and concepts of GIScience, a structured and productive means of collaboration between humanities scholars, natural scientists, and the Hawaiian community is possible. This is true simply because these dramatically different worldviews overlap in time and space.

Expanding digital humanities into the geographic domain

Very few digital humanities resources available now explicitly incorporate the geographic domain as part of the intellectual content provided. Promotional materials for the recently published *A Companion to Digital Humanities* suggests it is "a thorough, concise overview of the emerging field of humanities computing."[6] However, the applications selected for inclusion all demonstrate the focus of the field remains on textual analysis. A 2003 encyclopedia entry on *Humanities Computing* by Willard McCarty[7] begins with reference to an "intellectual map of humanities computing," yet there is nothing at all in this map that shows the geographic component.

Traditional humanities scholarship often involves working with a wide variety of materials in order to build a complex, interpreted reconstruction of events and processes.[8] While it has not been a significant, explicit component of research done in the disciplines focusing on the preservation, transmission, and interpretation of the human record, geographic information is nevertheless a fundamental attribute of much of the data collected and studied in these fields. Geographic references appear in texts, are part of the documentation stored with images of historic sites and events, and are used to differentiate between historic individuals with similar names.

Discovering items related by geography and exploring the geographic context of events, societies, or historical figures are important to unraveling an interpretation. Providing geographic context and relating things spatially, of course, are what GIS is designed to do. But with the opportunity come technical and scholarly challenges. In order to examine these objects of study in their geographic context, it is necessary to be explicit about location. In the humanities, location is often simply a place name whose geographic coordinates are poorly specified and whose name has changed over the centuries and through different linguistic communities. Boundaries are very fuzzy, in some cases may not ever have been explicitly determined, or, if they were, that information was lost in the intervening centuries of cultural upheaval.

Must humanities scholars accept that in order to "use GIS," they must put a dot or a line on a digital map precisely locating places for which the location is uncertain. How can geographically related digital information provided by other scholars be found automatically if everyone gives their dots different geographic coordinates? Are these different locations the same place? We need to find ways to make the technology sensitive to these kinds of locational uncertainties so that scholars can better employ GIScience and GIS technology – and better represent indigenous knowledge in a digital environment.[9] One challenge comes in integrating very different conceptions and representations of locations. Even within Western fields that adhere to the scientific method, the measurement and description of location is not carried out identically and locations recorded in different source materials are often difficult to co-register. Error, uncertainty, and representation, core concepts in GIScience, always create detachment between reality and what is recorded.

Geographically referencing Hawaiian cultural information engages a larger range of issues. First there are the obvious challenges with determining the historical location of long-obliterated places. Indigenous place names do not necessarily coincide with those currently used. However, the Hawaiian language is extremely descriptive, so literal translations of place names can provide clues to unraveling their locations on maps and landscapes currently defined in Western terms. Indigenous locations that cover areas, such as gathering sites, sacred sites, or cultivation fields may not have clearly defined boundaries, though their existence is well understood. Perhaps even more relevant to this project is determining the landscape to which legends or practices should be associated. What is the geographical footprint of a creation legend or a recorded narrative from a kupuna (elder)? How do we represent the integrated view of terrestrial and aquatic systems as in the *ahupua'a?* Despite georeferencing problems, a great deal of recorded Hawaiian knowledge provides clues as to how Hawaiians lived and flourished on the land that may be amenable to scientific study as a means of revealing the past and understanding current landscape processes. Consider a named cloud formation that the historic record in the form of legends and stories states appears daily over a local *pu'u* (cinder cone) in certain seasons. Western scientists might deconstruct this information about the cloud development and form, relating it to the landforms over which the cloud appears and explain its appearance using scientifically understood processes. Using GIS, this deduction can be used to identify and locate other places where similar clouds may have been observed and a further search in the historic record may uncover similar cloud references at those other locations. Importantly, if any of these clouds are no longer appearing on as regular a basis, this may be an important indicator of climate or other major landscape changes.

Ha'ahonua: the Hawai'i Island digital geocollaboratory

In 2007, funding from the National Endowment for the Humanities supported The Kohala Center, in Waimea, Hawai'i, and its project partners, including the Edith Kanaka'ole Foundation and the Redlands Institute, to lay the foundation for the development of a "digital geocollaboratory" to integrate traditional Hawaiian knowledge with Western scientific data. Humanities, science, technology, and indigenous Hawaiian experts participated in a collaborative workshop to frame the requirements for an applied research environment and a conceptual plan for the development of a geographically integrated knowledge management system at the intersection of Hawaiian culture and Western science. *Ha'ahonua* is the name given to the project at this workshop by Hawaiian cultural practitioner and project collaborator, Kekuhi Keali'ikanaka'oleohaililani.

Like many words used in Hawaiian, *ha'ahonua* can be translated to English in a number of ways. The Hawaiian language is complex: many words and phrases have multilayered interpretations, rich in knowledge and meaning. In the context of the effort discussed here, *ha'ahonua* can be translated simply as "the earth dances," with layers of meaning conveying a sense of humility and the richness of interpretations for *honua* as our home, whether it be the earth or a voyaging canoe.

In its grand form, *Ha'ahonua* is envisioned to be an interactive, online, geospatially referenced knowledge portal that provides scholars and the public with integrated place-based scientific and cultural information as well as modeling, visualization, and decision support tools. This system will demonstrate how indigenous knowledge can support and stimulate research and learning, enhancing scientific awareness of the largely untapped Hawaiian body of knowledge.

Building this vision unfortunately is not simply a technological challenge. An important separate, concurrent project that has begun to inform the conceptualization of Ha'ahonua is being carried out by scholars at The Edith Kanaka'ole Foundation. *Papakū Makawalu* seeks to construct a framework for teaching Hawaiian knowledge.[10] This framework articulates clearly the essential non-Newtonian character of Hawaiian knowledge. According to *Papakū Makawalu*, in Hawaiian epistemology relationships between entities, whether animate or inanimate in Western parlance, are strong, pervasive, and essential. An entity or event at a specific place on the landscape may be understood to be related to some very different entity in a very separate location. Thus simply putting a dot on the map to show where some knowledge about the landscape is evidenced does not necessarily mean that it is the environmental conditions at that place that determines its presence. The connections between place and entity are not simply environmental. This implies also that integration of Hawaiian knowledge and Western science cannot be completed simply by creating an ontology that combines two distinct world views. "Integration" may, in fact, not be possible. Indeed, the intersection of time and space may be a key way for Western and indigenous knowledge holders to begin to explore and experience radically divergent – perhaps parallel – universes.

Given the magnitude of the vision, the construction of *Ha'ahonua* is evolving through a series of related geospatial knowledge portal development activities. In this way, small achievable steps towards the full vision are made, each one prototyping specific aspects. Some current projects includeare described below.

- Design of a web-based GIS to support the work of volunteer trail stewards of the Ala Kahakai National Historic Trail which will run along a 170-mile corridor on the west coast of Hawai'i Island. Trail segments are delineated by *ahupuaa'a*.

Whenever possible, stewards will be lineal descendants of those who lived on the land prior to the arrival of Captain Cook. Using GIS as a knowledge management framework, each steward group will collect and manage information about their landscape, incorporating both cultural and scientific information. The GIS provides a means of connecting cultural information to the landscape so that it can be understood and explored alongside the scientific data. Each steward group will own their information and decisions to share with others will be at their discretion. However, by designing the system comprehensively, with standards for data reporting and georeferencing procedures, interoperability of individual components once shared, will be ensured.

- Implementation of an Agricultural Suitability GIS for the County of Hawai'i designed to incorporate crop suitability criteria articulated by both scientists and Hawaiian cultural practitioners and scholars.
- A multifaceted effort to collect, organize, and share scientific, historical, and cultural information about a single *ahupua'a,* from the coral reef up to the high mountain forests. This project will utilize both scientific and traditional knowledge about the integrated landscape systems in order to identify and explore indicators and impacts of climate change.

Unlike other Digital Humanities projects, *Ha'ahonua* goes beyond digitizing artifacts, developing engines for rapid cataloging and searching, and applying advanced visualization techniques. Instead, by applying GIScience and the associated spatio-temporal toolbox, the objective is to create an environment in which knowledge grounded in indigenous epistemologies can be brought together with knowledge grounded in the Western scientific epistemology. As such, it is expected that *Ha'ahonua* will provide benefits beyond the Hawai'i Island community, offering a framework through which indigenous environmental knowledge worldwide can stand next to traditional Western scientific knowledge, in ways that begin the process of mutual understanding.

Acknowledgment

The authors would like to thank the National Endowment for the Humanities for providing a start-up grant to support the initial efforts on this project in 2007–8.

Notes

1   T. Reichhardt, "Astronomers bargain for use of sacred site," *Nature* 410:1015, 2001.
2   D. Dibiase, M. DeMers, A. Johnson, K. Kemp, A. T. Luck, B. Plewe, and E. Wentz, "Geographic Information Science and Technology Body of Knowledge," Washington, DC: Association of American Geographers, 2006.

3 D. Mark, "Cultural Differences, Technological Imperialism and Indigenous GIS," *Directions Magazine* May 23, 2006.

4 W. McCarty, "We would know how we know what we know: Responding to the computational transformation of the humanities," Paper read at The Transformation of Science: Research between Printed Information and the Challenges of Electronic Networks., May 31–June 2, 1999, at Schloss Elmau.

5 J. Unsworth, "What is Humanities Computing and What is Not?" in K. E. F. J. Georg Braungart (ed.) *Jahrbuch für Computerphilologie*, Paderborn: mentis Verlag GmbH, 2002.

6 S. Schreibman, R. G. Siemens, and J. Unsworth, *A companion to digital humanities*, Malden, MA: Blackwell Publishers, 2004.

7 S. Schreibman, R. G. Siemens, and J. Unsworth, "Humanities Computing," in M. J. Bates, M. N. Maack. and M. Drake (eds), *Encyclopedia of Library and Information Science*, New York, NY: Marcel Dekker Inc, 2003, pp. 1224–35.

8 T. Becher, *Academic tribes and territories: intellectual enquiry and the cultures of disciplines*, Milton Keynes England; Bristol, PA., USA: Society for Research into Higher Education: Open University Press, 1989.

9 I. N. Gregory, K. Kemp, and R. Mostern, "Geographical Information and Historical Research: Current progress and future directions," in *Humanities and Computing* 13, 2001, 7–22.

10 Edith Kanaka'ole Foundation, "About Papakū Makawalu," Available on-line at www.papakumakawalu.org, retrieved August 7, 2009.

30

What do humanists want?
What do humanists need?
What might humanists get?

Peter K. Bol

I propose to address the intersection of the disciplines of geography with the humanities from three positions: as an historian, and then of China's intellectual and cultural history, as a participant in the China Historical Geographic Information System project, and as the director of Harvard's Center for Geographic Analysis, which serves all disciplines, including the humanities.[1]

On history and geography

Standing within my discipline and looking out, if I ask what historians care about that other disciplines do not it is, in overly simple terms, that historians care about change in society over time – how to describe it, how to understand its consequences, and how to account for it. We do not expect philosophers or scholars of literature to feel responsible for this line of inquiry, nor do we demand that historians of art or science, for example, be committed to the investigation of change over time beyond the confines of the domains they study. What is it then that geography offers that other disciplines do not? The answer that I find most useful as an historian is that geographers care about the location and distance, recognizing that when taken seriously location and distance are multivalent. It thus makes sense that cartography should be within the fold of geography, and that the application of computation to location and distance with geospatial analysis and geographic information systems has also found its most secure home within the geographic disciplines rather than history, literature, philosophy, sociology, political science, economics, landscape, design, or public health.

It is not, I hope, merely a rhetorical strategy to suggest that the two disciplines of Geography and History have similar relations to the humanities and that their basic modes of analysis face similar challenges. Both straddle the social sciences and humanities. To the degree that Geography – and here I have human rather than physical geography in mind – and History seek empirical knowledge about others that can be hypothesized and tested they are social sciences, but they also can be participants in the humanistic enterprise of reflexive cultural engagement. In simple terms, just as historians may study change over time and thus see the present in a new light, one might say that geographers study variation through space and thus illuminate our current location. There is more to it than this, of course, for some of the most widely read humanistic historians are less interested in an analytic understanding of change over time than in an intellectual and empathetic engagement with an historical subject, and there are humanistic geographers who achieve the same in reflecting on our involvement with place.

I am more interested in the similarities in how historians and geographers deploy their respective basic tools. The map is basic to thinking about spatial variation, just as the chronology is a basic tool for thinking about temporal change. The map is to geography as the chronology is to history. Neither discipline has ultimate authority over these basic tools, or over spatial and temporal thinking, but each discipline has contributed to them and has a certain responsibility for them.

The intelligibility of the map and the chronology depends on the choice of scale. To be effective they must be in some sense inaccurate, since the most "accurate" map and chronology would be useless – as big or as long as the area or time covered. Both are necessarily reductive; a choice must be made about what features to highlight. Each uses systematic ways of defining where something is in space and time, such as longitude and latitude or the calendar. In choosing scale and in representing things in space and time, both the chronology and the map can be as misleading in their effects as they are illuminating.[2] Both involve choices of what to show, choices from among multitudinous possibilities that are beyond our power to represent coherently *in toto*. It may be argued that geospace is finite whereas time in infinite, but historical time is certainly finite and, when one examines it on ever finer scales geospace seems quite inexhaustible.

History and Geography need each other; they need to be brought together.[3] To think about temporal change without attention to spatial variation has, it seems to me, the paradoxical effect of undermining history. For if we fail to see that change over time unfolds differently across space, we substitute a single history for the reality of multiple histories. We cannot, I think, fully divorce what humanists want from geography from what they want from history, nor should we. They want both place and time. Recent years have seen the extraordinary resurgence of the geographic disciplines; the question is what those outside of geography are going to make of it.[4]

Geography in the study of China's history

Rather than reviewing the role physical and historical geography have played in research on China's history I want to take note of ways in which a few China historians have made a geographic subject their own and argued the fundamental importance of spatial analysis to understanding of China's history.

China's cartographic traditions have been diverse, ranging from scaled representations of physical territory to images of particular landscape views, from overviews of the national field administration to detailed accounts of the administrative and physical geography of a particular administrative unit, from urban layouts to depictions of

building complexes.[5] Through the nineteenth century, before the government mandated Western cartographic practices, there was no one accepted standard for what
constituted a "good" map in practice nor was there a sustained discussion of what
should.[6]

A number of historians have taken an interest in the history of cartography, and
given Cordell Yee's authoritative English-language survey in the monumental Harley
and Woodward series, that interest should grow.[7] Their views divide into two groups.
One views developments from the perspective of the state – a majority of maps were
concerned with administrative entities – and see its use of cartography as a means of
asserting territorial and political claims.[8] The other looks at the actual production of
maps, for example in the context of a print culture dominated by commercial print-
ers, and challenges those who link the rise of cartography to the emergence of an early
modern state in East Asia.[9]

What does seem clear is that little has been done in the Chinese humanities to
make analytic use of location and distance in the manner that Moretti undertakes, for
example, in the study of the novel.[10] Instead, we find a fairly rich lode of writings on
space and place and on traditions of geographic knowledge, in both cases with an
emphasis on their sociocultural construction.[11] These works may not, as yet, have
made much of a methodological impact on how geographers practice historical geography, although it may be here that historically minded humanists and the humanists among geographers will find common ground.

Those scholars who have made location and distance integral to the historical study
of Chinese society have been social scientists such as the late G. William Skinner[12]
and "social science" historians.[13] For Skinner the advent of geographic information
system technology provided a means of systematically analyzing large amounts of
data and identifying temporal change and spatial variation in Chinese society; much
of his work during his last two decades was devoted to pursuing this. Skinner was
unique, I think, in his insistence that the variations in the organization of society and
change over time could only be explained in terms of spatial systems; in regional
systems analysis he saw the very structure of China's history.

Historical GIS and China

One need not share Skinner's conviction that spatial analysis is the true foundation of
historical understanding to think that a method for seeing change unfolding in many
places simultaneously broadens and complicates the study of the past. My second
position, one that is still uncommon among historians, is that of an advocate for "historical geographic information systems." The China Historical GIS is one of several
nation-based HGIS.[14] I shall not try to explain what is involved in creating an historical GIS aside from making the basic point that it allows the user to disaggregate
into layers the kinds of information that are joined together on a printed map.[15] But
it is the future, I think, for already major HGIS have established themselves as more
authoritative and informative than existing printed historical atlases.

What are the advantages of an HGIS relative to a printed historical atlas?

- Searchability
  - getting to the place directly
  - choosing the time period for oneself
- Visibility
  - it can be viewed at different scales
  - all aspects of appearance can be set by the user
- Analysis
  - different layers of data can be distinguished and turned on and off
  - new research can be added
  - GIS software can be used to analyze correlations between data
- Accuracy
  - source notes can be included for every location, name, and date
  - it can be continuously corrected without the cost of printing
  - and, of course, it can be shared internationally.

The China Historical Geographic Information System (CHGIS) project which began in 2001 is creating the authoritative common base GIS for Chinese history, from the creation of a unified bureaucratic empire in 221 BCE to the end of the dynastic period in 1911 CE. All CHGIS datasets are freely available for unrestricted educational use. The objective of CHGIS is to establish a common "base" GIS for Chinese history: a single, common basis which can be used to represent, analyze, and share all China's historical data with temporal and spatial attributes. To accomplish this we trace all changes in administrative units over time, for in the past and present data were collected according to a hierarchy of local administrative units overseen by officials sent from and reporting to the capital. The central government mandated the conducting of land surveys and household registration, held censuses, and compiled national administrative geographies, and local government units compiled more detailed local surveys. The point of creating a base GIS for China's history is to make it possible for users to represent their data – whether population reports, tax quotas, military garrisons, religious institutions, the journeys of poets, the social networks of philosophers – on a digital map that accurately reflects the administrative structure and settlements for which historical data exist. Ideally users will join their own datasets to it and create new data layers appropriate to their research, and, because CHGIS provides an internationally available and authoritative platform, share their data with others. Ultimately, the accumulation of layers using the same base GIS will create an ever-richer context for understanding the particular event.[16]

Cultural production, the dissemination of ideas, and religious networks all unfold in particular places. Poets and philosophers travel and communicate. Literary and historical works are replete with spatial references. Being able to determine where things happened and to see many things happening across space is something humanists find it easy to be interested in. GIS does impose a certain way of conceiving of space and place on the data – something those of us who study the past need to keep in mind and this fundamentally quantitative technology was not initially developed with the goal of tracing change over time. Such caveats should not blind us to the fact that geographic information systems are one of the new computational aids that enable us to make use of the ever-growing amounts of information in digital form that are becoming accessible to scholarship.

Historians have much to gain from spatial analysis, as Anne Knowles has shown.[17] Perhaps the challenge is to show historians in small ways and large that by making a small effort to take note of location and distance, historical possibilities emerge of which they would not otherwise have been aware. Two examples will illustrate how putting data from two different sources into spatial context with a GIS is forcing me to think about issues I had been unaware of before. The first has to do with religious sites in Yiwu county (Zhejiang province) as of 1480. The original list of 61 sites, from a prefectural history published in 1480, gives each site's name, religious affiliation, date of founding, and travel distance from the county seat along one of the eight compass headings. One's first inclination as an historian is to take note of founding dates and conclude that the majority of Buddhist monasteries, consisting of teaching monasteries (*jiaosi*) and meditation monasteries (*chansi*), were founded in the ninth and tenth centuries. But when we take location and distances into account we see something else.

With the exception of shrines (*ci*) devoted to deified literati notables, which were generally located in the county seat, a majority of religious sites (including the Buddhist monasteries, Daoist *daoyuan*, the popular religion temples labeled *miao*) were located near the county boundary and in the mountains, the two being closely

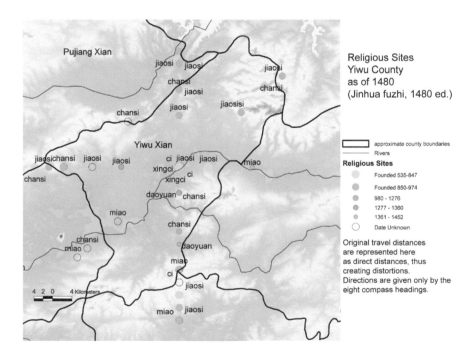

Figure 30.1 Religious sites in 1480, Yiwu County (Xian).

related. Note that the distances are travel distances, and thus a number of sites appear beyond the presumptive county borders. This suggests that I need also to explain why non-state-sponsored religious sites tended to maximize their distance from local government.

My second example draws on two complete lists of those who passed the civil service examinations in 1148 and 1256. The lists include information about the three generations of the patriline of the successful candidate, his age, number of brothers, and place of official residence. Hitherto scholarly focus has been on changes in the percentage of degree holders who had ancestors with official rank – thus to show that over time the examination system made increasing room for men who were not from families of officials.[18]

Again by including spatial distribution in our analysis we learn something we have not previously attended to: between 1148 and 1256 there was a shift in the regions that were represented in official ranks so that by 1256 men from the central Yangzi and far south were entering the bureaucracy. The spatial distribution of southern candidates in 1148 corresponds to the general trend prior to the thirteenth century as we can reconstruct it from other sources. It had been thought that it was only with the Mongol invasion from the north in the 1270s that elites fled south. Now we see that either this migration of literati elite families had already taken the place of

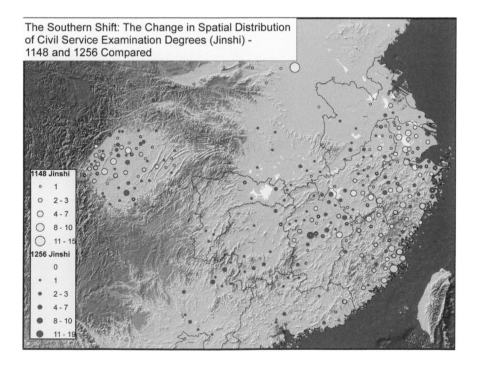

Figure 30.2 Distribution of degree holders, civil service examination of 1148 and 1256.

indigenous families in the far south and/or central Yangzi had already begun to acquire literati culture.

Both examples depend upon knowing where the places named in texts were, but they also depend upon historians paying attention to spatial information in their sources.

Serving the humanities: what do humanists want to practice?

This brings me to my third and final perspective, that of Harvard's Center for Geographic Analysis, which was established in the belief that GIS provided a common platform for bringing together information and knowledge from many disciplines in the humanities, social sciences, and sciences. Harvard University has not had a Geography department for over 50 years, since then-president Conant abolished it on the grounds that geography was not a university subject.[19] Having lost Geography, in particular those fields focused on human society rather than the physical environment, Howard T. Fisher ameliorated the situation somewhat with the founding of the Computer Graphics and Spatial Analysis Lab at Harvard's Graduate School of Design, but this too was discontinued, despite the objections of those involved.[20] The new Center, founded late in 2005, was not born out of an embarrassment over past mistakes but out of a belief that we were missing out on something of real importance to scholarship. At the time, although faculty in the School of Public Health, the Graduate School of Design, and in the environmental sciences recognized the value of spatial analysis, only one faculty member in all of Harvard had a PhD in Geography. Our purposes were quite pragmatic: to introduce GIS and related modes of geospatial analysis across the university, to support research, and to increase access to spatial data. We were able to build on efforts by the Harvard College Map Library, which has an outstanding collection of historic maps and an active acquisition program for both new publications and rare editions. The Map Library hired its first GIS coordinator in 1999 and began offering help to users who wanted to make use of GIS. In 2000 the library system began work on the Harvard Geospatial Library, Harvard's publicly accessible repository and finding tool for spatial data (georeferenced vector and raster data).[21] The Center has proven to be enormously successful and GIS has come to be seen as the most important new development in the way non-geographer humanists engage geographic information, theory, and knowledge. My sense of what humanists want is based on those we serve.

So, what do humanists want? The answer appears to be quite straightforward: they want spatial visualizations; they want maps. From the perspective of the discipline of Geography this is an interest of rather narrow band-with, one that favors cartography without understanding its real complexity and gives short shrift to the range of powerful spatial analytic methodologies. Yet at the moment this is what humanists want from geography.

This often takes the form of a request to map places mentioned in an article or chapter for illustrations of already completed research. Sometimes the request is bit more involved: for example, that a printed map be turned into a georeferenced digital map, information added, and made available to students or shared with fellow researchers through a mapserver. Too often the spatial representation is an afterthought, making us feel like the landscaper who is hired after the owners have spent all their money building the house. Sometimes the requests are more layered, and involve showing relationships between human settlements, communication networks, cultural sites, economic centers, and the landscape. For them – and their numbers are increasing – spatial analysis is an essential part of the research. They ask: What is the correlation between a certain kind of human activity and the places people live? What is the nature of the environment in which culture takes shape? What is the relationship between what exists today and what existed in the past? What do we learn from variations in the pattern of distribution?

We are now beginning to see long-term collaborative research projects initiated by humanists, in which an online geodatabase becomes the platform for multiple international collaboration and becomes a cumulative resource. Harvard's "AfricaMap" is an example of this.[22] AfricaMap combines raster, vector, and grid map data with spatially enabled datasets all served up at high speed. Moreover, it provides a spatial interface on which scholars can register the spatial footprints of their work. Finally, the system can be used for any place and discipline – currently adaptations are being used for social science research on a city and for teaching geology.

What do humanists need?

What seems at first glance to be requests for maps are, on closer examination, implicitly complicated requests to locate places on a visual representation of the earth's natural and cultural environments. Irrespective of the requests humanists make or the sophistication of their approach, we have found that at some point their projects entail the question "Where is this place?" And that, staff at the Center and Library have found, has been one of our greatest challenges. It is a challenge because although places are named, the names of places change. Names may appear in texts of a certain era or in maps for a certain date, but names are not fixed points on the earth's surface. (In China even the sacred mountains "move," that is, the name remains the same but gets attached to different peaks.)[23] Every new project involves an effort to find where past places might have been or discover the earlier names of contemporary places.

This brings me to a first answer to the question of what humanists need: they need a gazetteer. A gazetteer is in essence a dictionary of place names. To be useful it should tell us *when* the name was valid, *what* type of feature it was (a mountain, a settlement, an administrative unit, a religious site, etc.), *where* it was, and the source of this information. Ideally the *where* includes longitude and latitude and can be located on the surface of the earth, but there are many other solutions to "where" that allow us to identify a place in a useful way when the coordinates are unknown. This is particularly important as we move further back in time. For example, it is useful to know that the town of Laiguan was within the jurisdiction of the county of Longshui in the province of Zizhou in the Great State of Song (in modern Sichuan, China) in the eleventh century even if, at the moment, we do not know where exactly it was located. It is useful because since we happen to know the commercial tax quota for that town we can know something about the aggregate tax quota for the county. A gazetteer can tie an unknown location to the known, such as the record from 1480 which says that the Puji Monastery is south of the Yiwu county seat, a known location. In some instances we may only know that a place name appears in a text of a certain date, for example in the poet Basho's account of his journey to the interior of Japan in the seventeenth century, but registering that facilitates noting other mentions of the place. Ultimately the accumulation of references may lead to a more precise geographic location, but just as important is realizing that a place was of interest, that people wrote about it and attached meaning to it, and that meanings accumulated over time. One might think of a gazetteer as the geographical equivalent of the prosopographies of ancient Rome, Anglo-Saxon England, Byzantium, and China.

What would make a gazetteer for the humanities different from some of the massive gazetteers that do exist, such as the US Geonames database? It seems to me that there are at least five elements:

- It would be historical; that is, it would allow for various indicators of temporality such as start and end dates during which a feature is known to be valid, the date of a map on which a place name appears, or the date of a text or author in which it is mentioned. It would accommodate multiple scales of temporality, recognizing that historical periodization differs across space and time.
- It would be multilingual. The same place name exists in different languages; a gazetteer must be accessible to publics in different language communities; gazetteers need to be used to filter texts and maps in different languages; it must accommodate gazetteer data submitted in the vernacular. Lists of feature types exist in many languages, although we have yet to see a world standard with which all languages align themselves.
- It would incorporate the historical gazetteers being created as part of local and national historical GIS projects and it would allow for volunteered information. As the gazetteer grows it would be used to tag place names in texts, thus creating spatial footprints for historical texts while further populating the gazetteer.
- It would be open-ended and cumulative. At some point it would have fully mined the historical record, but the present is the future past, place names will continue to change and new places will appear. Ultimately a world-historical gazetteer would be integrated with the gazetteers that track the ever-changing boundary between present and past.
- It would be sustained. This is, of course, as essential as it is difficult. Designing a historical gazetteer, contributing to the formulation of international standards and conforming to them, and incorporating existing sources is not beyond us. But how do we sustain this?

A world historical gazetteer would serve as a geographical name authority and dictionary in its own right. It is a necessary resource for applications that extract geographic names from unstructured texts and for spatially enabling the plethora of datasets of historical information that already exist. Without a shareable gazetteer, every effort to broaden our understanding of historical data to include location and distance will require redoing the same preliminary work that we are now doing or it will limit us to those places and periods, primarily the last 200 years of human history, that have usable cartography. The compilation of a world-historical gazetteer ought to be led by those who know the most about how to think about location and distance: the geographers.

What might humanists get?

Culturally minded geographers have been critical of GIS in general, and historical GIS is open to all the same criticisms. As a student of China's history I am bothered that the world today, including China, has come to see the modern Euro-American ways of thinking about and representing space that are embedded in GIS as natural and universal. Criticism ought not, however, to put an end to experimentation and inquiry.

But what is possible? It seems to me that a cyberinfrastructure for the humanities and social sciences, the subject of a recent ALCS report,[24] will necessarily be composed of many systems managed by different groups. The Association of American Geographers (AAG), working with library systems, is the obvious group to take the lead in organizing a world-historical gazetteer interface that could query diverse online systems and harvest the work of individual and local research projects.

Given that Google is digitizing all the books in the world, who is going to be digitizing all the maps in the world and making them as searchable as the books? At the moment this may be too much to ask, but perhaps we could at least appeal for a national registry that would allow all those collections that are digitizing and georeferencing maps to inform each other and thus avoid the expense of reduplication. It may not be widely known that Harvard has digitized and georeferenced most of the Army Map Service 1:250K series for China.

It is not too early to start thinking about a world historical GIS – the European Union has already begun planning for an integrated European historical GIS and some American world historians will begin planning for a world HGIS next fall. We need a discussion of best practices, the active involvement of institutions outside of the United States, and serious reflection on compatibility issues and standards so that GIS datasets compiled in diverse places and languages can be shared and joined together.

I do not know, frankly, whether this approach to bringing history and geography together will do anything of value for geography. But it will, I am confident, change the way we do history.

Notes

1 For the history and current activities of the Center see www.gis.harvard.edu.
2 Historians might well write their version of Mark S. Monmonier, *How to lie with maps*, 2nd edn, Chicago: University of Chicago Press, 1996; similarly geographers

might well produce their own version of David Hackett Fischer, *Historians' fallacies; toward a logic of historical thought*, 1st edn, New York: Harper & Row, 1970.

3 It seems to me that thinking about the relationship between the two disciplines has come almost entirely from the geographers side, such as Alan R. H. Baker, *Geography and history : bridging the divide*, Cambridge, UK; New York: Cambridge University Press, 2003; R. A. Dodgshon, *Society in time and space : a geographical perspective on change*, Cambridge; New York: Cambridge University Press, 1998; Aharon Kellerman, *Time, space, and society : geographical societal perspectives*, Dordrecht; Boston: Kluwer Academic Publishers, 1989. Historical geography is a subfield of geography. I do not see a real equivalent within the field of history. Baker rightly, in my view, begins from the view that there is a divide to be bridged.

4 A search of articles with "geography" as a keyword in Google Scholar shows during the last decade a nearly ten-fold increase over the 1980s; even supposing that many earlier articles would not appear in such a search, the increase is dramatic. It is confirmed of course by other statistics such as the increase in the number of doctoral degrees in geography. I thank Dr. Wendy Guan for this information.

5 The best collection of historical maps in all their diversity is Cao Wanru 曹婉如 and *et al.*, eds, *Zhongguo gudai ditu ji* 中國古代地圖集, 3 vols., Beijing: Wenwu chuban she, 1990–94.

6 On the spread of Western cartography see Zou Zhenhuan 邹振环, *Wan Qing xifang dilixue zai Zhongguo : yi 1815 zhi 1911 nian xifang dilixue yizhu de chuanbo yu yingxiang wei zhongxin* 晚清西方地理学在中国：以1815至1911年西方地理学译著的传播与影响为中心, Shanghai: Shanghai gu ji chu ban she, 2000. On multiple modes of representation and the commercial publishers' willingness to reproduce outdated maps and maps of diverse standards, see Alexander Van Zandt Akin, "Printed maps in late Ming publishing culture: a trans-regional perspective" (PhD, Harvard University, 2009).

7 See Cordell D. K. Yee, "Cartography in China," in *Cartography in traditional east and southeast asian societies*, ed. J. B. Harley and David Woodward, *The History of Cartography, 2:2: Cartography in Traditional East and Southeast Asian Societies*, Chicago: University of Chicago Press, 1994.

8 Recent examples include Laura Hostetler, *Qing Colonial Enterprise: Ethnography and Cartography in Early Modern China*, Chicago: University of Chicago Press, 2001; Richard J. Smith, *Chinese Maps: Images of "All Under Heaven"*, Hong Kong: Oxford University Press, 1996; and Hu Bangbo, "Maps and Political Power: A Cultural Interpretation of the Maps in *The Gazetteer of Jiankang Prefecture*," *Journal of the North American Cartographic Information Society*, no. 34 (1999).

9 Akin, "Printed maps in late Ming publishing culture: a trans-regional perspective," Marcia Yonemoto, *Mapping early modern Japan: space, place, and culture in the Tokugawa period, 1603–1868*, Berkeley: University of California Press, 2003.

10 Franco Moretti, *Graphs, maps, trees : abstract models for a literary history*, London; New York: Verso, 2005.

11 Nicola Di Cosmo and Don J. Wyatt, (eds), *Political Frontiers, Ethnic Boundaries, and Human Geographies in Chinese History*, London: RoutledgeCurzon, 2003; Martin W. Lewis and Kären Wigen, *The myth of continents: a critique of metageography*,

Berkeley: University of California Press, 1997; Tang Xiaofeng, *From Dynastic Geography to Historical Geography: A Change in Perspective towards the Geographical Past of China*, Beijing: The Commercial Press International, Ltd., 2000; Jing Wang, ed., *Locating China: space, place, and popular culture*, London; New York: Routledge, 2005; Peter K. Bol, "The Rise of Local History: History, Geography, and Culture in Southern Song and Yuan Wuzhou," *Harvard Journal of Asiatic Studies* 61, no. 1 (2001); Timothy Brook, "Mapping Knowledge in the Sixteenth Century: The Gazetteer Cartography of Ye Chunji," *The [Princeton University, Gest] East Asian Library Journal* 7, no. 2 (1994); Bernard Faure, "Space and Place in Asian Religious Traditions," *History of Religion* 26, no. 4 (1987); Ronald G. Knapp, "Chinese Villages as Didactic Texts," in *Landscape, Culture and Power in Chinese Society*, ed. Wen-hsin Yeh, Berkeley: University of California Berkeley, Institute of East Asian Studies, 1998; and Knapp, ed., *Chinese Landscapes: The Village as Place*, Honolulu: University of Hawaii Press, 1999; Wen-hsin Yeh, ed., *Landscape, Culture and Power in Chinese Society*, Berkeley: University of California Berkeley, Institute of East Asian Studies, 1998.

12 Some of Skinner's seminal works are G. William Skinner, "Marketing and Social Structure in Rural China," *Journal of Asian Studies* 24, no. 1–3 (1964–65), "Cities and the Hierarchy of Local Systems," in *The City in Late Imperial China*, ed. G. William Skinner (Stanford: Stanford University Press, 1977), "The Structure of Chinese History," *Journal of Asian Studies* 44, no. 2 (1985). Skinner left his spatial analytic work, including spatial datasets, on China, Japan, and France to the Fairbank Center for Chinese Studies at Harvard, which is working to make it freely available to scholars.

13 See, for example, Robert Marks, *Rural revolution in South China : peasants and the making of history in Haifeng County, 1570–1930*; Madison, W1.: University of Wisconsin Press, 1984; R. Keith Schoppa, *Chinese elites and political change: Zhejiang Province in the early twentieth century*, Harvard East Asian series; 96, Cambridge, MA: Harvard University Press, 1982.

14 Of particular note are the HGIS for Great Britain (www.gbhgis.org), the United States (www.nhgis.org), and Germany (www.hgis-germany.de).

15 The best practical survey of historical GIS is Ian Gregory and Paul S. Ell, *Historical GIS: technologies, methodologies and scholarship*, Cambridge, UK; New York: Cambridge University Press, 2007.

16 For a brief account of the CHGIS project see Peter K. Bol and Jianxong Ge, "China Historical GIS," *Historical Geography* 33 (2005). For an extended discussion of the issues involved in creating this GIS see Peter K. Bol, "Creating a GIS for the History of China," in *Placing History: How Maps, Spatial Data, and GIS Are Changing Historical Scholarship*, ed. Anne Kelly Knowles and Amy Hillier, Redlands, CA: ESRI Press, 2007.

17 Anne Kelly Knowles, "Introduction to the Special Issue: Historical GIS: The Spatial Turn in Social Science History," *Social Science History* 24, no. 3 (2000); Knowles, ed., *Past time, past place : GIS for history*, Redlands, CA: ESRI Press, 2002; Anne Kelly Knowles and Amy Hillier (eds), *Placing history: how maps, spatial data, and GIS are changing historical scholarship*, Redlands, Calif.: ESRI Press, 2008.

18 The first to inquire into this was E. A. Kracke, "Family Versus Merit in Chinese Civil Service Examinations Under The Empire," *Harvard Journal of Asiatic Studies* 10 (1947).

19 The most detailed account of these events and the personalities involved is Neil Smith, "'Academic War Over the Field of Geography': The Elimination of Geography at Harvard, 1947–1951" *Annals of the Association of American Geographers* 77, no. 2 (1987).

20 For the history of Fisher's lab see Nicholas R. Chrisman, *Charting the unknown: how computer mapping at Harvard became GIS*, 1st edn, Redlands, CA: ESRI Press, 2006.

21 http://dixon.hul.harvard.edu:8080/HGL/hgl.jsp

22 AfricaMap is open-source and open-access; most map and datalayers may also be downloaded at http://africamap.harvard.edu/.

23 James Robson, *Power of Place: The Religious Landscape of the Southern Sacred Peak (Nanyue 南嶽) in Medieval China*, Cambridge: Harvard University Asia Center, 2009, Chapter 2, "Moving Mountains Nanyue in Chinese Religious Geography."

24 "Our Cultural Commonwealth: The Report of the American Council of Learned Societies Commission on Cyberinfrastructure for the Humanities and Social Sciences," New York: American Council of Learned Societies, 2006. Available at www.acls.org/cyberinfrastructure/OurCulturalCommonwealth.pdf. See also "Cyberinfrastructure For Us All: An Introduction to Cyberinfrastructure and the Liberal Arts," posted December 16, 2007 by David Green, in Filtered: The Academic Commons Magazine, at www.academiccommons.org/commons/essay/cyberinfrastructure-introduction.

Afterword
Historical moments in the rise of the geohumanities

Michael Dear

For as long as I can remember, there has been a constant two-way borrowing between geography and the humanities. So what justifies the call for a "geohumanities" at this time? I think the answer to this question lies in the nature of contemporary social change, a broadening of perspectives on the way we know things, and the responses of particular academic disciplines to these material and cognitive adjustments.

Altered material and cognitive worlds: a long view

Our world is in the midst of profound economic, social, political, and cultural change, including the tendencies toward *globalization*, understood as the emergence of a worldwide, integrated capitalist economy in which a relatively few "world cities" act as centers of corporate command and control; the rise of a *network society*, including unprecedented communications technologies and connectivity, also known as the "Information Age;" *environmental crisis*, a belated consciousness of the consequences of climate change and environmental hazards that threaten the viability of life on this planet; *social polarization*, the widening gap between rich and poor, among nations, ethnic, racial, and religious groups, genders, and those on either side of the digital divide; and *hybridization*, the fragmentation and reconstruction of identity and citizenship brought about by large-scale international and domestic migrations. Each of these five tendencies is causing radical adjustments in politics at global, national, and local levels. Now, some of these tendencies may find formal equivalence in previous eras, as in earlier manifestations of globalization, but never before have they appeared in concert, penetrated so deeply, been so geographically extensive, or overtaken everyday life with such speed and intensity. Taken together, they define an "Information Revolution" that is likely to prove as profoundly altering as the Agricultural and Industrial Revolutions of earlier times.

As the material world changes, concomitant shifts in our cognitive worlds occur. The history of Western Thought from pre-modern to modern times is characterized by several such adjustments. In a nutshell, the past 500 years have witnessed a shift from a belief about the divine basis of knowledge, through an understanding based primarily in science in which physical laws dominate explanations of the material world, to the current questioning of a science-dominated world-view. Four historical moments contributed to the formation of the modern world-view. In the fifteenth to sixteenth centuries, the Renaissance represented a rebirth of the traditions of classical learning that led eventually to the establishment of a secular education based on humanist values, contrasting the Christian view of humans as weak and sinful with a neoclassical sense of Rational Man's unbounded potential. The Reformation (sixteenth century) was a rationalization of religious beliefs and practices that shifted the emphasis from the sacramental experience to the study of texts, that is, the Word itself. The Protestant reforms of this era privileged personal judgment and autonomy over institutional authority; its focus on internal subjectivity became a cornerstone of modern thought. The Scientific Revolution (sixteenth to seventeenth centuries) located truth not in religious faith but in the material world, relying on what could be observed, measured, and systematically explained through natural laws. And the Enlightenment (eighteenth century) sought to extend the search for truth and laws to all aspects of human behavior; as the authority of organized religions declined, it offered the prospect of scientifically based reform for society as a whole.

Over time, the open-mindedness of the late-Renaissance generation of Montaigne – who himself regarded attempts to reach theoretical consensus as presumptuous and self-deceptive – was often stifled. Modernist practices could become as rigid and intolerant as that which they purported to replace; the scientific method, so extravagantly promoted for its purity and impartiality, was found to be time- and place-bound, its explanations contingent and mutable. Taste-makers such as Matthew Arnold helped establish a canon defined largely by European male aesthetics, philosophy, and method. However, during the last quarter of the twentieth century a variety of epistemologies emerged to challenge *all* kinds of discipline-based hegemony, emphasizing the importance of different ways of seeing. I refer to the "postmodern" as an umbrella term to encompass this variety, which includes post-structuralism, feminism, and post-colonialism. The *postmodern turn* was very much a product of its times, including (in the West) political battles for civil rights, gender equality, nuclear disarmament, and multiculturalism. Paralleling these political movements was a cognitive challenge that offered a revitalized relativism in place of the fictions invoked by proponents of any intellectual hegemony. Philosophers such as Isaiah Berlin took a further step, suggesting that the plethora of available ontological and epistemological stances were ultimately incommensurable, and that we had to learn to live with difference. This standpoint is, I believe, a major wellspring of the current interest in comparative transdisciplinary research. You can trace it through the humanities via the critical energies directed toward multiculturalism, transgender studies, postcolonialism, transregional studies, and cultural studies. Each of these moments has its own history that I cannot hope to summarize here. Instead let me focus specifically on how the convergence of material social change and cognitive adjustments became manifest in the case of geography.

Geography adjusts

The discipline of geography has evolved in response to the altered material and cognitive worlds I have just sketched. Recent geographical practice responded to two particular conditions: a revolution in geotechnology and a global environmental crisis. The former led to the rise of Geographical Information Science (GISci), and the latter to a revitalization of geography's longstanding interest in Environmental Studies. GISci is concerned with the creation, storage, manipulation, representation, and analysis of large spatially based data sets. In geography, the subfield developed along twin axes: as a science concerned with theory, hardware and software development associated with the creation, management, and utilization of multi-source geographic data (often in collaboration with computer scientists); and as an analytical tool dedicated to the application of GISci in specific knowledge domains (such as public health). Geography's participation in environmental studies is an organic outgrowth of the discipline's long-established traditions in landscape studies and physical geography, which includes biogeography, climatology, geomorphology, and hydrology. Geographers joined multidisciplinary teams in basic science explorations into environmental systems (such as biosphere sustainability, and climate change); they also increasingly worked in multidisciplinary applied environmental problem-solving teams (investigating recovery from natural disasters and pollution mitigation, for instance).

These two tendencies – geographic information technology and environmentalism – are occurring in the aftermath of a remarkable intellectual renaissance that characterized geography during the second half of the twentieth century. This was inspired by determined efforts from within the discipline to engage wider trends in the academy, which initially took the form of efforts to establish a social science basis for geography in what came to be called the "quantitative revolution." Apart from the intrinsic merit of this work, the quantitative turn – with roots in the 1950s recovery from the second world war – had the important effect of once more putting geography in touch with practices in a number of other social sciences that were simultaneously experiencing their own quantitative revolutions. It was not long, however, before cracks appeared in the edifice of quantitative geography. Early criticism surfaced from within the discipline as practitioners began calling for greater behavioral realism in model-building. In the late 1960s neo-Marxists launched their own attack on the false "science" of social science, insisting on a more critical approach to what they perceived as overly simplistic number-crunching. Subsequently, a number of prominent quantitative geographers switched to a Marxist view of scientific praxis that was as innovative and insightful as the original quantitative revolution.

The seeds of a distinct geohumanities were planted during the 1970s, when human geography was buffeted by a flood of fresh ideas from the social sciences and the humanities. Humanist geographers weighed in with a critique of Marxist geography for its excessive economic determinism. Cultural geographers opened an attack on geography's landscape tradition for its lack of a critical dimension. The first stirrings of a feminist geography began overturning the long traditions of a male-dominated discipline. By the early 1980s and the advent of postmodernism, these spirited awakenings had mutated into a dialogue with the entire realm of social theory, social science, and many humanities disciplines. Not long after, geographers began experiencing reciprocal expressions of engagement with place from disciplines as diverse as cultural studies, engineering, epidemiology, film studies, literature, painting, philosophy, photography, political science, and sociology.

It is unsurprising that geographers became active participants in an academy-wide shift toward transdisciplinary research and teaching. Indeed, the widespread acceptance of transdisciplinarity has been inspirational in geography. It has encouraged methodological diversification and experimentation with careful sequencing of suites of methodologies – commonly called "mixed methods" – by both large-scale collaborative research projects (typical of science) and by "lone" scholars (more typical of the humanities). These days, a geographer's methodological palette typically consists of quantitative, qualitative, textual, and visual methods. This is not a haphazard eclecticism, but instead represents a deliberate, creative methodological pluralism that is at the heart of transdisciplinarity. Such open-mindedness is one reason why the spread of GISci into other subfields of geography has inspired many practitioners to glimpse a common *methodological* ground for the discipline. In an analogous manner, the congregation of many disciplinary subfields around environmental issues has inspired glimpses of a common *substantive* ground that further integrates and unifies the discipline while at the same time sharpening its identity for outsiders.

Professional geographers in public and private practice, as well as those in the academy, are raising geography's profile as an activist, socially engaged discipline increasingly engaging society's hardest problems, such as immigration, sustainability, terrorism and war, homelessness, public health, poverty and inequality, discrimination, and environmental justice. The increasing public profile of geographers engaged in such research and applied work *requires* a multidisciplinary approach that furthers the outward orientation of geography. In addition, the nation's pre-eminent academic and professional organization, the Association of American Geographers (AAG), has adopted an aggressive agenda for interdisciplinary development and outreach, including initiating a cross-disciplinary dialogue in a "Geography and Humanities" conference that led ultimately to this volume.

Geohumanities emerge

What kind of geohumanities is emerging from the maelstrom of changing material and mental worlds? I believe this volume offers several signs that pertain to framing, precepts, and practice in the geohumanities.

Framing

The geohumanities that emerges in this book is a transdisciplinary and multimethodological inquiry that begins with the human meanings of place and proceeds to reconstruct those meanings in ways that produce new knowledge and the promise of a better-informed scholarly and political practice.

Place

A common analytical object in the geohumanities is *place*. It is an analytical "primitive," one of our principal "key words." Place is not the only relevant primitive, and the term has many meanings. Researchers commonly distinguish between *space* as an abstraction, and *place* as a social construct, that is, what humans create out of space. Another meaning is *landscape* which encompasses natural and cultural dimensions. The *production of place* refers to both material and cognitive processes and outcomes, and *representations* of place include the textual, visual, sculptural, quantitative, qualitative, perceptual, and oral dimensions. These varieties of meaning are themselves a source of riches that attract and facilitate interdisciplinary dialogue.

Embracing complexity: a non-exclusionary ontology

Human life is manifestly complex. Theory and practice inevitably involve degrees of simplification and abstraction in which some loss of complexity is unavoidable. In order to minimize information loss, a non-exclusionary ontology is preferred that, to the extent possible, retains variety, avoids reductionism or commitment to a single world view, and adopts a level of complexity appropriate to the task at hand. Such self-aware practice is linked to an ontological flexibility, that is, the ability to shift among analytical registers as required.

Transdisciplinarity: epistemological openness

After postmodernism, transdisciplinarity is unavoidable. The fullest knowledge is possible only when every epistemological alternative is included the geohumanities toolkit. Any *a priori* selection or rejection of a methodology must be regarded with suspicion, as prejudiced and partial. It follows that the geohumanities necessarily confronts the thorny issue of incommensurablity among epistemological alternatives. This is a heavy burden since no single discipline or individual can absorb all ways of knowing with equal facility. Hence transdisciplinarity, comparative analysis, and a critical self-reflexivity become prerequisites for successful theory and practice in the geohumanities.

Knowledge and action

Politics is personal, and political activism will not always be an explicitly acknowledged component of geohumanities research and practice. Yet the geohumanities in use simultaneously problematizes the production of knowledge *and* the practice of politics, just as ontological and epistemological awareness are together constitutive of the project. In practice, this involves consideration of the positionality of individual investigators as well as the relationship between knowledge and action.

Practice

The cross-boundary contributions in this volume demonstrate, time and again, that research and practice flourish in transdisciplinary contexts. The vignettes are proof that all grounded practice is inherently transdisciplinary. Still, most research practice remains dominated by a single discipline reflecting the training of the researcher. In contrast, our contributors – even when they begin from a disciplinary base – demonstrate a radically different practice in which *all disciplines adopt a supportive role*. The consequence of such self-awareness is a democratic intelligence that invigorates and extends our findings in ways that are demonstrably superior to research dominated by a single discipline. A collection of voices will deliver a closer approximation of knowing.

Readers and critics are better placed to judge the quality of the outputs from our geohumanities practice, but I will venture a few personal comments on what this practice has wrought. One concerns the presence and power of the *visual* in contemporary academic inquiry. Of course, the ubiquity of the visual in our contemporary world comes as no surprise, but new ways of seeing are infinitely enhancing our analytical capacities. Think, for instance, of the crowded exuberance of the vignettes; how a study of fiction is enhanced by a map; how a geography gains by being represented as fiction; and how geographical information science can help recreate psychogeographies of the past. Octavio Paz was right. In his momentous essay on the Mexican character, *The Labyrinth of Solitude*, Paz lamented the fact that "progress" had obliterated that half of the human character that "reveals itself in the images of art and love."[1] The contents of this book triumphantly displace that cognitive void by opening our eyes anew to the beauty and insight of the image.

Yet the geohumanities is not confined to the sublime; our contributors frequently return to the pervasive question of politics. Think of kanarinka's new cartography of Cambridge street names; of Paglen's detective work on the shoulder patches of "secret" agents who cannot resist flaunting their affiliations; or of Iglesias' subversive projection on the boundary fence between the two countries of a movie depicting the end of the Mexico–US border. Connecting again with intellectual history, this volume reestablishes one of the principal insights of the Frankfurt School (the critical theory of Adorno, Habermas, and Marcuse) that the cultural is political, and that political change may be forged through cultural practice.

This said, I readily concede that the engagement across geography and the humanities has hardly begun. For one thing, the range of humanities disciplines included in this volume is small, though it would be a relatively straightforward task to extend it to fields beyond architecture, history, literature, and visual studies (our principal interlocutors). In addition, our contributors, diverse as they are, began from disciplinary bases but did not engage in direct conversation; and we understand that textual propinquity is not sufficient to produce a community of inquiry. So those who would explore the geohumanities further are charged with inventing new vocabularies and attaining heightened levels of fluency across disciplinary boundaries. These tasks are daunting, but begin we must. Our text is offered as a first map of this challenging terrain.

Note

1 Octavio Paz, *Labyrinth of Solitude, and Other Writings*, New York: Grove Press, 1985, p. 226.

Index

T - #0626 - 071024 - C344 - 234/156/18 - PB - 9780415589802 - Matt Lamination